Converging NGN Wireline and Mobile 3G Networks with IMS

Converging NGN Wireline and Mobile 3G Networks with IMS

Rebecca Copeland

CRC Press
Taylor & Francis Group
Boca Raton London New York

CRC Press is an imprint of the
Taylor & Francis Group, an **informa** business

AN AUERBACH BOOK

Auerbach Publications
Taylor & Francis Group
6000 Broken Sound Parkway NW, Suite 300
Boca Raton, FL 33487-2742

© 2009 by Taylor & Francis Group, LLC
Auerbach is an imprint of Taylor & Francis Group, an Informa business

No claim to original U.S. Government works
Printed in the United States of America on acid-free paper
10 9 8 7 6 5 4 3 2 1

International Standard Book Number-13: 978-0-8493-9250-4 (Hardcover)

Library of Congress Cataloging-in-Publication Data

Copeland, Rebecca.
    Converging NGN wireline and mobile 3G networks with IMS / Rebecca Copeland.
        p. cm. -- (Informa telecommunications and media)
    Includes bibliographical references and index.
    ISBN 978-0-8493-9250-4 (alk. paper)
    1. Multimedia communications. 2. Convergence (Telecommunication) 3. Internet Protocol multimedia subsystem. I. Title. II. Series.

TK5105.15.C675 2008
004.6--dc22   2008012104

Visit the Taylor & Francis Web site at
http://www.taylorandfrancis.com

and the Auerbach Web site at
http://www.auerbach-publications.com

Dedication

To my father, who admired all knowledge and longed to see me in print

To my sons, who encouraged this voyage right up to the last sprint

To my husband, with so much to share

My deep love and thanks, for just being there

And care.

Contents

Foreword

At the publication of this book, almost every telecom operator in the world is planning to evolve his network toward IP-based infrastructure for advanced multimedia services. The IMS architecture is an essential component of the control layer of these networks.

In the operator's mind, this approach supports the main three aims: (1) cost reduction from consolidation, (2) faster time to market with new services, and (3) the ability to create access agnostic, advanced services without having to change the network each time. These aims lead inevitably to the IMS central idea of the layered approach, separating the session control from the access bearer network as well as separating the applications from the switching control.

Although developed initially for the Mobile world, France Telecom and BT and others realized quite quickly that the IMS functionality was also needed on fixed networks to allow users to log on anywhere and get the same service. The advent of WiFi and WiMAX only served to increase the similarity of the requirements. The result is that the IMS offers a real chance to bring together the architecture of the fixed and mobile worlds into one layered approach for all services — the converged NGN.

Against this background, this book by Rebecca Copeland is a very timely contribution to this field. It is suitable for both lay people wanting to understand the building blocks and drivers of IMS, as well as more skilled practitioners wanting to brush up on technical details and close some gaps in their knowledge.

Mick Reeve

Mick Reeve has a solid track record in technical innovation established during his 36 years at British Telecom (BT). Most recently, he was the BT Group CTO Chief Architect, responsible for BT's overall network architecture. He also led the BT Standards program, directing the architecture for BT's 21st Century Network and ensuring its alignment with global standards. In this role, he also served as a board member of the Tele-Management Forum, the WiMAX Forum, and ATIS. He has published widely in the telecom field, and he holds a number of patents in optical communications. Mick is a Fellow of the IEC and a Fellow of the UK's Royal Academy of Engineering.

Preface

In writing this book I aimed to provide an all-around view of the future network architecture and its main principles rather than a detailed description of merely core IMS standards and protocols. It is not easy to form an overall picture of the converged NGN and 3G Mobile network, where knowledge resides in the different camps (Mobile, Fixed/NGN, ICT, and Web services) and only few manage to straddle that divide.

I sincerely hope that this book will deliver to many readers the right scope and insight that I myself sought in my own journey toward gaining a deeper appreciation of the next generation of converged communication, with all its wonder and complexity.

The first chapter provides a full introduction to IMS, spanning from market trends and commercial drivers to the technology innovation that make it all possible, and migration issues. This introduction also provides a brief overview of subjects that are later expanded in the appropriate chapters, allowing the reader to appreciate how they are all logically linked. Chapter 2 contains descriptions of the standards sources and their particular areas of activities, including a description of the overall architecture for TISPAN NGN, 3GPP, and OMA.

The IMS advanced principles go beyond session control and multimedia handling. The innovation of IMS is also found in its ID management, Service Profiles, Event triggering, flow-based and event-based Charging mechanisms, Service-based QoS, and more. In addition, full appreciation of the new network requirements cannot be complete without a thorough understanding of network admission, security, border control, and blending legacy services. This book contains chapters dedicated to these important aspects, showing how they complement IMS.

Each chapter starts with a description of the main principles, from the generic technology to the new IMS principles devised to control that particular aspect. Where possible, this book not only points out convergence of the standards but also reflects separately the different specification and terminology for NGN TISPAN and 3GPP.

To make it a useful companion and a reference tool, this book includes many summary tables. A separate appendix (Appendix A) provides a brief introduction to the major IMS protocols, wherein a basic understanding of how the protocols work can explain some of the IMS features. The appendices also provide a comprehensive glossary of abbreviations (Appendix B), a list of terms and definitions (Appendix E), and a list of IMS reference points compiled from numerous standards documents (Appendix D).

Acknowledgments

I owe a debt of gratitude to those who took time to review and comment on the manuscript, especially to Maria Cuevas (BT), Gabriele Corliano (BT), Neill Wilkinson (Quartex), and to Mick Reeve (ex-BT GCTO Chief Architect), who wrote the Foreword.

About the Author

 **Rebecca Copeland** is a Telecom Consultant who has specialized in Next-Generation Networks, SIP applications, and IMS from the early days of concepts. At Marconi, Rebecca was the prime mover toward SIP-based applications, formulating the IMS vision for converging NGN with 3G, translating the emerging standards into a coherent set of IMS components.

Leaving Marconi in 2005, Rebecca was engaged by BT Group CTO to produce a strategic study where she outlined 21C IMS solutions for the Enterprise market, building on BT's plans for IMS.

Since August 2006, Rebecca has served as a senior IMS Consultant in Huawei's Core Network European Marketing, providing high-level expertise in support of strategic projects in Europe, articulating the Huawei architecture and product features to major telecom carriers.

Rebecca has been a frequent conference speaker for many years and has published a number of white papers and articles. Rebecca holds a degree in architecture and a degree in law (LLB hons), plus programming and business analysis certification.

Chapter 1

IMS Concepts

1.1   This Chapter

This chapter describes the concepts on which IMS is based and how these concepts have evolved from the original ideas as the scope of IMS widened. It describes the main requirements and the role of standards in IMS. It compares the IMS principles with both Internet principles and circuit network traditional designs. It also examines the new way of delivering an integrated environment for Information and Communications Technology. This chapter explains why IMS is necessary for the future network, especially for convergence of infrastructure and for enhanced multimedia services. It lists IMS benefits and risks, and how IMS makes the difference for delivering a desirable range of multimedia services.

1.2   Introduction to IMS

IMS was initially called IP Multimedia Subsystem — that is, the subsystem that manages multimedia services. This "subsystem" consists of the CN (Core Network) elements necessary for the provision of IP multimedia services (i.e., audio, video, text, chat, image, and a combination of them) delivered over PSNs (Packet-Switched Networks). It was soon realized that, in fact, this is the very center of the network, providing service control for multiple access networks. To reflect that, it is now referred to in standards documentation as the IM (IP Multimedia) CN Subsystem. However, the short reference of IMS remains the popular name for it.

The real value of IMS emerged when it became clear that this subsystem can provide an integrated control layer for many types of access, combining not only Internet services with communications, but also Mobile and Fixed networks. IMS provides service ubiquity for roaming or nomadic users. This brings opportunities for Visited Networks as well as Home Networks. IMS defines the underlying standards, including standards for security, quality-of-service, and inter-operator accounting.

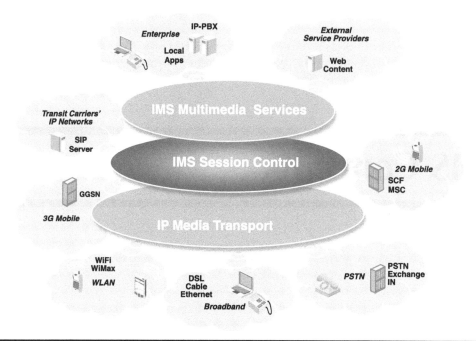

Figure 1.1   IMS connecting to multiple networks.

Figure 1.1 shows IMS with its three main levels: (1) Session Control, (2) Media/Transport Plane, and (3) Services Plane. It also shows how the Core IMS Network can interwork with multiple types of access networks — Mobile IP, Broadband xDSL, Wireless LAN (WLAN), and potentially Broadband IP Cable.

From the outset, IMS was conceived as evolutionary, not revolutionary. It builds on existing methods and protocols; it does not reinvent them. IMS adopts Internet principles, not just protocols, and combines them with the all-important mobility in addition to providing security, resilience, quality-of-service, and chargeability.

IMS also defines the standards for interworking with legacy networks, that is, CSNs (Circuit Switched Networks) — PSTN (Public-Switched Telephone Network) and GSM (Global System for Mobile) in particular. Interoperating with other IP network carriers is essential for today's globalization of services; therefore, IMS contains network border control between IP networks, as well as "breakout" to CSNs. For interim migration steps, IMS-based PSTN emulation has been specified, where the conversion of signaling from legacy phones occurs at the access gateways and IMS is used for session control and feature delivery.

1.2.1   The Origins and Evolution of IMS and 3G

IMS was conceived as the new Mobile communications session control over IP transport for 3G (Third-Generation) networks. It was meant as a subsystem within the 3G UMTS (Universal Mobile Telecommunications System).

IMS was first designed to work on upgraded facilities in the GPRS (General Packet Radio Service) servers: the GGSN (Gateway GPRS Support Node) and the SGSN (Serving GPRS Support Node). Therefore GPRS is referred to as the access network for UMTS (3G) as well as 2½G.

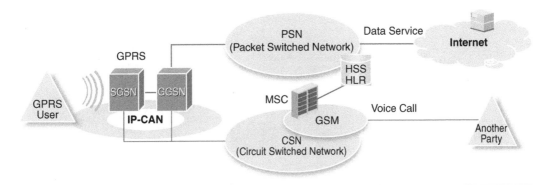

Figure 1.2 GPRS network model.

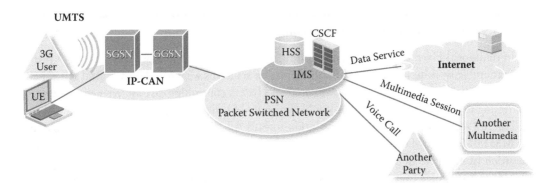

Figure 1.3 IMS UMTS network model.

In GPRS, as Figure 1.2 shows, IP Data services are entirely separate from Voice, and Voice is still delivered on CSN technology.

In Mobile IMS, both voice and multimedia services are delivered over the PSN, thus joining real-time and non-real-time services in a single framework, as shown in Figure 1.3.

UMTS is the full 3G Mobile network that provides integrated Voice and Data over Mobile Broadband. It consists of the air interface, the RAN (Radio Access Network), and the UTRAN (UMTS Terrestrial Radio Access Network). The UTRAN is the fixed network infrastructure that contains the facilities for the transmission to and from the Mobile users over radio, and relay signaling to the Core Network control. The base station that accepts the air signals for UMTS is called Node B, and the control node is the RNC (Radio Network Controller).

The IP-CAN (IP Connectivity Access Network) sits between the RAN and the IP CN, connecting access-side signaling to the service controls in the Core. The service control in GPRS (2.5G) and UMTS (3G), which are partially devolved to the GPRS nodes (GGSN and SGSN), should migrate to the new IMS in due course to fully benefit from the architecture.

IMS plays a crucial role in the SAE and LTE (Long-Term Evolution), where the IP-CAN is independent of Core IP and the Core Service Control. The HSS (Home Subscriber Server) unifies data for GPRS, UMTS, and LTE in terms of user profiles, service authorization, and mobility.

The emerging technologies of WLAN can provide alternative network access for voice that can reutilize IMS. DSL Broadband also makes use of the standard Core but in both cases, further adjustments to the standards are required.

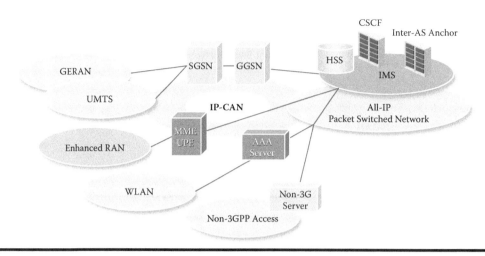

Figure 1.4   Integration of long-term evolution.

In traditional Fixed Line networks, transformation to PSNs is generally known as NGN (Next Generation Networking). Early PSTN transformation solutions were based on a central softswitch, MGCP (Media Gateway Control Protocol) or H.248 based, controlling distributed media gateways, connecting to fixed phones, analog, or ISDN (Integrated Services Digital Network).

Figure 1.4 shows the next step of evolution toward All-IP, with enhanced radio network and the MME (Mobility Management Entity) and the UPE (User Plane Entity) that will enable the transition from GPRS to the new architecture of SAE (Service Architecture Evolution).

Meanwhile, Voice over the Internet is gaining momentum. The Internet introduced users to the ability to log on from any location and access their own applications and settings, so Internet e-mail is available from any point that has access to the Internet. It also introduced Instant Messaging with Presence as a non-voice service, which is an alternative or a complementary service, enhancing the user's experience. Notably, it has established the concept of the user log-in that is equipment independent, unlike Mobile handsets.

For a while it seemed that there were two tracks of evolution toward packet networks: (1) Mobile (3G) and (2) Fixed (NGN, Internet VoIP). The merging point was only at the IP Transport layer. However, the lesson from the Internet meant that Fixed NGN communication now needs to achieve:

■ Service mobility, i.e., availability anywhere, with personalized settings
■ User mobility (i.e., enabling user registration) that is not equipment/access dependent
■ Equipment-agnostic service (i.e., equipment-independent applications on various devices)
■ Combination services, using various types of media

These attributes show that both strands of development are not dissimilar. "Tethered" but portable terminals (laptops) can register from different points of the IP network. They are considered nomadic — connected in different fixed locations but no service while physically on-the-move. Nomadic users need almost all the features that are designed for mobile users, who have continuous service on-the-go. The need to cater for mobility has, therefore, become paramount also for fixed NGN connections.

The advent of WLANs, with WiFi-ready laptops and dual-mode handsets (GSM and WiFi), made it a fast-advancing reality. Divorced from the way a terminal is attached to the network in the first place, given that all types of connections require mobility, it has become clear that session control is the same whatever the network — NGN, WiFi, and 3G. Thus, IMS has emerged as the unifying standard Core network architecture that can address all types of terminals and all types of access methods.

With the exception of the Internet and the Enterprise, IMS is now widely accepted as the reference architecture for all future public networks, wireless or wireline. This general acceptance of IMS is apparent not only in the mobile world, where it is considered as the next step from GPRS (2.5 G), but strong support comes from the fixed network side, where IMS promises to enable users' nomadism and enrich sessions with Web-based new applications.

1.3 IMS Principles

1.3.1 Unifying Layered Architecture

1.3.1.1 Separation of Responsibilities

The most distinctive characteristic of IMS is the layered architecture that separates the responsibilities among the different functions, unlike the circuit network. Normally, three main planes are defined:

1. Access/Transport carrying the media
2. Session Control and Service Enablers
3. Applications, Features, and Services Logic

However, other layers are sometimes added to distinguish, for example, the service orchestration and the data storage and management, which are also independent of the session control. This shows that the exact content of each layer is also not always agreed on.

The independence of the IMS functions is what makes the IMS architecture radically different from the approach of softswitch architecture, let alone circuit networks. It has far-reaching effects on how the whole network operates.

Where features are not embedded in the IMS session control functions, the Session Control layer is not disrupted when new features are added, for example. It also means that session control scaling up is divorced from the processing loading that is highly affected by features, and therefore bottlenecks can be addressed more accurately.

Applications may be closely related to particular types of terminals but can still use the shared IMS infrastructure for network admission, border management, security, QoS (Quality-of-Service), etc. This means that further application and terminal combinations can be introduced with little overhead and risk.

1.3.1.2 Multifacet Convergence

IMS is uniquely placed to unify different networks with a single service and session control infrastructure. This offers great cost savings for operators with both Mobile and Fixed networks, where consolidation can take place and services can be streamlined. This is realized by:

- Combining wireless and wireline breakout infrastructure (Media Gateways)
- Combining wireless and wireline in single session control
- Combining user identity and information (location, identities, single sign-in)
- Same applications on different handsets or clients, with a consistent user interface
- Sharing resources (HSS, Presence, media servers)

In most cases, the Fixed/Mobile Convergence (FMC) can bring far-reaching benefits and cost savings in equipment consolidation, lower maintenance costs, and common services implemented for all types of networks. It is the changing attitudes toward user requirements that drive this convergence. Instead of a set of network-centric services from which the user can choose, a service provider will offer a user-centric service package that combines multiple terminals and access methods and multiple applications. In this way, the services can be better integrated and closely tuned to the user's wishes.

IMS versatility goes beyond voice and multimedia. The same procedures and session control can be applied to real-time or non-real-time (data) services, whether they are interactive or user-server types of services. This enables services to integrate such sessions and enrich the user's experience.

1.3.1.3 Multiple Access

Although conceived as an evolution of GSM into packet network, IMS has now evolved into a core control system that is independent of the means of access into the network. Interfaces to three main Access Networks have been defined in the standards and are described here: 3G UMTS, WLAN, and Broadband DSL. Ongoing standards work continues to define further means of access, such as Cable, Metro Ethernet, and WiMAX.

The modular design of the 3GPP (3rd Generation Partnership Project) defined the IMS as a subsystem, positioned above the UMTS access network. When IMS was adopted by wireline networks, TISPAN (Telecommunications and Internet Services and Protocols for Advanced Networks) defined in more generic terms the Core IMS that is shared by all access networks. They also defined access-specific facilities needed for wireline access (e.g., the network attachment process).

Figure 1.5 shows the IMS three-layer model with all the interconnecting networks. These include the different IP-CANs through which the IMS terminal can gain access. Also included is the application-level access via Web Services or OSA (Open Service Architecture) gateways, which can also initiate sessions within IMS. In addition, Figure 1.5 shows connectivity to other IP networks and to the CSNs.

1.3.2 Openness and Interoperability

1.3.2.1 The Service Environment

The service environment is getting very complex, with old and new services still operating side by side. Further complexity arises from the sought-after combination of services — Fixed, Mobile,

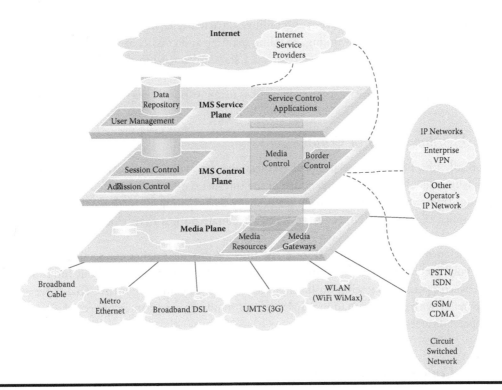

Figure 1.5   Multi-access IMS.

and Web. The cost of running a silo-style structure of services will mount up quickly unless the IMS approach is adopted.

IMS is designed with services in mind. It provides components and service enablers that can be shared and accessed via a standard format. For example:

- User identification and authentication
- Service profile data with feature-triggering filter criteria
- User location, presence, and dynamic status
- Service-based quality-of-service and policies

Having achieved that, a more flexible environment becomes available for multiple services to be trialed and deployed, quickly and cost effectively.

1.3.2.2   Open and Evolving Architecture

IMS is an open architecture. This means that it is built on published and agreed-upon standards between the network elements. Therefore it is easier to deploy products from different vendors who comply with these standards. IMS provides:

- Separate layers allowing for mix-and-match vendor equipment
- Clear separation of functions within these layers
- Use of open protocols, particularly SIP (Session Initiation Protocol)

- Open protocols to management and billing, particularly Diameter
- Full inter-operator interworking, enabling roaming and cross-charging
- Interworking with application servers on a standard open interface (ISC; IP-Multimedia Service Control)
- Enabling external entities to provide services using "exposed" network resources

1.3.2.3 Inter-Operator Connectivity

Historically, the PSTN achieved global connectivity gradually. Similarly, the Internet became universally available over time. The success of the Internet is often attributed to the ability to support huge traffic volumes moving freely across many participating network providers. Given such expectations, the new 3G/NGN network must facilitate no less than global connectivity from Day One.

While currently services of VoIP over the Internet provide neither connection between different VoIP providers nor any roaming capabilities, IMS sets out to bring about full interworking between operators, crossing business borders as well as country borders. To achieve service transparency with seamless roaming, IMS defines standards for IP interconnection, in terms of interoperability, security, and charging.

1.3.3 Concepts of Mobility

1.3.3.1 Mobility Definitions

Mobility is taken for granted when considering IMS as an evolutionary step for Mobile networks, but it is less obvious for wireline. The interaction between the Visited access network, intermediate networks, and the Home network should be well defined to ensure full interworking.

There are many ways of defining "mobility" because there are many aspects and modes for mobility. The ITU-T (International Telecommunication Union – Telecommunication) defines it as "the ability for the user, or other mobile entities, to communicate and access services irrespective of changes of the location or technical environment." See mobility definitions in Appendix E.

The concept of *Full Mobility* includes the seamless continuity of the session, even when crossing network borders. This means that a process of handover is taking place, not just between radio cells, but also between business entities.

Nomadism is mobility associated with devices that are "tethered" or wired when the session is connected, but this can be a wireless connection over a WLAN. This mobility notion is blurred even more when considering laptops connecting to Mobile 3G networks over a radio interface to gain access to e-mail. Therefore, nomadism is described as "Mobility with Service Discontinuity." The nomadic service does not accommodate a continuous service across different attachment nodes, such as Mobile cells. The nomadic user exits completely the previous service and logs on in a new network location, starting a session afresh, unlike the Mobile user who continues to receive connectivity service while on the move.

Terminal mobility refers to the ability of the equipment to operate when it is not tethered, and when it is able to discover a local network to which to connect. Terminal mobility is independent of the user and is concerned with the identification of the hardware or firmware rather than the user.

1.3.3.2   IMS User Mobility

A prime principle in IMS is *User Mobility*. As in wireless networks, users can appear in different spots on the IP network. The network must associate the IMS user profile with the current equipment being used by that user, and attach the user's equipment through authentication and authorization processes that assign an IP address and bind it to a user account.

The ability to receive service while roaming is taken for granted for Mobile handsets and, by extension, for terminals in NGN. Users of PCs with soft clients are deemed "fixed users" but they can log on from anywhere — that is, from any PC that provides the appropriate network access.

Moreover, several users can use the same PC to log on to the network. The fundamental difference from the Mobile handset, or from the circuit-switched line, lies in the fact that the User ID is independent of the equipment ID. Both the User ID and the equipment ID now must be recognized and managed, as well as dynamically or permanently associated. When laptops connect via WLAN, WiFi, or WiMAX, they are temporarily tethered to a port and therefore deemed to be in a Fixed network, while Mobile handset signals connect via cells that hand over control in mid-call.

Through IMS, both mobile and nomadic users can have their services made available when they log in from an external network. The mobility principle of Visited Network and Home Network is retained in IMS. IMS endpoints find a local proxy server via DHCP (Dynamic Host Configuration Protocol), as PCs do in a LAN (Local Area Network), or have such information preconfigured. This proxy is the first IMS node accessed by a terminal. The proxy is responsible for discovering the serving session controller that will manage the session. However, if the user is not within the Home Network, the proxy will ascertain the route to that network server, usually via a known border node in the user's Home Network.

1.3.3.3   IMS Service Mobility

Service mobility means that the same networked services, with their particular personalization and stored data, are available on alternative devices and access methods. Service mobility also means receiving services when roaming to another network.

Previously, service mobility has been limited. Traditional Fixed-line services are associated only with a particular circuit line. 2G Mobile phones also have services associated directly with the SIM card, not the user, and therefore offer no service mobility across devices. Any stored data kept locally on the equipment is not available elsewhere (e.g., personal address book).

Many services might be very useful when they become available in a variety of places and terminals. For example, a Pre-pay service can apply to 2G Mobile, IMS Mobile, laptop VoIP, and even residential lines, with a single credit account. Service interfaces and presentations can vary according to the terminal capabilities. For example, the user can be alerted to low credit by an audible tone, announcement, or by a message on the screen. However, the service functions operate in the same way for all terminals.

IMS provides tools for service mobility. The HSS enables user ID management that can support service delivery to multiple types of devices. The session controls and routing capabilities make it much easier to connect to the Home Network services. A unified session controller that can sustain a unified service supports the different access networks.

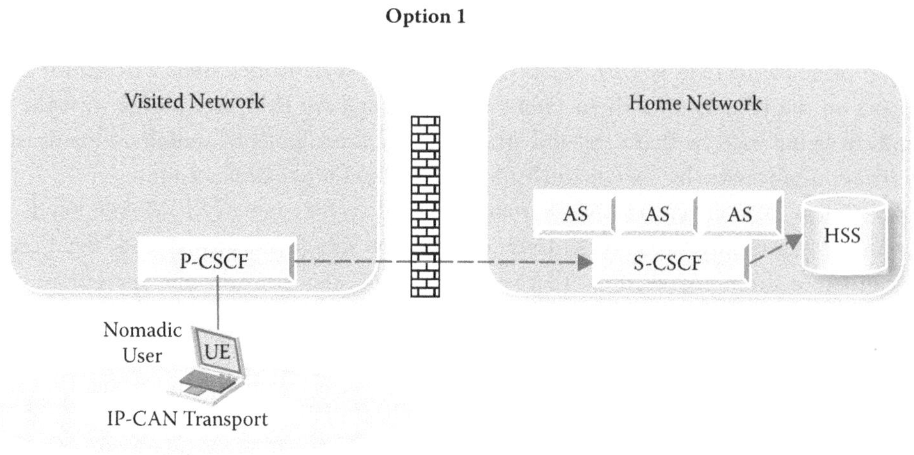

Option 1

Figure 1.6 Home network provides all services to nomadic user.

1.3.3.4 Delivering Services to Roaming or Nomadic Users

Service delivery to a nomadic user can be delivered via three options (1) home network, (2) visited network, or (3) as a combination of both.

(1) Normally when a user terminal is granted admission to the Visited Network, it can connect to the Home Network and perform an IMS registration. The user profile in the Home Network indicates which services should be activated and in what order of precedence. These services are performed remotely by the application servers in the Home Network, using data stored on the Home HSS or the Home Application Server. Session Control in the Visited Network passes control to the Home Network, where the normal services are activated remotely by the Home Application Server. Option 1 is shown in Figure 1.6.

(2) However, there are services that can be provided to all users regardless of their subscriptions. These services do not require any special settings or prestored data. These services may be offered to a Visited user from the Local network. In particular, Location-Based Services that detect the user's current physical location can provide services and information only available within the Visited Network, relevant to the locality. These services are therefore provided without the need to connect to the Home Network. Access to them may require only local authorization. As the users are not fully identified, they cannot be easily charged (unless they are on a Pre-pay service). However, local services can be based on advertising, calling party charging, or users' ability to pay in another way. With minimal local authentication, users are allowed to use Visited Network services that operate from a Local Application Server (AS). Figure 1.7 illustrates Option 2.

(3) A third option is a service that involves a local network AS. This AS may be contacting another AS in the Home Network to complete the service. The user's application data can be stored on the Home AS, and is reached via service interaction between the two ASs. This is typical of Distributed Messaging service and Internet-style mash-up services. A local AS is invoked but it connects to another AS in the Home Network. This option allows for regional variations while charging is still performed by the Home Network. In this scenario, there must be a close relationship between the operators and a standard interface between the applications, such as found in SMS messaging services. Figure 1.8 depicts Option 3.

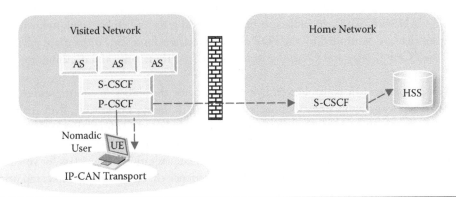

Figure 1.7   Visited network provides some selected services to visiting users.

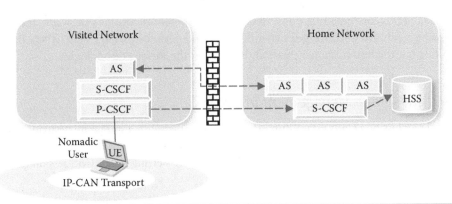

Figure 1.8   Cooperation of Visited network AS (proxy AS) and Home network AS.

1.3.4   Resourcing and Quality-of-Service (QoS)

1.3.4.1   Resource and Policy Administration

QoS control depends largely on the allocation of adequate network resources. As long as the IP network is over-provisioned with enough spare capacity, resources need not be monitored and controlled to deliver good service quality. However, with growing network congestion, and with the launching of many resource-hungry services such as Video calling IPTV (Internet Protocol Television), and content delivery, QoS cannot be assured without additional means.

Resources should be managed in different parts of the network — at the IP-CAN, the backbone transport network, and the Core network. Furthermore, full end-to-end QoS should allow negotiation of QoS across business borders between carriers. Even before all networks have this ability and can interwork correctly, QoS levels can be improved within each network and provide incremental improvements.

When a media flow is requested from an IP-CAN Bearer, it can be processed with SBLP (Service-Based Local Policy) or without it, i.e., by service-level control or from the transport network entirely. When using IP-CAN without SBLP, the Bearer is established according to the user's subscription, IP Bearer resources, admission control parameters, and roaming agreements. When using IP-CAN with SBLP, decisions are made according to service-level requirements and network rules that are controlled by the PDF (Policy Decision Function) or its more advanced successor, the PCRF (Policy and Charging Rules Function). These rules and decisions are applied to the Bearer.

As an access-agnostic control layer, IMS must provide the ability to control resources for sessions of all types and apply network policies according to the type of service. IMS enables operators to define a variety of network policies and service policies, and enforce these rules in real-time communications. Each request for session initiation involves the PDF that approves and reserves appropriate network resources. The services are classified, assigning a high class for voice and low class for text. Policies can also identify a level of precedence that gives, for example, top priority to an emergency call or a premium customer.

In 3GPP networks, the PDF is involved in session establishment for GPRS media flow within a PDP (Packet Data Protocol) Context. TISPAN defines a service-based policy decision function SPDF (Service Policy Decision Function) within the RACS (Resource Admission Control Subsystem) to control media flows in the network transport nodes.

The standards for these functions, despite their different definitions in 3GPP and TISPAN, can now be combined in converged network policy and resource management, where the controls are unified but the bearer implementation depends on the various technologies.

1.3.4.2 Resource Management in UMTS

Initiation of a session entails establishing the IP media flows with appropriate network resources as required for the session. The network resources are allocated as a result of negotiation of QoS with the network nodes (GGSN) and the UE (user equipment). Setting up an IMS session involves two levels of QoS negotiation and resource allocation:

1. *At the UMTS Bearer Service (BS) level.* UMTS-specific signaling, such as RAB (Radio Access Bearer) QoS negotiation and PDP Context setup, are performed at the UMTS BS level. When UMTS QoS negotiation mechanisms are used to negotiate end-to-end QoS, the *Translation Function* in the GGSN coordinates resource allocation between the UMTS BS Manager and the IP BS Manager.
2. *At the IP BS level.* This is the process that conveys QoS requirements at the IMS application level to the UMTS BS and performs any required end-to-end QoS signaling by interworking with the external network. This process is performed between the IP BS Manager at the UE and the GGSN. QoS signaling mechanisms such as RSVP (Resource ReSerVation Protocol) are only used at the IP BS level.

1.3.4.3 Resource Management in NGN

The scope of the Resource Admission Control Subsystem (RACS) has been defined by TISPAN. It defines the standards for requesting resources, approving and reserving them. Once agreed-on resources and charging rules, the subsequent flow of media is governed by network nodes that perform the Policy Enforcement Function (PEF). These functions are integrated into various network bearer nodes and access network nodes.

The SPDF is a logical policy decision element for service-based policy control. The SPDF makes policy decisions using policy rules defined by the network operator. The AF (Application Function), acting as a user agent, requests resources for an IMS session from the SPDF in the RACS. The SPDF chooses the local policy to be applied to the request based on the Requestor Name, Service Class, Service Priority, Reservation Class, and more. The SPDF maps the local policy into the parameters to be sent to the A-RACF (Access Resource and Admission Function) or the BGF (Border Gateway Function).

Chapter 6 provides additional details on QoS and resourcing.

1.3.5   IMS Security

1.3.5.1   Security Requirements

Addressing security in IMS networks is a critical factor for their success. The IP connection, compared with traditional circuit-based networks, is far more vulnerable to software attacks. The main control protocol in IMS is SIP, which was conceived as a simple peer-to-peer protocol. Its enormous success is due to its flexibile management of multimedia sessions. At the same time, SIP is also open to abuse, i.e., using it to bypass any network controls and avoid operator's charging.

3G Mobile is assumed to be reasonably protected because the radio link provides a level of protection, as the usage of the air spectrum is restricted and the interface is encrypted. That is not the case for Fixed IP NGN. It is more akin to Internet services, and therefore can be subject to the same security problems that plague Internet data services (such as e-mail and other laptop applications), i.e., viruses, worms, spam, spoofing, spy software, and so on.

Security measures should be deployed at different points in the network to provide a range of security aspects, including:

- ■ User authentication
- ■ Network-to-network authentication
- ■ Service authorization, according to subscriptions, roaming agreements, etc.
- ■ Charging policies and enforcement, preventing fraud, ensuring accurate billing
- ■ Resourcing policy enforcement, protecting against DoS or misuse of resources
- ■ Security key management for reliable identification
- ■ Data confidentiality and user privacy, protecting network data as well as the user's data
- ■ Data integrity, preventing database corruption or "poisoning"

1.3.5.2   Security within IMS Functions

The various security functions are built into the relevant IMS functions. Security is not achieved by a single box but rather through measures taken in various parts of the network and at various points of interaction, both in the NNI (Network-to-Network Interface) and the UNI (User-to-Network Interface).

The following table lists functions within IP communication that need protecting from various risks. It lists the type of security relevant for the function, along with examples of methods that can be applied to mitigate these risks.

Communication Functions	Security Requirements	Use of Methods and Security Functions
Network admission	Device authentication	IKE, ESP
Registration	User authentication, registration and re-registration	AKA, IKE, ESP, MAC, Digest, Anti-Replay (Call sequence, masking, etc.), Random challenge
NNI and UNI authentication	Mutual authentication	MAC, AUTN
Session initiation	Service authorization	IKE, ESP, SIP screening
Invoking AS	AS registration, third-party registration	HTTPS, ALG
Media payload	Media screening, packet filtering, tunneling	SA/IPSec, ALG, TLS
Charging	Charging rules enforcement	COPS/Diameter
Authorization of resources	QoS profiles, end-to-end QoS protocol	Encryption, IPSec
QoS enforcement	Traffic shaping (anti-DoS)	Secure COPS/Diameter
Inter-operator routing	NNI mutual authentication Border Gateway filtering	IKE, ESP, TLS SIP Screening, IPSec
Internal network elements and servers	Topology hiding, generic addressing	NAT/PAT, DNS queries for FQDN resolution
Service provider's access	Application authentication Filter/authorize operations	AKA/HTTPS Digest SIP screening, Applications gateway
User provisioning	Administrative privileges, data privacy	Encryption

Source: Compiled from multiple sources.

Chapter 4 provides further details about security at call admission time, and Chapter 7 describes security for the IP border.

1.3.6   Charging Facilities

1.3.6.1   Chargeability

IMS serves networks that can ensure a level of service quality that is predictable and chargeable. This is in contrast to Voice over the Internet. The Internet-style connection is peer-to-peer, where the media payload travels between the connected parties without any intervening servers. This makes it difficult to monitor and charge per usage.

IMS is being developed for privately managed networks, to deliver assured service control and resilience, which is controlled by media-bearing servers that obey network policies. In IMS, the routing is managed and the media flow is controlled by servers that check packet headers, filter unauthorized packets, and prioritize packets according to their class of service.

IMS is designed to provide the operator with:

- Chargeable services with information correlation
- Chargeability, online and offline, including events
- Control charging policies per user, service type, and session type

The charging mechanism has been extended to accommodate multimedia sessions, Mobile and Fixed data sessions, and WLAN connections, with additional types of contributing networks to follow. To deal with numerous streams of data from disparate sources, a charging gateway has been designed to funnel the information to existing billing systems.

1.3.6.2 Charging Rules

IMS includes charging mechanisms for online charging as well as offline charging. This involves IMS functions initiating, modifying, and terminating the recording of details on the Charging Data server, where each record identifies the element that produced it, but also the correlation of the session. Because IMS manages a wide range of services and connection types, more complex charging rules must be applied. For this reason, network-wide policies are defined for charging methods and procedures in the Charging Rules Function (CRF).

Following the convergence trends, similar functions are now merged in the PCRF. The PCRF allows the operator to apply such rules dynamically for each session, depending on a list of criteria applicable to a particular connection request.

Chapter 8 provides additional details about IMS charging.

1.4 IMS Services Environment

1.4.1 Service Environment Principles

More than anything, IMS is a multimedia service environment that improves service agility and flexibility, and reduces the time and costs of service introduction. By separating application logic from service enablers and session control, new services can be prototyped, trialed, and deployed much faster.

The IP Multimedia service delivery environment requires:

- A flexible, clearly defined architecture that can support any end-user device and network server, and encourages trying out innovative services
- An environment that allows fast deployment of IP multimedia applications across several types of networks, including the Internet, and can implement frequent changes cost effectively
- A set of network functions that have proven standard interfaces between them and toward the service layer, so that products from different vendors can interoperate easily
- Centrally managed data repository that can be shared across applications on separate servers (possibly from different vendors) and is available to the operator's internal systems, such as CRM (Customer Relationship Management) and other BSS (Business Support Systems)

IMS inherent principles provide for an ideal service environment that separates entirely session control from services, thus allowing greater service flexibility. Storing central common IMS data in the Home Subscriber Server/User Profile Server means that crucial user data can be easily shared across multiple application servers. Even dynamic information, such as the user's current registration and physical location, can be divulged to event-subscribing applications to create more advanced user-centric services.

IMS provides standard interfaces that enable interworking at both the network connectivity level and the service level. Defined interworking with legacy CAMEL (Customized Applications for Mobile network Enhanced Logic) Intelligent Networks and PSTN services, and interfaces to OSA for external services, ensure that IMS services can benefit from existing legacy systems. In addition, the ability for a variety of charging models enhances service capability.

1.4.2 Applications Environment

The primary purpose of IMS is to enable rapid service creation and deployment, where IMS provides a standard framework that supports a myriad of service enablers. IP multimedia applications are not standardized, as a matter of principle. This is intended to allow for operator differentiation because the service package is what attracts subscribers. Although the AS structure is not imposed, all the interfaces into IMS components must comply with the standards, so that any vendor's server can be installed into an existing service environment.

IMS can contribute to the application environment via the following functions:

- The mechanism for maintaining session state used in applications, detection of unreachable users that can trigger special services, and providing control mechanisms for session redirection or special routing
- Generic user ID management with extensive profile structure that can cover all aspects of user services
- Roaming facilities with exchange of information between Home Networks and Visited Networks, permitting roaming and nomadic users to access their services
- The security architecture that allows authentication of the user before the user is allowed to use network capabilities
- The mechanism for differential charging based on content (flow based and service based)
- Cross-charging between Home and Visited Network operators
- The mechanism for the network policy control to handle appropriately the traffic according to the user's contract and the application needs
- The mechanism for provision of services by third parties, which are required by the regulatory bodies in certain geographic regions
- Flexible media session negotiation and establishment, for voice, video, and combinations

Chapter 3 describes the management of application sessions by IMS.

1.5 Combining Technologies in IMS

1.5.1 Wireless IMS

Because IMS origins derive from the evolution of GSM, it is not surprising that it retained the concept of roaming into a Visited Network, using the VLR (Visitor Location Register) and HLR (Home Location Register) to refer back to the Home Network. The equivalent concept in wireline (Fixed line) is the Virtual Home Environment (VHE). Under VHE, the user's services and

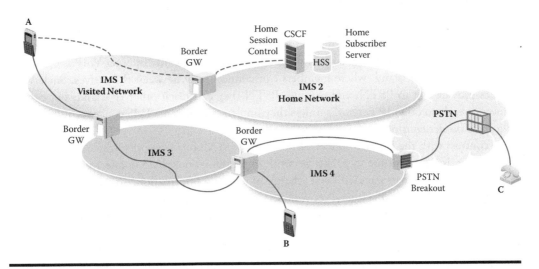

Figure 1.9   IMS inter-network connectivity.

applications are made available, regardless of what network is carrying the signals. Implementing this requires standard interfaces that allow signaling to cross over business boundaries toward the Home Network, much like the 3G Mobile network. Hence, only one converged standard (IMS) has gone forward.

Figure 1.9 shows Caller A in a Visited Network IMS 1. By agreement, Caller A is recognized as a subscriber of IMS 2, therefore, the signaling is forwarded to the Home Network for authentication. Connecting to Party B may require routing the media from the Visited Network IMS 1, via an intermediate network controlled by IMS 3, to the network where Party B is currently able to accept calls. Party B may be visiting IMS 4; therefore, the connecting call leg to Party B is performed in B's Visited Network.

Connecting Caller A to a destination C in PSTN follows a similar process of identifying User A in the Home Network, IMS 2, and connecting the media to PSTN through a breakout gateway. The media flows over IP intermediate network as long as possible to optimize costs, and emerges via the gateway nearest the relevant CS network.

1.5.2   Wireline and IMS

Wireline and transformed NGN networks are characterized by "tethered" terminals and nomadic users. Moreover, they assume that users may be using multiple terminals, unlike Mobile handsets that are closely associated with their owners. IMS caters for both types, for Mobile wireless and for nomadic wireline terminals. Moreover, IMS promotes the concept of user-centric service, where the user profile is kept centrally for multiple terminals in multiple domains. This is enabled by a structure of Public IDs (published known names or numbers) and their relationships with the Private IDs (network-specific identifiers).

User IDs can still be based on a packet network style of addressing, i.e., SIP URI, but a number, such as the E.164 Standard numbering scheme or Mobile International format, can still be used in certain circumstances. Therefore, the User ID authentication in IMS does not depend on the type of technology.

1.5.3 The Internet and IMS

The Internet is said to be open to all traffic. With the use of SIP (Session Initiation Protocol), it can facilitate user-to-user direct connectivity. However, to find each other, there must be an interworking agreement as well as gateway control via border servers. Typically, Internet VoIP is operating as a closed group, an island, where members can connect only between themselves. Such VoIP facilities provide exit to PSTN, and through PSTN they can reach other IP islands but cannot connect directly to other IP networks unless they implement the same IMS concepts of interoperability and border control.

In contrast with the Internet model, IMS is similar to Mobile networks, where cross-operator interworking is a prerequisite to packet network implementations. IMS connectivity is provided via its IP-to-IP border gateways with an exchange of mobility information. To achieve inter-operator routing and delivery of Home-Network services to visiting users, a great number of functions must be in place, such as addressing conventions and translations, inter-operator accounting, charging rules, etc.

From the Internet viewpoint, IMS is perceived as hierarchical and restrictive, in complete contrast to the Internet style of direct peering. The Internet is open for anyone to add a node, while IMS is a mesh of privately owned networks. Internet service providers can operate with virtually no network at all, while IMS operators are assumed to manage their own networks and take responsibility end-to-end. Internet services are offered "as is," while IMS operators promise guarantees of service level and compliance with regulations.

Despite the difference, it can be observed that many IMS concepts are built on Internet attributes:

- It supports distributed and federated servers and data.
- It is tolerant and uses flexible and extensible protocols (SIP, Diameter, XML) that are also used by the Internet.
- It allows adding servers and applications with relatively simple interfaces.

A major difference between Voice over the Internet and IMS lies in the treatment of the media path. Unlike the Internet, IMS media flows are inspected and managed by predetermined network policies, for both security and quality reasons. Media flows are therefore not transferred peer-to-peer, but they pass through media bearing nodes that filter packets according to operator policies.

1.5.4 IT and ICT

IMS is recognized as a unifying framework not only by wireline and wireless, but also by IT, or better phrased — ICT (Information and Communication Technologies). This is due to the fact that IMS is not concerned primarily with network connectivity and access mode, but with diverse service delivery.

Data services, such as e-mail and Instant Messaging, are an alternative means of communication. Their integration with IMS services is one of the major benefits delivered by IMS. This is apparent, in particular, in the ability to dial from a company directory, showing Presence, or a private address book, where the same directories are also used for e-mail. Another example is the use of a diary to initiate events such as schedule conferences and pre-call alarms. Workflow-type applications used in the Enterprise can now include interactive sessions triggered by an IT event.

Furthermore, OSS (Operational Support Systems) and BSS (Business Support Systems) are becoming essential ingredients of the service itself. The IMS HSS is not merely a database of user details, but also a central element in routing calls and service selection.

As the Services layer is separated from the Session Control, an application can interface into the network more easily than ever before. IT and Internet developers can now produce features without having to master the intricacies of network switching.

1.6 The Role of the Standards in IMS

1.6.1 Standards for Openness and Interworking

Interoperability is essential for wide adoption of new technology, especially in global communications. To achieve consistent implementation of a new technology, well-defined standards must be defined, along with universally available inter-vendor testing facilities. IMS is standards based, but the standards must be widely accepted and uniformly implemented. Having published clear and detailed standards, competing vendors can interwork with some ease, and ecosystems of collaborating vendors can be developed.

Chapter 2 provides more details about the IMS standards.

1.6.2 Global Collaboration of Standards Bodies

Historically, standards took shape within geographic territories. PSTN achieved global connectivity over SS7 (Signaling System No. 7), but each country had its own flavor, often dictated by a country's authority, local regulations, or the incumbent operator.

In addition, geographical blocs supported different standards organizations (e.g., ETSI in Europe and ANSI in North America). ANSI and ETSI developed different specifications for what essentially must be the same requirements. In addition, regulations in each country contributed to the creation of variations of SS7 signaling, hindering global connectivity. This became even more of a problem when Mobile handsets roamed into incompatible geographies, creating user demand for ubiquitous services.

In the case of IMS, existing organizations have rallied together to supplement the 3GPP effort rather than duplicate it. 3GPP is cooperating closely with the IETF (Internet Engineering Task Force) on extending Internet protocols, in particular extensions of SIP and Diameter. Other organizations develop extensions to the IMS architecture, for example, the OMA (Open Mobile Alliance) is defining service interworking standards, such as PoC (Push-to-Talk over Cellular). Most notably, TISPAN (ETSI) is producing wireline interfaces to incorporate into 3GPP.

Figure 1.10 details the various standards organizations that are collaborating in the specifications of IMS. In North America, the 3GPP2 group is merging the North American WCDMA (Wideband CDMA) standards with the GSM-led 3GPP, and the ATIS (Alliance for Telecommunications Industry Solutions) is contributing to interworking of wireline there. Other organizations and workgroups, such as the GSM Association (GSMA), various ITU (International Telecommunication Union) groups, and the W3C (World Wide Web Consortium), also are contributing by addressing certain issues but not duplicating existing work.

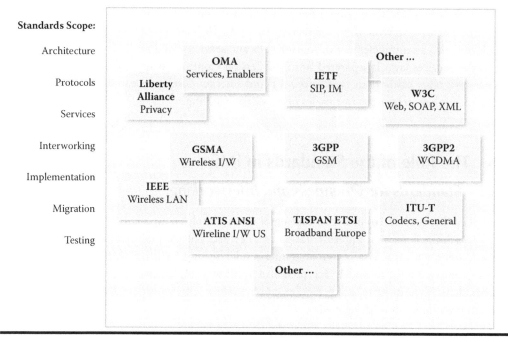

Figure 1.10 Collaborating standards organizations.

Work from the different organizations is merged through a process of obtaining mutual endorsement, which is documented and published as part of their standards documentation. The endorsement documents list what has been accepted and any exceptions that are not the same and need slight modifications.

1.6.3 The Standards Protocols

IMS is, in fact, a set of standard protocols and specified reference points between defined functions. In the face of expanding requirements for further sophistication and enhanced capabilities, the Standards bodies are determined to keep the number of protocols low and the protocol specifications simple. At the same time, they must attempt to produce accurate enough definitions that will lessen ambiguity and improve the chance that different vendors will be compatible.

The predominant protocol for session control is SIP in collaboration with SDP (Service Description Protocol), which provides a wide range of headers and parameters to enable complex exchange of information. Equally important for IMS is Diameter with its extensible range of AVPs (Attribute Value Pairs) that are used for interrogating the network's central data storage and exchanging charging information.

H.248/Megaco (Media Gateway Control protocol) is the protocol used to control media gateways which has evolved from the MGCP (Media Gateway Control Protocol). It is a master–slave type of protocol, as used in circuit networks. It is used by the IMS Media Gateway Controller to manage media gateways for media connections, playing tones, playing announcements, and other TDM-compatible media operations.

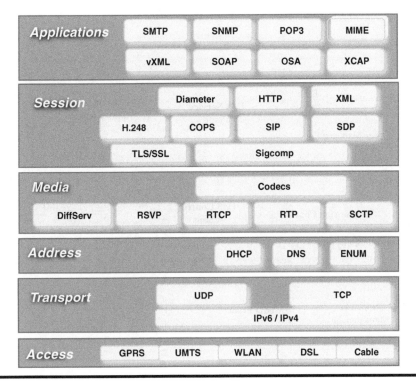

Figure 1.11   The IMS main protocols.

Recently, Diameter is preferred instead of COPS, to negotiate Qos according to network policies and charging rules. The policy descisions are enforced by protocols such as DiffServ and RSVP at the bearer layers.

The Internet protocols HTTP (HyperText Transfer Protocol) and XML (eXtensible Markup Language) are still used in applications for session-related functions, for direct interaction between applications and the endpoint client, where complex service data and formats must be transferred.

Figure 1.11 depicts the main protocols that play a part in IMS-based network, along with their relative position in the layers of intelligence. As IMS essentially defines the Core Network session control, the significant IMS protocols are found at the Session layer, where SIP and Diameter are the predominant protocols. IMS also mandates media handling by selecting TCP and SCTP for media flow control.

Chapter 11 describes some of the IMS protocols.

1.7   Introduction to the IMS Architecture

1.7.1   The IMS Layered Model

The IMS architecture is based on a layered model. These layers can be seen in relation to network layers models. When discussing layers in a network, there can be more than one definition of these layers.

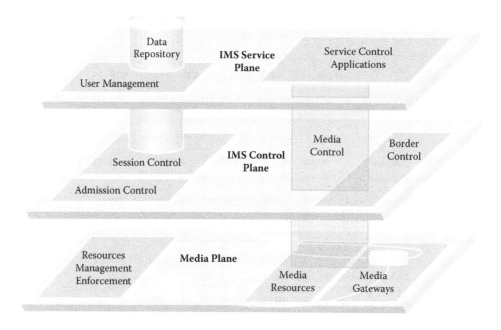

Figure 1.12 IMS three-layer model.

Figure 1.12 depicts the IMS three-layered model, with the Control layer in between the Service layer and the Media layer. The Media control is shown here as a vertical plane linking the applications with the Media Plane. This reflects the fact that media (text, image, and video, as well as voice) is often defined and controlled by the applications, and media formats, content, and other requirements can be conveyed to the endpoint terminals directly, rather than through the session control nodes.

When discussing the structure of IMS and the role of certain protocols, the ISO layered model falls short of clarifying how elements interconnect to each other for all types of links. In particular, what is missing is the segregation of the media bearing elements from those that control the session (that is, the signaling). The IMS control layer as well as the IMS services are regarded as the Application layer for OSI, while the Media Plane (or the "User Plane") encompass the rest.

IMS distinguishes between the Media layer, the Control layer and the Services layer to allow for rapid development of new applications and services on the one hand, and linking them to the ever-increasing types of access and media technologies on the other hand. The decoupling between these layers is therefore a cornerstone design principle and a marked deviation from traditional voice networks. This layered model also defines clear interfaces between Access Nodes and Control Servers, with admission control and authentication procedures that can be adjusted to suit differences in terminals and access methods.

1.7.1.1 The Session Control

The heart of IMS is in its Control Plane, containing the control of the session initiation and flow and controls of ingress and egress of the network, as well as the control of the media flow. The Control Plane components include:

- *Session Control:* initiation, routing, tear-down, and also registration.
- *Admission Control:* authentication, allocation of bandwidth, control priorities.
- *Border Control:* selection of breakout point, interworking for IP connections.
- *Media Control:* session media connection, Media Server, media gateway to CS networks.

Data Management can be considered a sublayer that contains User ID Management and User Mobility Management, including data storage for IMS. The position of the HSS in the Services layer is sometimes disputed. The HSS is a data repository for the network, containing user profiles and session data, supporting admissions, registering, and locating the active session controller for the subscriber. As such, it is regarded as part of the Session Control. However, the HSS also contains subscription details, service data, and service selection filters, and therefore could be seen as an element of the Service layer.

1.7.1.2 The Service Plane

Service logic and applications are actually not considered part of IMS, although certain features can exert control over the sessions. The Service Plane is defined in IMS by the interfaces to the AS, by triggering and orchestrating services, and by the interaction with the Data Control elements, the HSS (Home Subscriber Server), and its equivalent in NGN, the UPSF (User Profile Server Function). The Service Plane provides three types of IMS interfaces to application servers: (1) the SIP application server, (2) the IN CAMEL server, and (3) the Parlay/OSA server.

The separation of features from Session Control must be established cleanly to realize the benefits of this architecture. Since all features are served from the AS and are *not* embedded in the CSCF (Call Session Control Function), the network control is not impacted when features are changed, as it had been in TDM and in softswitches. This means that new features can be introduced much faster, at lower cost, and without network disruption. For this reason it is important to resist the temptation to build session features into the session controller server, despite some advantages at the start.

1.7.1.3 The Media Plane

The lower layer is the Media Plane, or the User Plane, representing the Transport layer. This plane handles the media separately from the signaling, using different protocols to control the flow. Here, media is carried over, manipulated and converted (CS to PS), monitored and inspected when crossing over the border.

The following are the main functions performed within the Media Plane:

- The Media Gateway performs media conversion.
- The Media Server provides media processing, i.e., storing and playing media content (announcements, music), bridging (conferencing), and transformation between media types.

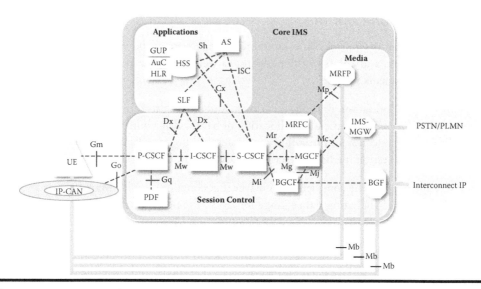

Figure 1.13   Core IMS in 3GPP.

■ Border media bearing nodes perform media filtering and deep-packet inspection functions.

1.7.2   The 3GPP Approach

IMS was designed as the subsystem that enables delivery of multimedia services to Mobile 3G, and was based on Mobile phone access. It was originally concerned primarily with connectivity over packet, with User ID and location management and with roaming capability. The architecture is focused on the ability to roam and making user data available to applications — hence the careful definition of the session controller (CSCF) and the HSS.

The strength of this architecture lies in its inherent distributed nature, and in the common central data. These attributes make it easily scalable and well suited for global networks as well as small overlay implementations.

Figure 1.13 shows the main elements in Core IMS as described by the 3GPP. The Session Control elements detail the roles within the CSCF and the breakout to PSTN via the MGCF that controls the media gateways. The Data Management (HSS/SLF) is shown together with the links to the AS.

Naturally, the main concern at the early stage was connectivity to Circuit Switched Networks, via the MGCF controlling the Media Gateways. The procedures for network accessing are relatively simple, relying on the Mobile handset firmware to provide identification, and on airway mechanisms to provide a high level of security. This architecture therefore lacks details of generic network admission and strict border patrol functions.

The original, architecture grew more complicated as the scope of IMS expanded to cater for additional requirements of wireline access and WLAN technologies. Concerns over security and quality of service also have dictated more elaborate solutions. In addition, some compromises were necessary to allow for early implementation with gradual network migration. In particular, the assertion that IMS will only support IPv6 addressing had to be relaxed. The initial assumption that IMS will force through a global adoption of IPv6 gave way to the realization that IPv4

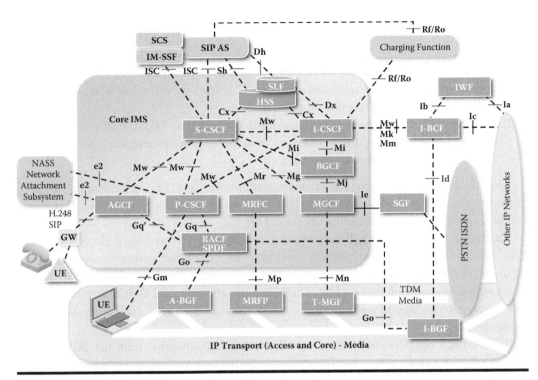

Figure 1.14   TISPAN definition of Core IMS.

will remain in use for some years. Therefore, IPv4 interworking with IPv6 had to be worked out through Border Gateways and Network Address Translation (NAT).

Chapter 2 provides further details of the 3GPP standards.

1.7.3   The TISPAN Approach

TISPAN is an ETSI group that evolved from the Fixed network standards (TIPHON) to define IP standards for NGN (Next Generation Networking) and migration from Fixed Line networks. TISPAN work drove forward the idea that Mobile and Fixed connections can be converged not only with common transport, but also with common core components.

IP-based wireline terminals include not only IP phones, but also PCs, laptops, pocket PCs, and PDAs. Laptops and PDAs are not stationery but can be attached at different points of the IP networks and behave more like Mobile devices than like a fully wired connection. Therefore, many of the 3GPP principles of IMS were recognized as relevant for the new generation of wireline networks. For that reason, TISPAN did not set out to reinvent another type of IMS, but adopted whatever possible from the 3GPP specifications — namely, define the Core IMS that is common to 3G Mobile and the TISPAN-defined NGN.

Figure 1.14 represents the TISPAN early definition of Core IMS, with elements that have been defined already by the 3GPP and endorsed by TISPAN, but also the functions that have been added by TISPAN, such as AGCF (Access Gateway Control Function), A-BGF (Access Border Gateway Function), RACF (Resource and Admission Control Function), and others.

The TISPAN architecture extends the architecture for tethered connections. It specifies the two principal connection types: (1) analog Fixed phones interfaces and (2) IP terminals with advanced multimedia capabilities. These types of terminals require two types of feature management:

1. *Emulation.* For analog phone to be connected to the Core IP network, TISPAN developed emulation methods that retain almost all the PSTN features, and subscribers may not even notice that they connect to a new IP network.
2. *Simulation.* For other terminals with multimedia capabilities, TISPAN defined the simulation method, where the same services can be delivered in a different way, with a different format, and with different user interfaces.

TISPAN developed additional subsystems to complement Core IMS. In particular, the NASS (Network Attachment Sub-System) and the RACS (Resource and Admission Control Subsystem) are instrumental in the multi-access capability of IMS.

The role of the NASS is to facilitate network admission for NGN terminals. The NASS UAAF (User Access Authorization Function) is responsible for access-level authentication, maintaining authentication information in its PDBF (Profile Database Function). The NASS provides local configuration and allocation of IP addresses to devices (e.g., via a DHCP server). The CLF (Communication Location Function) in the NASS keeps location data that tracks the physical connection.

The TISPAN definition of resource management differs somewhat from the 3GPP definition. The RACS enables resource management driven by operator policies. The SPDF (Service Policy Decision Function) in the RACS must be session aware to make service-based policy decisions. The RACS negotiates resource allocation with the bearer nodes, which then will enforce the agreed QoS level.

TISPAN also has developed further the procedures at the network border, both on the access side and on the network interconnection side. The control of the border point is the role of the IBCF (Interconnection Border Control Function) that instructs the BGF (Border Gateway Function) to enforce security and routing across the border. The I-BGF (Interconnecting BGF) connects to other IP networks. The C-BGF (Core BGF) connects the IP-CAN to the Core network. The A-BGF (Access BGF) is a border gateway at the access side that applies border control between the CPE (Customer Premises Equipment) server and access networks.

Network Address Translation (NAT) has always been an important issue with NGN where the connection is between elements that maintain different addressing plans. Interworking between private and public NAT zones and between IPv4 and IPv6 are facilitated at the border servers as well as security filtering of packet headers and payload. The IWF (Interworking Function) has been added to smooth out incompatibilities between implementations of SIP or translation of IMS signaling to non-IMS communication (e.g., H.323).

Chapter 2 contains additional information about TISPAN.

1.7.4 The OMA Approach

The OMA (Open Mobile Alliance) concentrates on service delivery capability, rather than session admission and session control. As such, the OMA recognizes IMS type of services and non-IMS services. Non-IMS services can be Web services, data-centric services, near-real-time services, or machine-to-machine services. The OMA defines service enablers for both types of services.

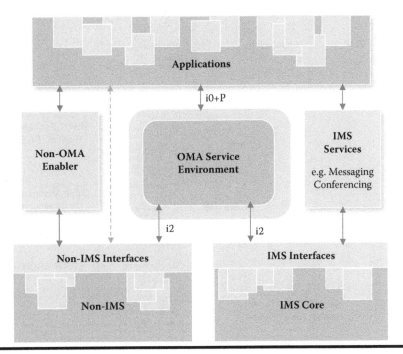

Figure 1.15   OMA and IMS

As shown in Figure 1.15, the applications can access these enablers directly, for exchange of information that is outside the scope of IMS. The OMA service environment manages the interaction between IMS services and non-IMS services.

The OMA architecture and standards are described in Chapter 2.

1.7.5   The Converged Architecture

The concept of converging 3G Mobile and NGN networks has great appeal to operators who can consolidate their networks with the help of IMS. The concepts of IT and data services integrated with communication is reflected in the demand for more sophisticated communication services, such as dynamic dial-up directories and group management, which in turn injects requirements to the session control functions and central data management.

The process of converging Internet and communication is yet to unfold fully, using Web 2.0 service technologies. The Internet may well act as another access network, alongside 3G Mobile carrier networks, Enterprise IP VPN networks, and NGN IP networks. Fixed IP networks include several access methods, namely xDSL, cable, or Metro Ethernet. Fixed wireless LANs are also in this category, with WLAN WiFi and WiMAX. The future integrated network must support all these access methods.

Mapping out requirements from all these areas is no small task. The definition of the shared Core IMS components must be refined, and some components may have special requirements for a certain network access. Other components remain network or device specific, and can be implemented only for a specific use.

Figure 1.16 is an abstract depiction of a network based on IMS at the core, supported by the functions necessary to enable it to operate efficiently.

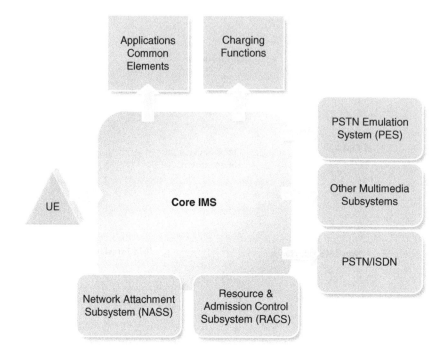

Figure 1.16   Core IMS with subsystems.

Core elements definitions are reasonably stable now and able to process session requests from different access networks. Convergence work is continuing in various subsystems interfacing to Core IMS, in particular the charging functions and resource control.

1.8   Evolving Services

1.8.1   New Addressable Markets

As the market reshapes with the consolidation of Fixed and Mobile areas, new addressable markets are opening up opportunities for operators to increase their customer bases. Subscribers in the Western world are becoming more and more discerning when selecting their communication tools, and seek both greater convenience and new Internet-style services. Subscribers in emerging economies, especially in the Asia-Pacific region, have also exhibited demand for advanced broadband services and the latest technology.

Operators of Mobile networks, Fixed networks, both networks, and service providers with no networks all have different perspectives of the opportunities with IMS:

■ Mobile operators with a saturated market can now also pull in Fixed phone users.
■ Fixed operators can offer a combined service through MVNO (Mobile Virtual Network Operator) agreements.
■ Operators with both networks can merge them, with significant savings.

■ Network-less service providers can offer a raft of user-centric services through wholesale relationships. This could include Internet giants such as Google and Yahoo.

1.8.2   New Services

1.8.2.1   Cross-Domain Services

IMS makes it easy to create new nuances for services to fit users' personalities and preferences. Personalization is a powerful motive on which successful applications often rely. IMS ID management allows for a single user account to support multiple profiles and multiple devices, yet have, for example, the same preferences. This enables creation of user-centric services — not network-centric or technology-domain-centric.

The separation of services from session control facilitates easy modification and roll-out of services. The flexible structure of the user data central repository in IMS enables storing user-specific service triggers and parameters. The ability to merge in Web-based services and utilize user-generated content allows for infinite variations.

Having installed a unified network that enables easy introduction of blended services and full multimedia, controlled by agile service logic, services that are not imagined yet can become feasible. Potential new revenues can come from many services, for example:

■ Push-to-Talk, Push-to-Video, Conversational Video
■ Rich call involving multiple sessions and multimedia
■ Personal media displayed at session setup time, push or pull
■ Combined account or Pre-pay for any type of access (WiFi, Fixed, and Mobile)
■ FMC Centrex with WLAN, for business

These services are not new ideas but have not fulfilled their potential so far, due to implementation difficulties and the lack of usability. New IMS services will arise from the ease of combining capabilities, allowing for new ideas to flourish. New ideas will also come from the fusion of domains that can integrate together: entertainment, information, and communications.

1.8.2.2   Broadband Services

While certain services with video streaming and video clips messaging are already on offer, they still use relatively narrow bandwidth. IMS supports full multimedia environment, with mixing media, real-time two-way video, and higher-complexity services. In fact, IMS realizes the potential of broadband to the fullest.

Full broadband is also the vehicle for IPTV, over DSL as well as radio mobile networks. IPTV, on any terminal can benefit greatly from reusing the IMS service enablers, such as authentication, security, QoS control, policy and resource management, and especially billing (including flow-based charging). This is another example of how IMS can support varied and innovative services, regardless of the access network and bearer, and add value to the operator.

1.8.2.3 Harnessing Internet Power

Accessing the Internet from a communication device is now commonplace. Mobile operators have deployed the "Walled Garden" model (where only the operator's chosen Web sites can be linked to), which is frowned upon by the Internet world. Needless to say, NGN using the laptop as a communication device cannot adopt such strategies. With the advent of open handsets, new Mobile devices will resemble PDA-like handsets that act as both Mobile phones and laptops. They may herald the end of an era for the "Walled Garden."

As communication facilities, such as SMS, appear on Web sites, the opportunities for Telcos to offer network services to ISPs grow. Combining Telco knowledge of users and their locations with web surfing information can provide novel services and attract advertising revenues. The new web technologies allows for mashup, stringing several applications to provide powerful variations.

The power of the Internet, enhanced by contributions from the general public, is an immense engine for innovative services. No single operator can implement such a variety of services. Instead, the aim is to harness this power for the user, while ensuring privacy, fraud prevention, security protection, and convenience. This is achievable by opening the network capabilities for external service providers to gain access to the IMS network and its service enablers.

1.8.3 Blending Services

1.8.3.1 CS and IMS Combination Services

At the early stages of IMS penetration into the market, there is high scope for services that combine the proven and available TDM mechanisms and existing voice quality with new multimedia capabilities of data services, thus enriching existing services prior to the full facilities of broadband multimedia coming online. Such combination services make it easy for the user to escalate a simple voice call to multi-session interaction, without having to initiate separate calls. Advanced Combination Services for IMS make use of IMS session control, as well as other IMS facilities, such as user profiles and locations.

Chapter 9 provides a description of CSICS (Circuit-Switched IMS combinational services).

1.8.3.2 Merging of Functions in ICT

The term "ICT" was coined to express the fact that not only Mobile and Fixed networks are converging, but there is also another convergence force in evidence — the coming together of IT systems with Telecom. IT systems are generally based on data services such as e-mail, directories, calendars, and other support systems. These systems are not real-time (i.e., not time-critical) because they involve man-machine interaction. Telecom, by contrast, delivers voice/video as interactive services in real-time.

In the new Telecom world, there is a blurring of these borders. For example, text chatting, dynamic (Presence-based) address book, and e-mail access on handsets mix and match time-sensitive services with less time-critical functions. This means that the services are converging, the markets are converging, and so are the involved vendors and service providers that previously specialized only in one area of functionality.

IMS provides the means of threading-in communication services within traditional IT, through session control that can equally manage telephony or data services. Their combination is just beginning to be realized and will spawn yet-unimagined services. These can be merely enhanced usability of tools — for example, embedded communication links in documents that can initiate a call back to the author, or a multi-party conference scheduled and activated according to a workflow process when certain preconditions are met.

1.8.3.3   Blurring Private and Business

As users become more sophisticated, the demand for business-like services in their private lives continues to grow. For example, private users now want constant access to e-mail, address book or directory dial, and Internet portals, as they do in their corporate lives. In addition, flexible working hours and home working are more and more commonplace, especially in developed countries. This means that home life and work life become intermingled and blurred. There is also in evidence a shift toward self-employment and small private companies. Such SoHo (Small office Home office) and SMB (Small to Medium Business) create a growing demand for similar services for both private and business use.

IMS can satisfy this need too. IMS supports the concept of a subscription with multiple user profiles that can use the same services but with a separate set of preferences and parameters. For example, one user profile can support a call screening list for work and another for home, and a private networked dynamic address book as well as business group management for the Enterprise. Certain types of calls, at certain times and dates, can be forwarded to a personal voicemail, but at other times to a hunt group of colleagues.

While such services were available previously, IMS enables setup and operation much more easily and quickly.

1.8.3.4   Incorporating Games and Entertainment

IMS can be used as the underlying session management for games, particularly interactive real-time games, and other types of entertainment. IMS will take care of user registration and location, authentication, network admission, policy and resource management, charging rules, and charging online and offline. New games have a ready-made environment to support them, where only the game logic varies. This facilitates fast turnaround of new ideas, easy trials, and, most importantly, low-cost failures.

Merging IMS capability with entertainment is receiving a special boost in the definition of IMS-based IPTV, where IMS will provide session support and user management. This can lead to interesting combinations of chats, Push-to-Talk, messaging, and address-book dial on TV screens via a digital set-top box, where IMS combines the sessions and provides the same service enablers for all applications.

1.9   Implementation Issues

Implementation of IMS is not without risks and problems. Issues span from the initial business case approval, through technical worries over performance, to staff and skills shortages. These issues must be addressed to mitigate possible effects. The following is a selection of some of these

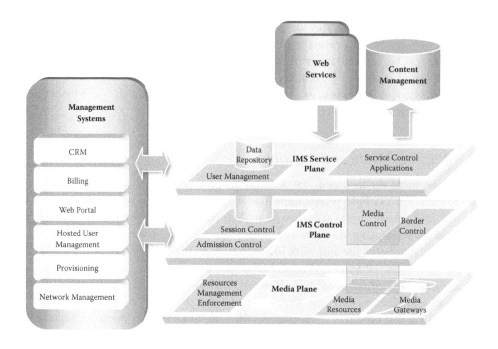

Figure 1.17   MS auxiliary planes.

issues, which is meant to recognize difficulties and give a balanced view, but is by no means an exhaustive list.

1.9.1   Supporting Functions and Services

There are further functions that affect the session, where standard interfaces are just as essential. These are the management systems, content management, and Web services, as shown in Figure 1.17.

In advanced communications, management elements play more central roles, often in mid-session — not only after the event. Real-time Billing is a service in its own right and can dictate ending the session or opening an associated session that allows topping up, for example.

Web portals are used not only to enter personalized parameters, but also to initiate calls. Web sessions or special content can be involved as part of session initiation, e.g., as a welcome message or ring-back service. Therefore, it seems that the old borders between what is a service and what is a management function are no longer so clear.

As a result, a service "wrapper" — its provisioning style, charging, linking services together, Digital Rights Management, etc. — should be defined at concept time. The IMS goal of enabling fast service deployment, not merely fast prototyping, could not be accomplished if charging, provisioning, and management were not established too. IMS scope covers only functions that feed information to the management functions — namely, session and event information. The standards for other management systems are developed in parallel by the TMF (Tele-Management Forum) and other bodies.

1.9.2 Implementation Options

Previously, the building blocks of the architecture were aligned very closely with network entities. In 3GPP, the functions are entities that can be combined together in products where it makes sense for network topology and scalability. Such implementation decisions are left to product vendors. For example, many vendors combine all types of CSCF in one product, but operators may prefer to install a separate P-CSCF, as the network entry point, and perhaps co-locate it with an access-side border gateway and with NAT service.

Only a few vendors combine Media Gateways (MGW) with the Media Server (Media Resource Function, MRF), although both use the same codecs and media manipulation processing. This may be for historic reasons, where the Media Server capability was perceived as part of an IVR (Interactive Voice Response). Other vendors prefer to co-locate the MGW with the Signaling Gateway (SGW) to provide a trunking gateway for network breakout. Some Media Server capabilities, e.g., DTMF (Dual Tone Multi Frequency) tones and system announcements also have been integrated into Edge devices and Access Points.

Another notable option is the PDF (Policy Decision Function) and the AF (Application Function). In early 3GPP documents, the PDF was viewed as a module within the P-CSCF or even the GPRS nodes. TISPAN, on the other hand, describes the RACS (Resource and Admission Control Subsystem) that contains SPDF (Service-based PDF) and A-RACF (Access Resource and Admission Function) as separate entities, and the AF as the requesting agent module, which is external to them and can co-reside in the P-CSCF as well as in other Access Gateway Controllers (AGCF).

1.9.3 Technical Issues

1.9.3.1 Standards Maturity

In the early days, SIP was said to be very simple, a lightweight signaling protocol to replace the complex TDM SS7 and associated protocols. However, SIP, as it was then, could not provide unambiguous interfaces due to its permissiveness and flexibility. This meant that extensions had to be defined and the enhanced SIP is no longer simple to implement.

As the scope of IMS increased to envelop all types of access networks, existing IMS specifications were extended and new functions needed integration. Not surprisingly, widening the scope slowed the progress, delaying the publication of globally approved specifications.

Vendors have to work around these pitfalls, delivering their own extensions where the standards are not available yet. At the same time, more than ever, operators are fully aware of the importance of adherence to standards and tend to reject proprietary or semi-proprietary solutions.

1.9.3.2 Performance

Many of the early technical problems, such as echo cancellation, jitter, and latency, already have been largely resolved for IP trunking — carrying voice for the long haul over IP networks. However, full SIP solutions need to be proven for high performance, adequate quality levels, and large-scale networks.

Performance in IMS can be affected by heavier processing of longer session descriptions in SDP (Session Description Protocol) and lengthy addresses, and by multiple hops between several servers. This is sure to occur in roaming, and also occurs within the same network. Devolving features to separate application servers provides greater operability and scalability, but may insert delays and extra processing when the session is forwarded to other servers.

In response to these concerns, many labs are set up to prove and improve performance, with tools that can analyze the weakness points. Optimizing performance can be helped by computing advanced technologies, such as parallel processing, by carefully thought through network topology and co-locating functions that interact intensively.

1.9.3.3 Database Issues

Multiple database dips often cause performance issues. Several network nodes (applications and session controllers) need to retrieve data from the central data storage, the HSS, and can create another bottleneck. To mitigate this, some caching of data in local memory should be designed, with an efficient data replication mechanism that retains data integrity.

While the concept of a single data repository is very attractive, reaching this goal is not easy. Many current systems need to be upgraded to support the IMS interfaces (Diameter). Existing data stores, with various proprietary extensions, may contain data structures that are not accommodated. Existing data may contain data corruption — errors, duplications, and omissions that are not visible until the data is converted. Therefore, data conversion is also an exercise in a complete database overhaul that demands additional skills and human resources.

1.9.4 Security and Privacy Issues

The IP/Internet space increases the danger from debilitating Spam, ID theft and fraud, etc. Because IP Multimedia communication involves exchanging increased amounts of information about the user, the issues of ensuring user privacy, confidentiality, and data integrity are becoming more significant. Telecom services over IP are no longer immune from the kind of threats that affect the Internet. Sensitive data must be secured even from the operator's own staff, as well as from external hackers.

However, security measures such as encryption, decryption, and deep packet inspection increase processing loading, slow down the performance, and add to the overall costs. Constructing a good IMS-based IP network requires a finely balanced assessment of the actual risks compared with the impact of security measures.

1.9.5 Migration Issues

Migration issues are, of course, specific to each network, within the particular business type, demographic and geographic attributes, existing equipment and existing applications, profile of current customer base, and type of addressable market.

Worries with regard to IMS practical implementations often include issues at several levels:

Migration steps via a softswitch or to IMS directly:

■ It may be easier to migrate first to an interim solution that delivers packet voice based on H.248, and not to full IMS multimedia based on SIP. This will *not* revamp the network, as the architecture follows the same principles of TDM silos but will allow more gradual evolution that can be perceived as safe.

■ Choosing a temporary solution (softswitch) or a long-term solution (IMS-based PES) to support those subscribers who would prefer to keep their traditional phones while the transport network migrates to IP. IMS-based PES provides for POTS (Public-Owned Telephone Service) accessing a SIP network via a gateway and receiving IMS-based session and feature control in the new architecture. Telephony services for IMS-based PES will remain in force as subscribers move to IMS terminals.

Migration procedures and connectivity:

■ Interworking with the complex and varied legacy signaling while moving to SIP and IMS, where linking to further networks can also occur, i.e., mixed signaling protocol environment during the migration period.
■ Ensuring that successful revenue-earning services continue to be delivered, despite moving the Core to IMS, e.g., call centers, premium rate games, or location-based advertising.
■ Network engineering when experiencing difficulties to anticipate traffic volumes due to unpredictable speed of service take-up and as a result of the increased length of the variable signaling messages.

Investment and cost:

■ Justifying new investment that is still worthwhile to extend the life of existing services connecting them to IMS or to advanced enablers (e.g., Presence). This, however, can drain financial resources and prevent moving onto the target architecture.
■ Stretching successful services over the new access-agnostic IMS enablers for additional access methods (e.g., WiFi, WiMAX, or PBT) requires new investment and further decisions as to which access methods to adopt.

Service prioritization decisions:

■ Differentiating feature-rich terminals that will attract users and provide hybrid technologies and multiple access facilities at reasonable cost and high resilience.
■ Offering new value that can fund the new equipment rather than merely replace voice features.
■ Weaving the mass of new services, such as IPTV and Web content, into coherent communication packages.

Impact on existing business:

■ Introducing IMS without cannibalizing existing services, while fending off diminishing revenues from bare voice.
■ Keeping loyal users and avoiding prompting them to look elsewhere if they are not inclined to do so yet, i.e., continuing to provide identical, undisrupted service while moving to IMS.

1.9.6   IMS Terminals

The availability of good-quality IMS terminals is another issue that has slowed the pace of IMS implementations. IMS terminals should have the ISIM (IMS SIM) client, whether on an UICC (UMTS Integrated Circuit Card) or as a soft client. The ISIM provides the IMS identities and authentication material necessary for network admission and registration.

Mobile handset technology is evolving with competing technologies running side by side. The handset is becoming more and more like a full computer or PDA, with a wide range of built-in services and supported clients.

There are still many open questions regarding the most suitable technologies:

■ Operating system: Mobile Window, Symbian, Mobile Linux, J2ME
■ Open browser with downloadable applications or "Walled Garden" approach
■ Means of authentication and security

To benefit from IMS capability, ISIM clients should enable access from various networks. This means that the appropriate access network authentication is enabled. Furthermore, to enable seamless handover between technologies, the Mobile handset should have dual-mode or multi-mode, with a growing number of required access protocols and service parameters. This increases the complexity and the testing time.

As the handset becomes more valuable, more powerful, performing sensitive functions (e.g., electronic purse), users may worry more about handsets being lost or stolen. Further security technologies are becoming available (e.g., biometric signatures), but also network-based services and data backup are needed to retain service availability in such situations.

Battery life must be improved because multimedia is more demanding and depletes power much more rapidly. At the same time, battery size must not increase the size and weight of the terminal, despite higher functionality.

The ergonomic design of the handset is also not settled. This includes choice of size and layout for the keyboard, size and shape of the screen, programmable keys, touch-screen or light-pen, etc. The use of the devices for video means that associated Bluetooth devices become necessary to enable viewing the screen while talking and listening. Interference that may occur between Bluetooth and WiFi or WiMAX must be resolved.

Performing conversational video reliably over the air interface is yet to be established and may require further compression and other RAN improvements. However, video streaming (i.e., not in real-time) is possible and can provide a rich range of new services.

Soft clients will also mature. Many soft clients merely "pretend" to be a handset, with a numeric pad. While they provide a familiar environment, IMS can deliver far more than VoIP. Soft clients that interface with IT systems are available in the Enterprise space for the private sector, and should be on offer in the public sector with IMS. This will enable integration of e-mail, chat, and directory services with IMS sessions.

Chapter 2

IMS Standards

2.1 This Chapter

This chapter examines the standards that define IMS. Following the standards is more important than ever in an environment that consolidates disparate technologies globally and merges mobile and fixed networks, Web technology, Enterprise solutions, WLANs, etc. To achieve the goals of manageable and versatile network with fast service deployment, all parties must agree on standards that will ensure interoperability.

This chapter explains how the standards are defined and the interaction between them, through several layers of network, each governed by suitable protocols. It describes the various contributions to IMS from the collaborating standards organizations and consortia around the globe and their respective areas of activity. Each of the main contributors to IMS — namely, the 3GPP (Third-Generation Partnership Project), TISPAN (Telecommunications and Internet Services and Protocols for Advanced Networks), IETF (Internet Engineering Task Force), and OMA (Open Mobile Alliance) — are described here, along with their architectures and main concepts.

Finally, at the end of this chapter there is a useful list of all the IMS components, according to both the 3GPP and TISPAN. The list includes the standard component names with brief descriptions of their functionality.

For a better understanding of the underlying protocols, see Appendix A. Appendix E provides a list of terms and definitions, Appendix B provides abbreviations and acronyms, and Appendix D is a list of standard Reference Points.

2.2 The Standards Scope and Uses

2.2.1 Standard Architecture

The Standards address a high-level architecture for a network that contains multiple functions combined to form a framework. This architecture also addresses the external connections to other networks. In general, the architecture is a tool to make sense of complex information in a coherent layout. The standard architecture may show all the possible functions, although these functions are not always implemented in every solution. The theoretical architecture can be modified according to the existing network's requirements and precise usage.

2.2.2   Functions

The Standards aim to define all the required functions for the type of connectivity. The scope of each function can be small or large, and can constitute a module within a product or a separate network entity. The granularity — that is, the level of detail that makes a single function — is not defined anywhere, nor what is considered part of the same function or what is defined as another entity. It is therefore not surprising that some earlier functions have undergone major changes and morphed in later releases. A good example for this is the PDF (Policy Decision Function), which was first embedded in GPRS nodes, then thought to be part of P-CSCF (Proxy Call Session Control Function); and now considered a separate element, also containing charging rules (PCRF: Policy and Charging Rules Function).

Defining a function individually does not decree that it should be implemented in a separate element, but it does ensure that:

■ Each function is fully understood with its scope and relationship to other functions.
■ Each function can be implemented separately (to scale up) or jointly with other functions.
■ Each function interfaces to other functions via a standard interface, even when it is co-located.

The interpretation of the standard functions into products is the responsibility of the product developer. The way functions are grouped together can have a considerable impact on their scalability, maintainability, network positioning, and price. Therefore, such product determinations provide differentiation for Telecom vendors.

2.2.3   Reference Points

Previously, an interface to a function was defined by mapping all activities to and from that function. The modern approach is to define "reference points" between two functions. A reference point is defined as one interaction at a time, always between two functions and belonging to both functions. On occasion, new reference points are defined between the same functions to describe additional functionality. In other cases, reference points can merge with others when it becomes clear that they do carry exactly the same interface. This approach is more flexible and can cope with later introduction of additional features.

Reference points define in detail not only what protocols are deployed between the two functions, but also data structures, headers, parameter values, and behavior. Where there are no reference points, there is no direct interaction. Where the reference point is between two nodes with the same function in the network, it describes interaction between multiple instances in a distributed network.

Reference point definitions help avoiding ambiguity in the description, which often causes problems of interworking between vendors and between networks. Although different interpretation of standards cannot be entirely eradicated, following the reference points definition closely can aid coming closer to the ideal of "plug-n-play."

For a list of reference points from both TISPAN and 3GPP, see Appendix D.

2.2.4 IMS Protocols

2.2.4.1 Protocols and Protocol Profiles

Protocols define formats of dialogs between elements. Such formats are then used with further definitions of parameters, dialogs, and data structures of the exchanged information to define case-by-case behavior for various scenarios.

Protocols have emerged from various Standards organizations over the years to cover particular sets of requirements. The type of interaction may dictate what kind of protocol would be suitable. In choosing a protocol, one can trade off flexibility against efficiency and assured compatibility, or simplicity against clarity of definition.

Despite its renowned flexibility, SIP is not always adequate, as believed by staunch SIP proponents. Protocols have an inherent assumption of the type of dialog they manage. Therefore, session signaling protocols may not be the same as media flow protocols or authentication protocols. For example, while SIP is perfectly suited for session control, dialogs for exchange of parameters, retrieval of dynamic information, or establishing network rules seem to need something else, such as the Diameter, to provide data improving facilities.

A "protocol profile" defines the behavior that fits a particular use case. A profile for a protocol is a standardized interface between communicating parties that exchange protocol messages. While the protocol description specifies all that is possible to convey by this protocol and the entire range of values for the parameters, each protocol profile describes the message sequence with the permitted parameter values at every stage. A profile specification can contain a dialog that uses the protocol commands and message types, with specific headers and their parameters, and with their anticipated responses. It also can contain dependencies on other profiles and conditions. By following a given profile, developers are guided more precisely and therefore can achieve a higher level of compatibility.

2.2.4.2 SIP, Diameter, and Megaco with IMS

The main IMS protocols are SIP for session control, Diameter for authentication, QoS resourcing, and charging, and H.248 for interaction with media gateways and breakout to CSNs. Appendix A provides brief protocol descriptions.

a. SIP

SIP is the main IMS protocol and is a relatively new protocol. It originated from the desire to provide voice and multimedia over the Internet. It was originally intended solely for peer-to-peer communication. It assumes that the endpoint is intelligent and capable of determining the session requirements.

SIP has rapidly become popular due to its simplicity and versatility, and its ability to negotiate multimedia sessions. SIP was established as the prevailing session control protocol for the new packet network when it was adopted by the 3GPP.

However, SIP lacked many facets that public network communication required, especially regarding inter-network cross-charging, security, policy, quality of service, resource management, network address translation, orchestrating service features, and more. Such extensions to SIP have been added through the normal process of proposing draft RFCs through the IETF. SIP is now rich in capability but no longer simple.

b. Diameter

This IETF protocol is another important protocol in packet networks that is emerging as the pre-ferred protocol for IMS data exchange, for queries of databases in real-time, for provisioning user data and network policies, and for the transfer of accounting information.

Diameter is extensible, through the assignments of AVP (Attribute Value Pair) that can be added to define more data exchanges, queries, and responses. As such, it is flexible and can support future, as-yet unspecified requirements. Diameter supports the IMS requirements of added secu-rity in roaming and the proxy/server architecture. This allows for more secure authentication in the Home Network as enhancement of the currently popular RADIUS Protocol.

c. H.248

The H.248 Protocol evolved from the MGCP and other similar protocols, including Megaco, that were designed primarily to manage media streams. MGCP, which is popular with many softswitch implementations, was superseded by H.248, which has wider applications, including media server (conferencing, multimedia, etc.) as well as media gateway (voice media conversion, tones and DTMF) functionality.

The H.248 Protocol is designed to address master–slave relationships between a call server or session controller and a media resource. H.248 is deemed more precise in the control of media streams, as compared to SIP, but too restrictive when it comes to providing multimedia session control. H.248 therefore is utilized between the media controller and the media gateway or the media server.

2.2.5 Network Layers Models

The use of layered models enables abstractions of the services that each layer provides and build-ing up the total service from selected elements in each layer. Thus, upper-layer protocols need not worry about lower-level techniques and do not overlap with their capabilities.

A notable example for this is the IP protocol that need not consider transmission reliability, which is defined by additional protocols. A lower-layer protocol can be chosen to provide reliability to suit the type of connection. The UDP (User Datagram Protocol) provides data integrity but does not ensure the delivery of all packets. TCP (Transmission Control Protocol) provides both but is less efficient and may create delays as a result of error correcting retransmission, while the SCTP (Stream Control Transmission Protocol) provides better media management in a reliable protocol.

2.2.5.1 OSI Layers Model

The ISO/OSI Network Model defines a standard model for networking protocols and networked applications. This model was produced by the International Standard Organization (ISO) while defining the Open System Interconnect (OSI). The model contains seven network layers:

1. *Layer 1 — Physical layer.* This layer involves the hardware medium (e.g., cable optic fiber, unshielded twisted pair, or card). It defines the transport of the raw bitstream. Example standards for this layer include IEEE 802 and ISDN.
2. *Layer 2 — Data-Link layer.* At this layer, bits are formed into units, packets, and data frames. In this layer, connections are controlled between network elements using, for example,

Ethernet address or Media Access Control (MAC). Example protocols include IEEE 802.1 or 802.2 link control.

3. *Layer 3 — Network layer.* This layer is responsible for network addressing and routing, managing congestion and packet switching, and translating between the physical address and logical address. The main protocol for packet networks is IP — Internet Protocol — including IPv4 or IPv6 and the conversion between them.

4. *Layer 4 — Transport layer.* This layer reforms the data units and manages their flow, with or without error checking and error correction and retransmission. Transport protocols include Transmission Control Protocol (TCP) and User Datagram Protocol (UDP), as well as Real-time Transport Control Protocol (RTCP) to support RTP (Realtime Transport Protocol).

5. *Layer 5 — Session layer.* This layer deals with "stateful" connections. It establishes a session, monitors it, and tears it down. In this layer, services are synchronized, parties are allowed to connect, and lengths of sessions are determined.

6. *Layer 6 — Presentation layer.* This layer transforms data between the various applications and the network unified format. In this layer, encryption, protocol conversion, character conversion, and data compression are performed.

7. *Layer 7 — Application layer.* This layer relates to the user interface and provides access to network services. Mail, File Transfer (FTP), telnet, and DNS (Domain Name Server) are all examples of network applications.

Figure 2.1 shows the ISO/OSI seven-layer model. The lower three layers refer to media, while the upper four layers deal with session control signaling. The OSI model is applicable in different networks with a number of protocols within each level.

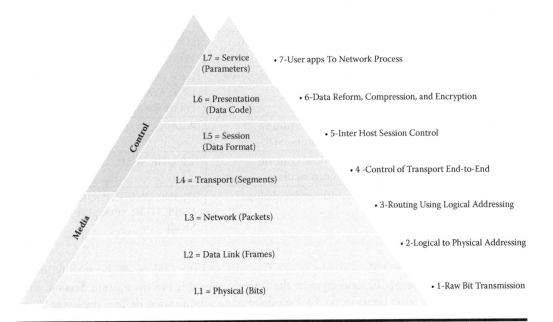

Figure 2.1   The ISO/OSI model of seven network layers.

It often is easier to understand the scope of a certain protocol in the context of the layer to which it belongs. The following is a table of common protocols within the seven OSI layers to help associate protocols with functional layers:

Protocols per Seven-Layer ISO/OSI Model		
	Layer	*Description*
1	Physical	E1/T1, SONET/SDH, G.709, 10/100BASE-T,
2	Data-Link	Ethernet, 802.11, L2TP, PPP, ISDN, Frame Relay
3	Network	IP, OSPF, ICMP, X.25, RIP, IPSec
4	Transport	TCP, UDP, RTP, SCTP
5	Session	TLS, SSL, NetBIOS
6	Presentation	NCP, XDR, ASN.1
7	Application	SIP, HTTP, NFS, SMTP, XMPP, DNS

2.2.5.2 The Internet Layer Model

The Internet model, or TCP/IP model, was created in the early days of the Internet, and the structure of the Internet is still closely reflected. Initially it had just four layers that were deemed simpler and more flexible than the OSI model.

This Internet model has never been enshrined in the standards; therefore, many variations can be found with different references to the layers. This Internet model later evolved into a five-layer version that splits Layer 1 into a Physical layer and a Network Access layer, corresponding to the OSI Physical layer and Data-Link layer, as shown in the table below:

Protocols per Seven-Layer Internet Model			
	Layer	*Description*	*Protocols*
1	Layer 1 Physical	Wiring connection, plugs and circuits	ISDN, SONET/SDH, G.709, WiMAX
2	Data-Link	Management of packets or service data units	GPRS, 802.11, EVDO, HSPA, PPP, L2TP, ATM, Ethernet, FDDI, Frame Relay
3	Internet (Network)	Interworking and routing, navigating packets /SDU	IPv4/IPv6, IPSec, OSPF, ICMP
4	Transport	Host to host, flow-control, connection protocols	TCP, UDP, SCTP
5	Application	Processes and service dialogs	SIP, SDP, DNS, RTP/RTCP, TLS/SSL, XMPP, HTTP, SOAP, IMAP4, MIME

The differences between the Internet IP/TCP model and the OSI model lie mainly in the upper layers, where the Application layer in the Internet model is broken up into Session, Presentation, and Application layers in the OSI model. For communication design, the lower layers (i.e., Layers 1, 2, and 3) are the most significant, so there is little to distinguish between these two models now.

2.3   The Standards Organizations

The 3GPP group has been leading the effort to define the architecture for IMS for Mobile networks. The TISPAN group (at the ETSI) has been leading the effort to define fixed broadband networks — NGN, as well as PSTN and ISDN replacement, as extensions of IMS.

These groups have decided that they will not respecify new protocols, but adopt existing ones and, where necessary, will extend them to fill in gaps of functionality. As a result, these influential Standards bodies have been working closely together and with the IETF and ITU.

2.3.1   *The Organizations and Their Work*

The Standards organizations are generally funded by the main industry players, and their active members come from the main carriers, service providers, and vendors. Interest groups can be formed when a certain topic seems to require special study. These groups may make their recommendations then dispersed. Alternatively, such groups may merge, for example, the Wireless Village/Liberty Alliance and OMA.

The table below summarizes the main contributing organizations and their activities associated with IMS, but is by no means exhaustive. The table shows what focus each group has and the standards for which it is responsible:

Organization	Focal Areas and Topics	Examples of Standards
IETF — Internet Engineering Task Force	All IP networks Internet protocols	SIP, SIMPLE, ISCCOPS, Diameter, PPPXML, HTTP, HTTPS, MegacoIP, and Internet Security
3GPP — Third Generation Partnership Project	Universal Mobile Telecommunications Service (UMTS), Wideband Code Division Multiple Access (W-CDMA) European focus	GSM, GPRS, etc. IP Multimedia Subsystems (IMS) LTE, SAE
3GPP2 — Third-Generation Partnership Project 2	CDMA2000 mobile networks North America focus	Multimedia Domain (MMD), an equivalent to IMS for North America
ETSI — European Telecom Standards Institute TISPAN NGN subgroup	Next-Generation Networks (NGN) wireline migration to NGN European focus	NGN Sub-systems: NASS, RACS/SPDF, I-BGF, etc., Enterprise, IPTV PSTN Renewal: PES, IMS-based PES, and PSS
ITU — International Telecommunication Union	Next-generation wireline networks Next-Generation Networks (FGNGN) Global focus	H.248 Protocol Service descriptions (REV, MLPP, etc.)
IEEE	Wireless LANs	802.11x and more
OMA — Open Mobile Alliance	Service architecture Enablers and services for mobile networks	Open Service Environment (OSE) Capability Exposure (CEF) Mobile services (e.g., Push-to-Talk over Cellular, etc.) Device enablers
GSMA — GSM Association	Mobile networks implementations European focus	GSM implementations and migration to IP, testing Interworking, GPX, etc.
CableLabs	Cable IP networks PacketCable 2.0 Project	DOCSIS, EuroDOCSIS

Note: The itemization of standards gives relevant protocols as examples, not an exhaustive list.

2.3.2 *The 3GPP Group*

3GPP is an independent group for defining wireless network architecture and interworking interfaces. It publishes its work in documents, classified as a:

■ Technical Report (TR), which identifies the general requirement
■ Technical Specification (TS), which provides actual specifications

The Technical Specification Group (TSG) is responsible for reviewing the output from the various subgroups, leading to a formal TSG approval. The TSG may modify the contents of the documents and re-release them.

2.3.2.1 *3GPP Releases*

3GPP Release 1999. 3GPP Release 1999 laid the foundation for packet networks with specified requirements for a CS system with packet network data services. It was functionally frozen in December 1999. It included concepts of UTRAN (UMTS Terrestrial Radio Access Network) Iu, Iub, and Iur interfaces; AMR (Adaptive Multi-Rate, Mobile) speech codec; and USIM (UMTS SIM) definition for a UTRAN terminal.

This release also included service capability in Location Services (LCS), Multimedia Messaging Service (MMS), GSM supplementary services, Customized Applications for Mobile network Enhanced Logic (CAMEL) and Open Service Access (OSA).

3GPP Release 4. 3GPP Release 4 was functionally frozen in March 2001. It is a further functionality for GSM and CS-based services as well as packet-based services. This includes a collection of enhancements, Bearer-independent CS Core Network, CAMEL enhancements, and OSA enhancements. Significantly, Release 4 also developed the IP data service capability, with IP transport of Core Network protocols.

3GPP Release 5. 3GPP Release 5 was functionally frozen in March 2002/June 2002. This was the release that introduced IMS. It also defined bearer enhancements and is essentially the release of IMS for UMTS, including:

■ High-Speed Downlink Packet Access (HSDPA) and Wideband AMR
■ IP transport in the UTRAN, with Iu for GERAN and Gb over IP
■ Global text telephony
■ Location services enhancements
■ UTRAN sharing in connected mode
■ Security enhancements

3GPP Release 6. 3GPP Release 6 established IMS Phase 2 and made IMS deployable. Most of the specifications involve the Core IMS or peripheral functions. This release brought in the Fixed IMS and defined the xDSL access method via endorsing TISPAN standards. Among the main features are:

■ IPv4-Based IMS Implementations
■ Interoperability and Commonality of IP-connectivity Networks
■ Multimedia Broadcast and Multicast Service Plus Packet-Switched (PS) Streaming Services
■ WLAN-UMTS Interworking Rel-6

- Subscription Management and Generic User Profile (GUP) Rel-6
- Presence Capability
- Security Enhancements: Network Domain Security (NDS) and Subscriber Certificates
- 3GPP enablers for Services like PoC
- Support of Push Services and MMS Enhancements
- Location Services Enhancements 2
- Speech Recognition and Speech-Enabled Services
- Digital Rights Management
- Priority Service and QoS Improvements
- Network Sharing
- Charging Management

3GPP Release 7. Release 7 provides enhanced IMS functions, refinements of interfaces with TISPAN elements and with OMA standards. But most importantly, it established the WLAN access mode. This release contains:

- IMS Phase 3 and IETF Protocol alignment
- Multimedia Telephony Capabilities and Telephony Service for IMS
- Combinational Services, Stage 3 CSI Terminating Session Handling
- IMS Emergency Sessions and Transferring of Emergency Call Data
- In-band Modem Solution
- Access Class Barring and Overload Protection
- Advanced Global Navigation Satellite System (A-GNSS) Concept and Performance
- Enhancement of E2E QoS and QoS provisioning over 3GPP/WLAN Interworking
- IMS Transit
- The Mp (between MRFC and MRFP) Interface
- Evolution of Policy Control and Charging
- Diameter Gi interface to GGSN and the PDG Wi interface
- Liberty Alliance and 3GPP Security Interworking
- WLAN Interworking, with UMTS Phase 2
- Private Network Access from WLAN 3GPP IP Access
- GUP Phase 2
- Voice call continuity between CS and IMS (including I-WLAN)
- Trust Requirements for Open Platforms in 3GPP
- MBMS Enhancements and MBMS User Service Extensions
- GRUU (Globally Routable User Agent URI) in IMS
- IMS Service Identification and Service Brokering enhancements and IMS Local services
- IMS Support of Conferencing and Messaging Group Management
- 3GPP System Architecture Evolution (SAE) Specification
- ISIM API for Java Card
- Selective Disabling of UE Capabilities

3GPP Release 8. While Release 7 is still delivering results, work on Release 8 is already ongoing. Release 8 concerns enhancing the broadband over the air interface and creating a unified all-IP network that can handle multiple technologies. Among the topics addressed are the following:

- All-IP Network (AIPN)

- Registration in Densely Populated Areas
- Packet Cable Access to IMS
- Consumer Protection against Spam and Malware
- 3G LTE and interworking with GERAN
- Local Charging Zone Requirements
- Video Telephony Enhancements to Bearer Service
- NDS Authentication Framework Extension for TLS
- Location Services for 3GPP Interworking WLAN
- 3GPP System Architecture Evolution (SAE) Specification
- Personal Network Management (PNM)
- eCall Data Transfer Requirements
- IMS Service Brokering Enhancements
- Network Composition
- Multimedia Priority Service
- Support of Customized Alerting Tone Service
- Charging and Security Enhancements for IMS

2.3.2.2 3GPP Architecture

The main achievement of the 3GPP is to define the entire structure of the future architecture, which takes care of sessions from the initial registration and roaming relating to a particular user and device, through the admission controls, session attributes, addressing, and rerouting, up to service logic and billing.

The 3GPP architectural components cover the following:

- Subscriber Profile Management
- Identity Management
- Service Control
- Service Orchestration
- Network Policy and Policy Enforcement
- Single Sign-On Authentication
- Media Authorization and Allocation
- Charging Correlation
- Conferencing Support
- State Security (Lawful Interception)

Figure 2.2 shows the 3GPP architecture with its main elements, interfacing to three types of access networks: (1) GPRS/UMTS, (2) xDSL Broadband (NGN), and (3) WLAN. As shown, in a high-level view, IMS components serve all three IP-CANs.

2.3.2.3 3G Reference Points

To standardize not only the functions in the network, but also how they interface with each other, the links between the architectural elements are specified in reference points identified by a unique identifier. Each interface between the functions is a different reference point, although several of them may be using the same protocol, or extensions to that protocol. For example, Mw and ISC (IP Multimedia Service Control) are both SIP, while Dx and Cx are both Diameter. However, the

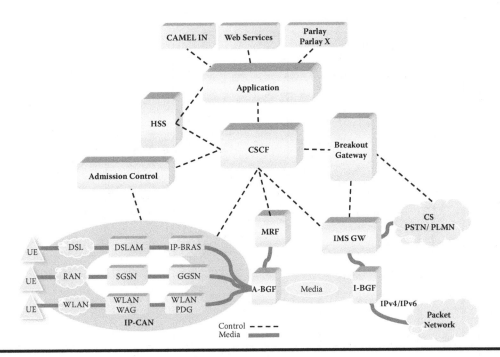

Figure 2.2   Overview of 3G model with three types of IP-CAN.

reference point can be defined further with a *Protocol Profile* that describes particular behaviors for use cases, with parameters and data formats for the particular dialog. Protocol interfaces with different Protocol Profiles are maintained as separate references.

The IMS model was first published by the 3GPP, as shown in Figure 2.3. This model intended to highlight clearly the reference points between the IMS components and what protocols they use.

This model shows the three CSCF functions — Proxy, Interrogating, and Serving session controllers — connecting via SIP to other control elements: Border Gateway Control Function (BGCF), Media Gateway Control Function (MGCF), Media Reserved Control Function (MRFC), and via Diameter to the HSS and SLF. The interface to the AS is also SIP, although it contains extensions and therefore is referred to as ISC. The MGCF interfaces to the Media Gateway by H.248, as does the MRFC to the MRFP (Media Resource Function Processor). The Ut interface between the UE and the AS is using secure HTTP for transferring service configuration data in XCAP (XML Configuration Access Protocol).

2.3.2.4   3GPP Overall Components

The 3GPP Pre-IMS Network consists of the Packet Data Network (PDN) for pure data services, interworking with legacy Mobile Switching Center (MSC), and with SMS Center, etc. Some preexisting elements, such as GGSN or MSC, have been enhanced by 3GPP specifications to accommodate IMS.

The GPRS nodes are used for 2½G as well as 3G terminals, but the WLAN terminals are connected via a separate IP-CAN, with a separate authentication and accounting process, using the

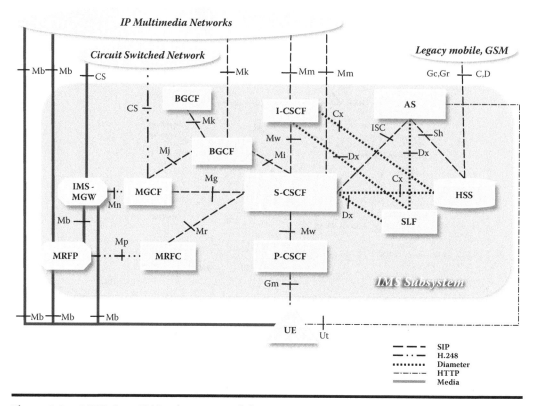

Figure 2.3   3GPP Core IMS reference model.

3GPP AAA Server (Authentication, Authorization, and Accounting Server). For a more detailed look at WLAN, see Chapter 10.

Figure 2.4 is a detailed reference diagram that includes the existing GSM elements and the Data Network used in GPRS to show how they integrate with the new IMS. GSM interaction with IMS is via a special IW-MSC (Interworking MSC). Integration with SMS and MMS services are also shown. Below the IMS elements, the access via the WLAN is shown, integrated with the Data Network, which already delivers data services through GPRS. Figure 2.4 also shows WLAN interworking with existing GPRS and Data Network as I-WLAN (Interworking Wireless LAN). The 3G/UMTS IP-CAN is also shown, interfacing directly to IMS.

2.3.3   TISPAN

TISPAN (Telecoms and Internet (converged) Services and Protocols for Advanced Networks) is a standardization body of the ETSI, specializing in Fixed Networks and Internet convergence. It was formed in 2003 as a result of the amalgamation of the ETSI bodies of (1) Telecommunications and Internet Protocol Harmonization Over Networks (TIPHON) and (2) Services and Protocols for Advanced Networks (SPAN).

While 3GPP defined IMS for pure Mobile Networks, TISPAN perceived the potential of a single sharable CN, supporting a rich service environment for both wireless and wireline, regardless

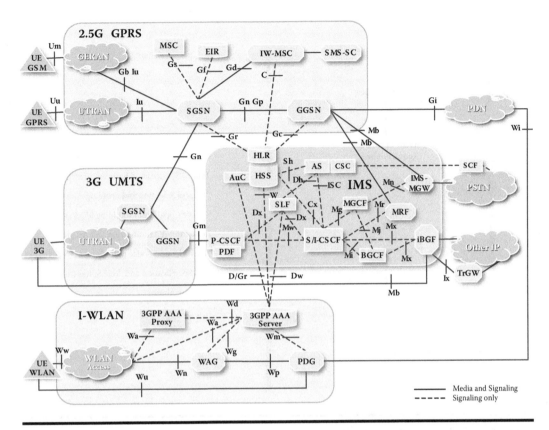

Figure 2.4   Overall 3GPP elements map.

of the means of access. Therefore, TISPAN set out to define generic functions as far as possible, while providing enough detail to enable realization in products.

TISPAN adopted the main 3GPP IMS standard framework and added the necessary sub-systems for the extended application to wireline. The first subsystems to be defined manage the wireline subscriber's network attachment and authentication, including network addressing, and the dynamic requests of network resources per session, enabling the negotiation of these resources according to network rules and availability.

2.3.3.1   TISPAN Scope and Aims

The focus for TISPAN was originally to define the NGN, i.e., the evolution of Fixed Networks toward packet-based IP networks. As an ETSI workgroup, the scope for TISPAN was essentially European but it attracted interest and participation from other regions outside Europe.

Figure 2.5 is the TISPAN structure, in terms of activities that are based on the Goals, Definition, and Specification phases. TISPAN activities for NGN definition center on the following areas:

■ Wireline access aspects and the impact on the CN
■ Service aspects for wireline access and their interfaces to the CN components

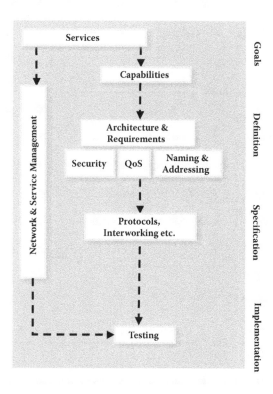

Figure 2.5 TISPAN areas of activity.

- Service-level interworking with CSN
- Communication services that are access-agnostic and services that can utilize a heterogeneous environment independent of any specific underlying network technology
- Communications services that can interwork with existing circuit switched and packet-based networks
- Adaptation of existing protocols for packet-based wireline and access-agnostic services
- Public network, government and corporate network requirements and solutions
- Closing the gap between Fixed and Mobile Networks in terms of regulatory requirements, security, lawful intercept, emergency communications, quality-of-service, numbering and naming, etc.
- Integrated telecommunications network management
- Predictable level of quality for packet-based communication, per session and per service type
- Ensuring global connectivity from the new NGN- and IP-based services

2.3.3.2 TISPAN Working Groups

The TISPAN organizational structure consists of a single technical committee, presiding over several focal areas with core competencies, and their Working Groups and Project Teams. Working Groups have specific responsibilities, scope of work and focal technology space, and a program of activities to meet their objectives. Project Teams are responsible for driving the work forward within a defined set of time scales and objectives.

TISPAN has several Working Groups responsible for the main focus areas, including Architecture requirements, Security requirements, Naming and Addressing, Protocol Selection and Profiling, Network and Service Management (NGN OSS), and Test Specification and Test Suites (to facilitate interoperability).

2.3.3.3 The TISPAN Architecture

TISPAN describes a heterogeneous network that is a combination of subsystems, with IMS at its core. The components of Core IMS were taken from the 3GPP, with the aim of adopting them as they are, or with minimum change.

TISPAN has modeled the NGN network with multiple access technologies. The network Attachment Subsystem can provide support for PSTN Emulation, IMS, and other subsystems such as IPTV. The shared data repository is an important element of this architecture. It is where user data is stored and made available to all subsystems, including the applications. Also considered in this model is an array of different terminals, including various access gateways, nomadic terminals, and mobile handsets for any IP-CAN.

The IMS system is regarded as a subsystem, not a higher layer of control, connecting to CS PSTN/ISDN as well as PSTN Emulation. IMS connects to both a common application layer and to a common IP Transport layer. Other subsystems, such as IPTV and PSTN Emulation, are perceived as stand-alone independent systems with their own control. The emergence of IMS-based PES (PSTN/ISDN Emulation Sub-system) solutions and IMS-based IPTV change this perception, as the IMS subsystem becomes the session controller for multiple services in other subsystems.

The subsystems of NASS and RACS can be used together with the other subsystems. The NASS is responsible for enabling admission of fixed terminals into the network with an allocated IP Address and logged location. The RACS is responsible for resourcing according to network policy rules and service requirements, for both Access Network (AN) and Core Network (CN).

Figure 2.6 is a schematic representation of the high-level TISPAN subsystem architecture. It contains two distinct layers: (1) the Transport layer, with transport functions and media bearing nodes; and (2) the Service layer, with session control, features, and applications.

Figure 2.7 is a more elaborate view of the TISPAN network solution for a converged network, showing how different access networks, including legacy CSN, make use of the common subsystems. This model distinguishes between the IP-CAN and the Core Transport IP Network, thus enabling these IP-CANs to co-exist with their different requirements. Note the distributed data function between the subsystems, which is still a central repository.

2.3.3.4 The TISPAN Integration with IMS

The elements of the TISPAN subsystems must be fully integrated into the 3GPP definition of functions in IMS. This process requires cooperation between 3GPP and TISPAN, gaining an understanding of the common parts, regardless of the initial different set of concepts and terminology.

To that purpose, TISPAN defines "Core" IMS functions which remains the same as in 3GPP while other functions are added or modified with further TISPAN definitions of reference points and protocol profiles.

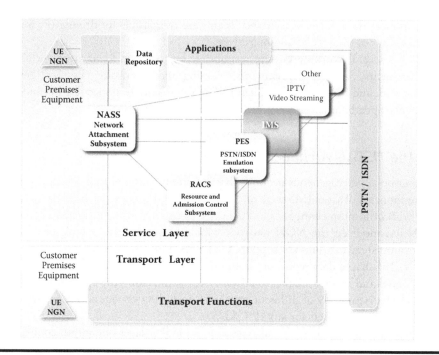

Figure 2.6 TISPAN subsystem architecture.

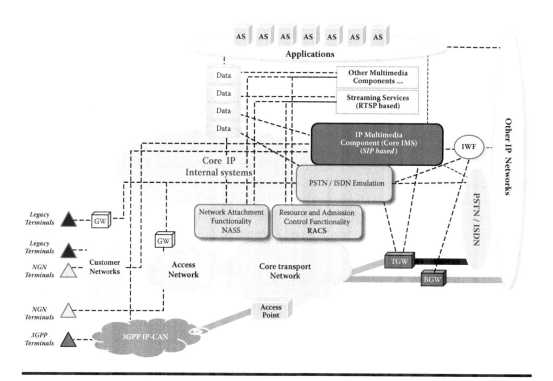

Figure 2.7 TISPAN subsystems architecture.

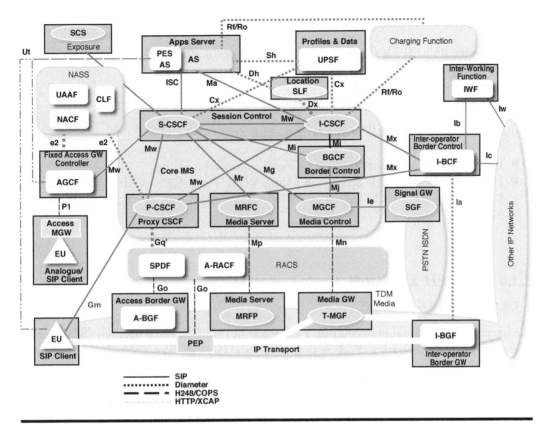

Figure 2.8 TISPAN Release 1 reference model.

The TISPAN components include:

- The UPSF (User Profile Server Function) as the data storage (equivalent to HSS but without the HLR [Home Location Register])
- The UAAF (User Access Authorization Function), NACF (Network Access Configuration Function), and the CLF (Connectivity session and Location repository Function) is used in the NASS for network attachment
- The AGCF (Access Gateway Control Function) is used for connecting IMS to the H.248 local gateway

Figure 2.8 is the IMS reference model, showing the interaction of IMS components with the TISPAN components, through the TISPAN defined reference points.

2.3.3.5 TISPAN Releases Plan

The first TISPAN release was published in December 2005.

Release 1 deals with Multimedia services, including:

- Nomadic users and user-controlled roaming
- xDSL access
- The RACS with service-based policy
- Access Network Attachment Sub-System definition

Release 2 plans to enhance access resource usage, video subsystem, Enterprise, and PES:

- Video streaming IPTV (Internet Protocol-based Television)
- RACS enhancements
- Enhancements to PSTN/ISDN support
- Enterprise-hosted services and trunking requirements

Release 3 will deal with full inter-domain nomadism:

- Internetwork domain nomadic user with user-controlled roaming
- Higher bandwidth access services including VDSL (Very high speed Digital Subscriber Line), FTTH (Fiber to the Home), and WiMAX (Worldwide Interoperability for Microwave Access).

2.3.4 The IETF Standards

2.3.4.1 The IETF and IMS

The Internet standards and protocols play a major role in IMS. The IETF develops and promotes Internet standards, cooperating closely with the W3C and ISO/IEC standard bodies. It deals with standards of the TCP/IP and the Internet Protocol suite (HTTP, XML, etc.).

The IETF work pioneered the ideas of voice over packet and the protocols that can support multimedia communications over the Internet — in particular, IETF has developed SIP that has been widely adopted for VoIP and Multimedia and is the key protocol for IMS.

2.3.4.2 The IETF SIP

The IETF SIP Working Group is responsible for the SIP evolution. The main tasks of the SIP group involve approving the SIP standard and proposed extensions that arise from strong generic requirements. However, this group will not explore the use of SIP for specific environments or applications. This group aims to maintain the basic model and architecture defined by IETF SIP, according to the following principles:

- Services and features are provided end-to-end whenever possible.
- Standards-track extensions and new features must be generally applicable.
- Simplicity is essential.
- Reuse of existing Internet protocols and integrating with other Internet applications is crucial.

The SIPPING Working Group, which is involved in inter-operation requirements and testing, is the main contributor of new requirements for consideration by the SIP Working Group. Additional requirements are produced by other IETF Working Groups using SIP, including the SIMPLE WG, which is using SIP for messaging and presence, and the XCON WG, which is using SIP for centralized conferencing.

In addition, the group is dealing with known issues related to SIP, such as security and privacy mechanisms, including:

- Secure expression of identity in requests and responses
- Secure request of services delivery by non-terminal elements ("end-to-middle"), for example by the AS or any Back-To-Back-User-Agent (B2BUA)
- Guidelines for the use of existing security mechanisms with SIP, such as TLS (Transport Layer Security), IPSec (Internet Protocol Security), certificates, and SAML (Security Association Markup Language).

2.3.4.3 The IETF Organization

The IETF is now formally an activity under the umbrella of the Internet Society. During the early 1990s, the IETF changed from an activity of the U.S. Government to an independent global activity associated with the Internet Society. The IETF is an open, all-volunteer organization with no formal membership. It consists of multiple Working Groups and BoFs ("Birds of a Feather" — interest groups), each dealing with a specific topic, with the intent to shut down when the work is deemed complete.

The Working Groups are organized into "Areas" by their topics. An Area has an Area Director who appoints a Working Group Chairperson. The Area Directors together with the IETF Chairpersons form the Internet Engineering Steering Group (IESG), which is responsible for the overall operation of the IETF. The IETF Working Groups strive to achieve rough but wide general agreement. There is no voting, but rather a consensus-based decision-making procedure that ensures future adoption of the standard. The Working Groups are open to all who want to participate and have much of their processes performed through low-cost e-mail and Web-site media.

2.3.4.4 The RFC Process

An "Internet Standard" in the IETF is a specification for a new internetworking technology or methodology that the IETF has ratified. Through the Internet Society, engineers and computer scientists can publish and discuss any technical idea or proposed standard as an RFC (Request For Comments) Memorandum.

RFCs that contain drafts of standards must go through a process of ratification. In due course, the IETF may adopt some of these proposed RFCs as *Internet Standards*. The process involves posting an Internet Draft, which is open to general peer review by the engineering community. The RFC Editor allocates a unique serial number to each RFC document. Amendments are published as revised documents, thereby rendering previous ones obsolete but retaining the old documents as a track record.

The IETF RFC ratification process is fundamentally different from that of standards organizations such as ANSI (American National Standards Institute) and ETSI. It is an experience-driven, after-the-fact standards definition, which is accomplished by individuals or small workgroups rather than elected committees. Internet Drafts can be submitted by anyone, thus encouraging innovation. Only the IETF, represented by the IESG, can approve an RFC as an Internet Standard.

Internet Drafts are working documents that are not intended to be cited or quoted in any formal design document. They include drafts that have not been reviewed or undergone any hardening, and therefore cannot be relied upon. Unrevised documents placed in the Internet Drafts directories have a maximum life of six months. After that time, they must be updated or they will be deleted. Some of the interest areas have also been removed because all the drafts under these areas have become STDs or were deleted.

The RFC ratification process, called "standards track," follows three maturation stages: Proposed Standard, Draft Standard, and full Internet Standard:

1. Proposed Standard specification must be already stable, with resolved design issues and well-understood functionality. It should have gone through extensive engineering scrutiny and proven to attract sufficient interest from developers. Such proposed standards can undergo changes, or even be retracted.

2. Draft Standard is the next stage, when its specification has been the basis of at least two independent and interoperable successful implementations. A Draft is not expected to change in any significant way, so it is commonplace for vendors to go ahead and use it in their advanced products.

3. Internet Standard specification status is achieved when a Draft has been proven through significant implementations and successful operational experience. An STD is characterized by a high degree of technical maturity. It should be widely considered a protocol or a service that provides significant benefit to the Internet community.

This process of maturing the drafts by seeking widespread community review is deemed fair and resistant to pressure from the larger vendors The fact that it requires practical implementation means that the theory is hardened into a realistic specification before it can be fully qualified but delays the final approval.

At the same time, this is criticized by many developers for lacking milestones and roadmaps, and for having no clear plan of publication. Therefore, it is an unpredictable process that cannot be relied on in a program of development. Because the approval process is lengthy and requires real implementations, vendors often have to implement an RFC long before it becomes a confirmed standard, taking the risk that a particular RFC may never be fully accepted as an STD. As such, the RFC process lacks the clarity of other organizations and may create pressure by vendors who already have committed to immature drafts. The IETF process also fails to resolve opposing opinions and rival methods that may have been proposed as different RFCs, and hence promotes uncertainty, fragmented market, and duplication of effort.

2.3.4.5 The IETF Areas and Working Groups

The IETF acts as an open forum for discussions of any new technology idea. The IETF Working Groups are engaged in evaluating many new Internet Drafts prior to becoming RFCs and many that never become RFCs. These Working Groups within their Area of Interest may be closed down when they conclude the evaluation of the drafts. For example, Concluded Areas include the IP Next Generation Area, Operation Requirement Area, OSI Integration Area, Sub-IP Area, and User Service Area.

The following are the current Working Group Areas with examples of their topics:

◾ *Applications Area:* These are draft standards for applications-related items. The applications can be Calendaring and Scheduling, Messaging, and E-mail. Such items can be, for example, Internet Message Access Protocol Extension and Language Tag Registry Update. Functions related to Internet E-mail include, for example, E-mail Address Internationalization,

Enhancements to Internet E-mail to Support Diverse Service Environments, and Sieve Mail Filtering Language.

- *General Area:* This includes, for example, Intellectual Property Rights.
- *Internet Area:* This area includes issues related to the IP protocol, such as IP over IEEE 802.16 Networks, IPv6 and IPv4 Mobility, and Handoff Optimization. It also includes topics about DHCP, DNS Extensions, Extensible Authentication Protocol (EAP), Layer Two Tunneling Protocol (L2TP) Extensions, Point-to-Point Protocol (PPP) Extensions, etc.
- *Operations and Management Area:* This area deals with protocols for Management — for example, RADIUS EXTensions and Diameter Maintenance and Extensions. It covers general issues of Network Configuration, Benchmarking Methodology, Internet and Management Support for Storage, and Packet Sampling. Operational issues also include IPv6 Operations, Global Routing Operations, IP Flow Information Export, IP over Cable Data Network, Operational Security Capabilities for IP Network Infrastructure, etc.
- *Real-Time Applications and Infrastructure Area:* This area includes session control protocol issues, i.e., SIP, and related topics such as SIP Services interoperability and SIP for Instant Messaging). Also in this area are real-time interaction issues, for example, Multiparty Multimedia Session Control, Peer-to-Peer Session Initiation Protocol, IP Telephony, Media Server Control, and more. In addition, communication issues such as Telephone Number Mapping, Geographic Location Privacy, Emergency Context Resolution, and Centralized Conferencing are also in this area.
- *Routing Area:* Routing issues include, for example, Inter-Domain Routing (secure and not secure) and Protocols, Layer 1 Virtual Private Networks, and Virtual Router Redundancy Protocol. Methods of routing include Bi-directional Forwarding Detection, Protocol-Independent Multicast, Multi-Protocol Label Switching (MPLS), Open Shortest Path First (OSPF) Interior Gateway Protocol, and Path Computation Element.
- *Security Area:* Many security spheres are included in this area, for example, Multicast Security, Domain Keys Identified Mail, Better-Than-Nothing Security, and Simple Authentication and Security Layer. This area includes security protocols such as TLS, S/MIME (Secure/Multipurpose Internet Mail Extensions) for e-mail security, EAP Method Update, and Public Key Infrastructure. Also considered are Handover Keying, Integrated Security Model for SNMP, Provisioning of Symmetric Keys, Kitten (GSS-API Next Generation), Kerberos, Long-Term Archive and Notary Services, Network Endpoint Assessment, and Security Issues in Network Event Logging.
- *Transport Area:* This area includes traffic management such as Congestion and Pre-Congestion Notification, Datagram Congestion Control Protocol, Behavior Engineering for Hindrance Avoidance, and Reliable Multicast Transport plus IP Performance Metrics. This area also includes management of the transport nodes and storage, including the FEC Framework, IP Storage, Middle-box Communication, Network File System Version 4, Remote Direct Data Placement, Robust Header Compression, and Reliable Server Pooling.

2.3.5   The OMA

2.3.5.1   OMA Goals and Activities

The OMA was established in June 2002 by a group of influential mobile operators and major equipment manufacturers. The OMA has amalgamated several industry fora, including:

- The Wireless Village (focused on instant messaging and presence)
- The WAP Forum (focused on browsing and device provisioning protocols)
- The Location Interoperability Forum
- The SyncML Consortium (focused on data synchronization)
- The Mobile Games Interoperability Forum
- The Mobile Wireless Internet Forum

The OMA was created to gather these initiatives under a single umbrella, making sure that there is no overlap or incompatibility. The OMA set out to provide open standard-based enablers for the intelligence layer of the new integrated network. In doing that, the OMA is ensuring that services will interoperate across business borders from the outset and prevent market fragmentation as has occurred in the past. Although started as a closed group, the OMA later stated its commitment to openness in an effort to achieve broad industry adoption.

The OMA ensures that the service enabler specifications provide interoperability across different devices, geographies, service providers, operators, and networks, and encourages consolidation of standards to decrease operational costs. In its work, the OMA considers all parts of the value chain, including content and service providers, information technology providers, mobile operators, and wireless vendors.

2.3.5.2   The OMA Model

The OMA architecture establishes the service environment in relation to the session control (i.e., Core IMS), treating IMS as a "black box," not to be tampered with. The architecture clearly identifies the enablers in both the endpoint client and the application server. The UE interacts with Core IMS in the normal way (using SIP) to activate a session and a service from the AS. There is also a direct interaction between the UE and the AS for non-IMS services and for an exchange of service configuration data. To provide connectivity to service enablers, there are clients in both the UE and the AS:

- ETI: Enabler Terminal Implementation
- ESI: Enabler Server Implementation on the AS

The ETI and ESI ensure that the UE can interface to enablers and to applications making use of these enablers.

The OMA services make extensive use of Web technology and interfaces to the Internet. They contain many elements regarded as data services or Web-based services that are not governed by IMS. Therefore, the OMA model describes separately service enablers for IMS and non-IMS elements.

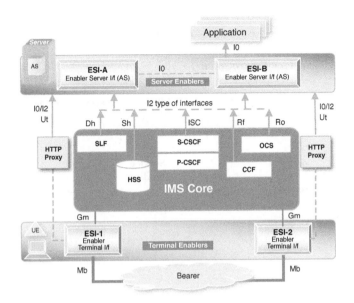

Figure 2.9   OMA model for IMS and non-IMS services.

2.3.5.3   OMA and IMS

The OMA concentrates its efforts on service delivery. Multimedia services rely on the client in the handset or laptop on the one hand, and the server with the application on the other. The OMA defines enablers that operate over the OMA-defined reference points. These enablers must be implemented on both the terminal and server sides.

Figure 2.9 depicts the interfaces to IMS from both the UE side with its ETI, and from the AS side with its ESI. The normal IMS reference points (the Dh and Sh interfaces) link the ESI to the data store on the HSS. The ISC channels session control, and Rf and Ro transfer charging information, as per IMS. The OMA has introduced direct link between the AS and the EU, now generally accepted, which is the Ut interface that carries HTTP XCAP information, such as Presence status.

2.3.5.4   The OSE

The OMA Architecture Working Group is developing further the OSE (OMA Service Environment) architecture, which is becoming more complex with multiple business models, greater openness, a wider and richer variety of types of services, along with their seamless integration. To that end, the OMA defines an OSE that encompasses the service enablers and their interfaces to the Session Control layer, as well as provisioning, life-cycle service management, links to business systems such as ERP (Enterprise Resource Planning) and CRM, etc.

The OSE contains a Policy Enforcer that interacts with the application to enable application-aware QoS control and other network policies. The Policy Enforcer architectural element supports enforcement of policies that call for using, for example, special authentication and authorization or reassigning common facilities at the Service layer. The Policy Enforcer may use enablers to evaluate and enforce the policies that have been specified for the domain or target enabler. The

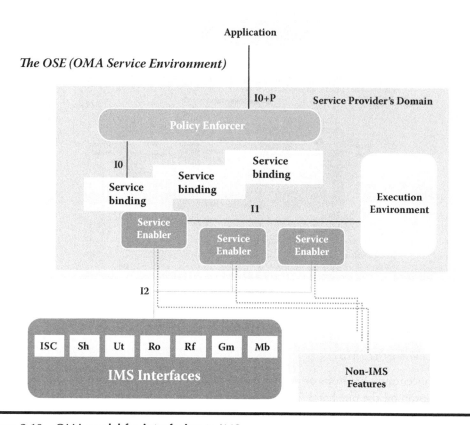

Figure 2.10 OMA model for interfacing to IMS.

Policy Enforcer also can be used to compose enablers into higher-level functions, although these have not been defined yet by the OMA.

Figure 2.10 shows OMA links to IMS interfaces, such as SIP (ISC, Gm), Sh (Diameter to HSS), Ut (HTTP directly to UE), charging interfaces (Ro and Rf), and the media (Mb). Non-IMS nodes also can interface to the OSE. This OSE diagram also depicts the Policy Enforcer function that enables performing service-dependent policies. Such service policies are maintained separately from the network policies defined in the PCRF, but operators can choose to combine these functions in the future and coordinate their effects.

2.3.5.5   OMA CEF

The Service Architecture as defined by the OMA includes a Capability Exposure Function to allow access to third parties wishing to use network capabilities in a wholesale business model. This function provides a framework that regulates access to network resources by external service providers and monitors their SLA (Service Level Agreement) to ensure appropriate allocation of resources.

Parlay/OSA has already addressed some of the issues of enabling external service providers, including corporate networks as well as network-less service providers, to use the network resources and to manage the relationship with the network carrier. Now OMA is taking it further, adjusting the thinking to accommodate Web services, multimedia, and content delivery.

The Application Environment consists of:

■ *The Capability Exposure Function (CEF):* These functions provide the public interfaces used to access the network component in a form suitable for a particular class of application layer. An example CEF may be a Parlay gateway or a simple Web services interface.
■ *The Application Environment:* This is the software infrastructure in which an application runs. Such an environment can be situated locally, at the operator's CN (hosted applications), or externally, on a service provider's premises (third-party applications). Examples of application environments could include IBM Web Sphere or the Microsoft .Net enterprise servers.
■ *The Service Provider Framework:* This function allows application service providers to gain access to the CEF. It provides for security, account management, SLA administration, and discovery of CEF servers.

While OSA exposed network operations that allowed call management, the OMA is concerned with the exposure of a wider range of network capabilities and service enablers that can be utilized by external entities. OSA/Parlay provides a gateway to CSN by defining APIs (Application Programming Interfaces) that are translated into INAP (Intelligent Network Application Part) and CAMEL. OMA provides a similar function to IP networks with SIP and IMS session control.

Figure 2.11 shows the SP (Service Provider) creating applications that are "untrusted" external applications. These applications can utilize the "exposed" network capabilities via the provided function, having passed through the framework. The common capabilities, including the IMS session control, HSS data sharing, location information, authentication procedures, etc. are available for use for both internal applications as well as external network carriers.

The OMA also defines an SP Framework that allows the SP to be recognized and authorized to use the network and to manage the SLA that governs and monitors usage. Such an SLA for multimedia services brings in different considerations than those used for CSNs, for example, treatment

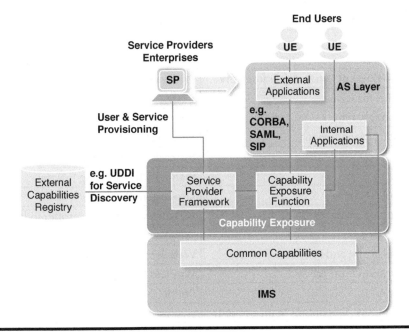

Figure 2.11 OMA Service exposure.

of long-duration sessions, DRM (Digital Rights Management), variable quality of services, wider bandwidth for certain services.

An external registry can be set up for users to discover the service independently and automate the procedure, allowing for service collaboration, Internet style. The SP can use an exposed API to activate user authentication, even Single-Sign-On, that may be provided by the network carrier.

2.3.5.6 OMA Specifications and Enablers

The OMA maintains a number of specifications, including:

- Browsing facility, previously known as WAP browsing, now based on HTML Mobile Profile
- MMS (multimedia messaging)
- DRM
- IMPS (Instant Messaging and Presence Service) for mobile phones (formerly Wireless Village)
- SIMPLE IM (Instant messaging based on SIP-SIMPLE)
- Specification for Data Synchronization (DS) using SyncML
- Specification for Device Management (DM) using SyncML
- Specification for Mobile Broadcast Services (BCAST)
- Specification for Push to Talk Over Cellular (PoC)

The following table provides a list of enablers found in OMA publications to help product development:

Browser Protocol Stack	On-Board Key Generation
Billing Framework	Mobile Location Service
Browsing	Multimedia Messaging Service
Client Provisioning	Online Certificate Status Protocol Mobile
Client-Side Content Screening Framework	Presence Simple
Data Synchronization	Push
Device Management	Push to Talk Over Cellular
Digital Rights Management	Secure User Plane for Location
DNS	SyncML Common Specification
Download	Standard Transcoding Interface
E-mail Notification	User Agent Profile
External Functionality Interface	vObject Minimum Interoperability Profile
Games Services	Web Services
Instant Messaging and Presence Service	Web Services Network Identity
Mobile Location Protocol	Wireless Public Key Infrastructure
IMS in OMA	Profile
XDM – OMA XML Document Management	

These enablers assist in the development of multimedia services. The implications on the Core IMS functions are conveyed to 3GPP and TISPAN, to ensure full compatibility.

2.3.5.7   The OMA Specified Services

Although the OMA primarily is concerned with generic service environment and service enablers, it also has produced detailed specifications of certain services, especially services that can be used as enablers for further functions. In particular, the OMA has been involved in specifying the following:

- *PoC:* defining a service that enables multi-way conferencing in a similar fashion to a "walkie-talkie," where only one person has the "floor" having pushed a button to speak, but the whole group is listening. PoC is now expanded to video and image as well.
- *Advanced Messaging:* investigating multimedia messaging technologies while leveraging the OMA architecture, basic and advanced messaging, plus content handling is specified.
- *Presence and Availability:* addressing presence and availability requirements from all technologies, the OMA develops shared presence and availability specifications to avoid duplications and overlaps. The OMA aims to make the presence and availability technology protocol-neutral in its effort to encourage reuse of the technology by multiple services.

2.4   The Converged IMS Standard Components

2.4.1   The Converged IMS Component Map

The IMS standards are maturing now with a wide range of functions and network elements. The next effort is to amalgamate and simplify some of these functions as convergence efforts progress.

Some areas of functionality lend themselves to merging for all types of access and for "tethered" (Fixed) and "untethered" (Mobile) communications. The process of merging functions still continues. Therefore, the diagram in Figure 2.12 is only a suggestion of the combined architecture. It shows the range of IMS functions with the three access networks (xDSL, WLAN, and UMTS) and the range of IMS products that facilitate the convergence. The following is a list of the standard functions that make up a converged Fixed-Mobile IMS network, as defined by the 3GPP and TISPAN, grouped by their functions.

To summarize the current standard functions, a list of them is given here, containing both NGN and 3G Mobile functions, classified according to their roles.

2.4.2   Session Control Components

The *CSCF* is the session controller, the nerve center of Core IMS. It contains subcomponents that are defined independently.

The *Serving CSCF (S-CSCF)* acts as a switching center with access to full user details. It connects sessions, maintains session state and links to appropriate applications, and subsequently produces charging records. It is responsible for the triggering of the subscribed applications in the right sequence and for management of mid-call events as well as orderly call termination.

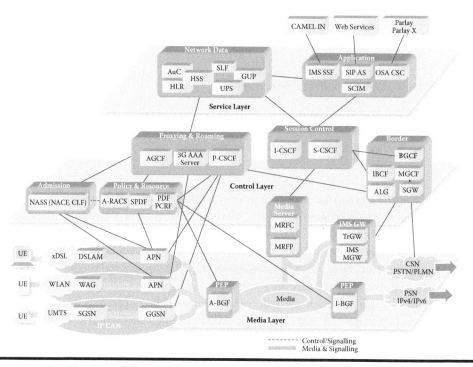

Figure 2.12   Layers and components of converged IMS.

The *Interrogating CSCF (I-CSCF)* acts as a forwarding agent and a topology-hiding server. It interrogates the HSS for locations of S-CSCF for users and routes to Home Networks.

The Proxy CSCF (P-CSCF) is described below.

2.4.3   Proxies and Access Gateways

The *Proxy CSCF (P-CSCF)* acts on behalf of the SIP user as a proxy for sending messages to network servers, assisting in admission control, authentication, and resourcing. It fields all signaling messages, such as registration, re-registrations, and session INVITE messages to the appropriate network servers. It also plays a key role in the Visited Network, routing messages between the user access side and the network border.

The *AGCF (Access Gateway Control Function)* manages the access for analog and H.248 devices communicating to IMS via the PSTN Emulation System (PES). It facilitates signaling between the analog devices (via residential or other access gateways) and the SIP-based Core IMS functions, appearing to Core IMS as another P-CSCF. It initiates the authentication of the "line" and binding it through the NASS or retrieving line profiles from the NASS location database (the CLF) before requesting registration at the S-CSCF. It also controls the access-side media gateway functions for tones, announcements, etc., and creates charging identifiers per session. The PES features are provided via a PES AS connected to the S-CSCF.

2.4.4   Data Management

The *HSS (Home Subscriber Server)* is a data repository and retrieval processing function. It is central to the IMS subsystem and is an essential component that ensures sharing of information between multiple network servers and functions. It is important in maintaining data integrity and reducing the costs of user management. The HSS contains ID management, user profiles, and service triggering data, as well as integrating Service layer authentication.

The *UPS (User Profile Server)* is the equivalent of HSS in the TISPAN specifications. The UPS (or UPSF) contains user profiles and service subscription data similar to the HSS, but excludes the HLR location management. The wireline location/line configuration and authentication data is held on the CLF (Communication session Location and repository Function) and PDBF (Profile DataBase Function), respectively.

The *SLF (Subscription Locator Function)* is a database that contains knowledge of a distributed HSS system, including the IP Addresses for the HSS and other database particulars, such as regional servers, server hierarchy, and distribution parameters. The SLF is only required where there are multiple HSS servers.

The *HLR (Home Location Register)* is the location manager in CS Mobile Networks. The equivalent function for PS (Packet-Switched) Networks is in the HSS. Early implementations of IMS interface to the HLR to manage user mobility between CS Mobile Networks and PS Networks.

The *VLR (Visitor Location Register)* function is the proxy element of the HLR in a Visited Network that enables roaming users to sign in with their Home Network and receive their services outside their own network.

The *AuC (Authentication Center)* is a subset of HSS information that is responsible for authentication and security. The AuC operates together with the HLR in the HSS.

The *GUP (Generic User Profile)* is a server that facilitates retrieval of distributed data from multiple data repositories and authenticates the requestors of the data. It was introduced in Release 6 as a generic data retrieval system, and the RAF (Repository Access Function) for the HSS was specified in Release 7.

The *EIR (Equipment Identity Register)* was introduced to GSM as the central store of International Mobile Equipment Identities (IMEIs), which identify the mobile handsets when roaming. IMEIs still can be used in IMS in some cases for early methods of device authentication.

The *ENUM (Electronic Numbering) and DNS (Domain Name Server)* are servers already in use for IP routing for Internet and IP data services. They are used to resolve domain names into actual server addresses and translate telephone numbers via the ENUM (Electronic NUMbering) algorithm to routable IP Addresses.

2.4.5 Media Server

The *MRF (Media Resource Function)* represents a pool of media resources for IMS that provides support for bearer-related services, such as bridging multi-party sessions, user announcements, or bearer media transcoding.

The *MRFC (Media Resource Function Controller)* is the signaling controller element that interacts between the media processor and the requestors of media services.

The *MRFP (Media Resource Function Processor)* is the equipment that provides the media resources, that is media connection, media mixing and bridging, media transcoding, recording, and playing or broadcasting stored media.

2.4.6   Admission and Resource Management

The *NASS* (Network Attachment Sub-System) provides the mechanism for attaching users to the network. This function allows nomadic users, logging in via soft clients on PDAs or laptops, to receive an IP Address, to bind the User ID with the location and address, and to be authenticated.

The *UAAF (User Access Authorization Function)* is the NASS component that authenticates wireline terminals and allows wireline users to gain access to the network. This element uses Diameter to perform authentication against prestored security information in the PDBF (Profile DataBase Function).

The *CLF (Connectivity session Location and repository Function)* is the server that keeps location information for NASS users, that is, it keeps the path information, including the physical nodes used to connect a Fixed Line terminal to the IMS network. The location is stored when users register through the NASS. The CLF is interrogated by the P-CSCF and I-CSCF to route incoming calls.

The *AF (Application Function)* is a component in either the P-CSCF or in AGCF that requests resources from an IP bearer and helps ascertain the charging mechanism.

The *PDF (Policy Decision Function)* in 3GPP analyzes requests for resources against a predetermined set of rules and parameters, and decides on resource allocation to a session that has been requested by User Equipment (UE) or an application. Note that this function is now superseded by the *PCRF (Policy and Charging Rules Function),* as the combined PDF and CRF (Charging Rules Function) (see below).

The *RACS (Resource and Admission Control Subsystem)* is defined by TISPAN as a subsystem responsible for controlling NGN admission (setting up sessions or non-sessional communications) and reserving resources (such as bandwidth or ports) for that communication. The RACS contains the SPDF and A-RACF as described below.

The *SPDF (Service Policy Decision Function)* is defined by TISPAN as a similar role to that of the PDF, but with emphasis on service-based resource allocations rather than bearer flows. It vets the requests according to predetermined rules and makes decisions regarding prioritization and level of resources to allocate.

The *A-RACF (Access – Resource and Admission Control Function)* is the element in the RACS that performs the reservation of resources toward the access network, committing these resources and releasing them, according to instructions from the SPDF. It also sends notifications of events if requested by the SPDF.

2.4.7   Breakout to CSN

The *BGCF (Breakout Gateway Control Function)* is the signaling server that determines where to exit the current network. The BGCF can be set up as an exit contact point on various internal network servers, and connects to several IMS nodes when the destination requires a PSTN breakout or connection to another IP network. The BGCF determines the next hop for routing the SIP message, for example, to a media gateway controller.

The *MGCF (Media Gateway Control Function)* is the server that enables IMS to communicate to and from PSTN or ISDN by determining the media path and media management requirements. The MGCF is responsible for controlling the media channels in an IMS-MGW (IMS Media Gateway) and the call state that relates to that media connection. It determines the next hop for the media path, depending on the routing number for the legacy networks. It also performs protocol conversion between ISUP TCAP and the IMS call control protocols.

The SGW (Signaling Gateway) performs signaling conversion (both ways) between SS7- based CS networks and IP-based signaling, that is, between SIGTRAN SCTP (Stream Control Transmission Protocol) and IP and SS7 MTP. The SGW does not interpret the application layer (e.g., MAP, CAP, BICC, ISUP, INAP) but may have to interpret the underlying SCCP (Signaling Connection Control Part) or SCTP (Stream Control Transmission Protocol) to ensure proper routing of the signaling.

The *IMS MGW (IMS Media Gateway)* is responsible for converting CS communication to PS networks and vice versa, to enable interworking between the CS domain and the PS domain. It terminates bearer channels from both CS networks and media streams from a packet network (e.g., RTP streams in an IP network). It also handles resources such as echo cancelers and supports various codecs for media conversion.

2.4.8 IP Border and Media

The *IBCF (Interconnect Border Control Function)* forwards the session signaling to the destination network, applying the operator's security policies and methods, as received from the PDF. The IBCF provides application-specific functions at the SIP or SDP protocol layer to perform interconnection between two operators' domains. These include network topology hiding and screening of SIP signaling information.

The *TrGW (Transition Gateway)* is located within the media path and is controlled by the IBCF. The TrGW function provides network address/port translation and IPv4/IPv6 protocol translation for media packets as well as for signaling.

The *IWF (Interworking Function)* provides the functionality necessary to allow interworking between the IMS network and any other network running different protocols or variations of them. It includes interworking to H.323 session control and VPNs (Virtual Private Networks), as well as different "SIP Profiles" that can vary in different implementations.

The *SEG (Security Gateway)* is responsible for enforcing security policies between network segments and in the borders with external IP networks. The security is provided via security association for IPSec but may also include filtering policies and firewall functionality. The SEG is an IPSec tunnel responsible for enforcing the security policy of each IP security domain, and links with other SEGs. All IP traffic must pass through an SEG before entering or leaving a security domain.

The *PEF (Policy Enforcement Function)* is the generic function that enforces the decision by setting up the bearer's packet flow as instructed. It can restrict which IP destinations can be reached by the IP-CAN bearer, according to a packet classification process. PEF nodes are often incorporated in relay network nodes or network access point nodes, such as GGSN. Note that this is a generic function that is delivered through the border media gateways.

The *BGF (Border Gateway Function)* is a gateway between a network supporting GPRS/UMTS and an external inter-network backbone network, or with other operators' packet networks. The Border Gateway performs the same function between NGN network and interconnecting IP networks, such as GPRS or VPNs.

2.4.9 Application Server Types

The *AS (Application Server)* in IMS is from where all the features are served, as the Session Controllers are dedicated to session management, not services. All IMS-AS interfaces to the session controllers via ISC, which is extended SIP, and to the HSS data repository via Diameter. IMS supports three types of application servers, in terms of their interfaces to the Session Control components, based on SIP, OSA, or IN/CAMEL.

The *SIP Application Server* is any platform that provides SIP-based applications. Typically, this type of platform also interfaces to Web services.

The *OSA Application Server* is a platform that provides Parlay/OSA-based services. The OSA framework manages external service providers. Parlay/OSA is used to interface to softswitches and TDM switches using INAP and CAP, but some platforms are evolving by providing Parlay-to-SIP conversion, namely the ISC interface to interwork directly with IMS session elements.

The *CAMEL IM-SSF (IP Multimedia Service Switching Function)* is the module that enables traditional CAMEL platforms to interface to IMS session elements. This module converts CAP commands to SIP messages compatible with the SIP ISC interface.

The *SCIM (Service Capability Interaction Manager)* is also known as service orchestration, service choreography, or service broker. This function ensures that services are triggered in the right order, using input from the previous service when applicable. It supports non-IMS services as well. It is required where complex service interaction may occur and where non-IMS services are used in the IMS session.

The *PES AS (PSTN/ISDN Emulation Service AS)* is a type of SIP AS that contains service logic to emulate PSTN/ISDN services, working together with the AGCF primarily, but also via a VoIP Gateway and P-CSCF. A *PSS AS (PSTN Simulation Services AS)* is the same as any other SIP AS, but is dedicated to Telephony features.

2.4.10 WLAN Components

The *3GPP AAA (Authentication, Authorization, Accounting) Server and Proxy*, also referred to as AAA Server and Proxy, perform the main attachment process for WLAN-enabled UE to access IMS. Where roaming is detected, according to the WLAN UE's realm ID, the local AAA Proxy provides the initial access service and relays requests for service authorization to the Home AAA Server. The AAA Server authenticates the WLAN user at access level for initial network attachment, provides it with an IP Address, and establishes roaming needs and charging regimes for the session.

The *WAG (WLAN Access Gateway)* is a gateway for the local WLAN access nodes through which the UE gains admission to the IMS network. In the case of roaming, the WAG resides in the Visited Network. The WAG provides filtering, policing, and charging functionality for the traffic between WLAN UE and 3GPP or TISPAN NGN networks.

The *PDG (Packet Data Gateway)* provides media-level access to PS-based data services from the WLAN. The process of authorization, node selection, and subscription checking determines whether a service will be provided by the Home Network or by the Visited Network. It performs address translation and mapping, encapsulation and decapsulation for tunneling, authentication and ensuring the validity of the WLAN UE, as well as charging reports.

2.4.11 Charging Components

The *OCS (Online Charging Subsystem)* responds to Credit Control Requests to produce real-time credit and charging reports. The OCS includes the Bearer Charging Function (BCF) for bearer nodes, the Session Charging Function relating charging to controlled sessions, and the Event Charging Function (ECF) relating to chargeable network events.

The *IMS GWF (IMS Gateway Function)* provides interfaces between the CSCF and the OCS, smoothing out incompatibilities. The OCS can be connected directly to IMS elements, where implementations permit it.

The *OFCS (Offline Charging Subsystem)* is the system that provides charging information during and after session termination, that is, not in real-time. It contains the CCF (Charging Control Function), which consists of the CDF and CGF.

The *CDF (Charging Data Function)* part of the OFCS is where Changing Data (CDRs) are generated. The CDF receives requests to start recording, modify, or terminate the recording of charges from various IMS elements, such as S-CSCF, I-CSCF, P-CSCF, and MRF. The records identify the requesting IMS element for later reconciliation of the information. The CDF can also receive modifiers in mid-session, and finally, a request to cease recording. At this point, the CDR is completed and closed.

The *CGF (Charging Gateway Function)* part of the OFCS is an intermediate node that simplifies the interface to the billing system. It acts as the central point for data collection from the numerous network nodes involved in the charging process. It can perform some preprocessing, such as a consolidation of CDRs designed to reduce the load on the billing system and to optimize its operations, as well as to shield it from frequent changes.

The CRF (Charging Rules Function) acts as a policy controller for flow-based charging functionality. The CRF selects and provides the applicable charging rules to be applied to the GGSN and WLAN, and potentially other bearers. The CRF is now part of the PCRF (see below).

The PCRF (Policy and Charging Rules Function) is the integrated function that includes the CRF functionality with an extended range of network policies. Such policies can be used for resource allocation in conjunction with charging rules, therefore combine PDF/SPDF and CRF in one element.

2.4.12 User Equipment (UE)

UE refers broadly to the user endpoint device that allows a user to access network services. In NGN, this can be, for example, a residential gateway, laptop, or phone. In 3G, it is a mobile station with an IMS client (ISIM).

The 3GPP UE is the element that interfaces to the network over the radio interface. This is referred to as the MS (mobile station). It comprises the mobile equipment (ME) and the subscriber identity module (SIM). The ME contains the mobile termination (MT), which, depending on the application and services, may support various combinations of terminal adapters, and the terminal equipment (TE) functional services. These groups can be implemented in one or more hardware devices.

The ISIM is the IMS client within the UE. This client contains the IMS identifiers (public and private IDs), with security materials and procedures to allow authentication.

The USIM (UMTS subscriber identity module) is the UMTS client version of the SIM, used for authentication of the data services provided by GPRS, alongside the GSM normal authentication.

The IMS-UE is a UE with ISIM (IMS SIM card) on board or an IMS client in a laptop. The IMS client is compliant with IMS standards for UE identification and authentication and other IMS signaling that provide QoS and security.

The WLAN UE is the WLAN User Equipment, that is, UE (equipped with a UICC card including (USIM or ISIm) that is utilized by a subscriber capable of accessing a WLAN.

Chapter 3

Session Control

3.1 This Chapter

This chapter describes the various ways that convergence IMS delivers session control, dealing with a variety of IMS session processes. The focus in this chapter is on the various aspects of session control for different types of sessions and scenarios, while further details are given in the following chapters. Rather than spelling out SIP messages and call flows, the varied types of IMS sessions, session flow logic, and how applications are interlaced are described here, to provide full appreciation of the session control process. In particular, note that Chapter 4 describes separately registration and network admission, and Chapter 6 describes allocating session resources.

After a short introduction to the main functions of the session controllers and their respective responsibilities, this chapter gives an overview of session processes, including session initiation, session routing, managing application triggering, session termination, and mid-session processing including event-driven sessions, third-party call control, and session handover with voice continuity. It also describes how media is managed and how media resources are engaged by both session controllers and applications. This chapter also includes the session control aspects of special sessions such as Emergency calls, Immediate Messaging, Number Portability, and Presence.

3.2 Session Control Elements

3.2.1 The Session Control (CSCF) Components

At the heart of IMS is the CSCF, the Call Session Control Function. This is the function responsible for the control of user registration, initiating sessions and managing them, linking to applications, and providing information to session support tasks such as billing and applications. The CSCF, unlike CSN switches, deals only with signaling. Although it is instrumental in defining the media type and media path, the media payload does not traverse CSCF servers. This principle has a profound effect on network topology, as the CSCF is placed centrally but the media-bearing elements are distributed along the network perimeter.

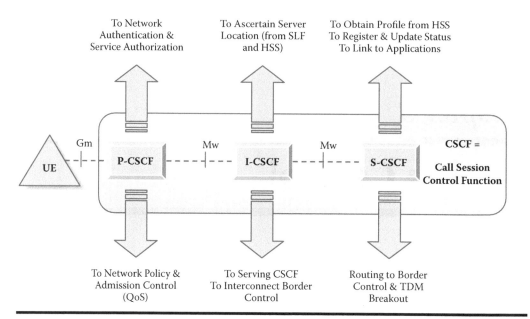

Figure 3.1   CSCF components and links.

As shown in Figure 3.1, the CSCF consists of three functions that can co-reside or occupy separate servers as desired for a particular network size and shape. The three CSCF modules, as described below, are:

1. *P-CSCF (proxy CSCF):* acting on behalf of the user as a proxy for sending messages to network servers, assisting in admission control, authentication and resource allocation, as well as routing roaming user's messages to the Home Network.
2. *S-CSCF (serving CSCF):* acting as a switching center with access to full user profile details, it connects sessions, maintains session state, links to appropriate applications, and subsequently produces charging records.
3. *I-CSCF (interrogating CSCF):* acting as a forwarding agent and a topology hiding server, it interrogates the HSS (Home Subscriber Server) for locations of serving CSCF for users and routes to Home Networks.

3.2.2   The Proxy CSCF

3.2.2.1   The P-CSCF Role

The P-CSCF is primarily a network-based user agent proxy. It behaves like a proxy as defined in RFC 3261 and its updates. It accepts user requests, analyzes them for routing purposes, and then forwards them. In this role, the P-CSCF does not modify the UE-requested SIP message but forwards it on. For terminating calls, the P-CSCF delivers the messages to the destination user via an access-side border gateway where they are implemented.

The P-CSCF can also act on behalf of the user in the role of a user agent, which means that in some circumstances it can terminate and independently generate new SIP transactions. In

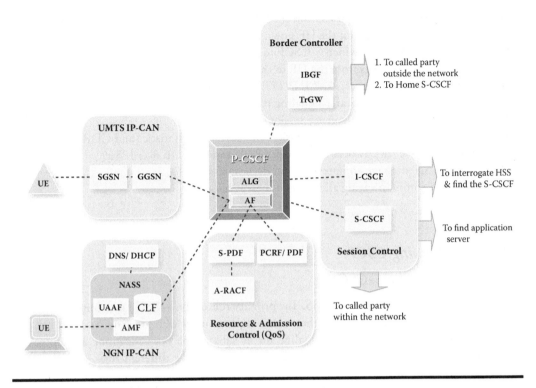

Figure 3.2   Multi-access P-CSCF element and its connections.

performing such a function, the P-CSCF can act as a B2BUA (Back-to-Back User Agent). This can happen for security, routing, and roaming purposes.

The P-CSCF responds to user requests and handles all interfaces to the AN and related subsystems such as NASS. It is the first contact point for the UE with the Core IMS network, having traversed the IP-CAN through its appropriate security gates. In the case of roaming, the P-CSCF forwards session control signaling through the Border Control to the Home Network. Therefore, the P-CSCF enables session admission to the network, with the appropriate addressing, and ensuring interworking between IPv4 and IPv6, QoS parameters setting user/device network-level authentication, and security filtering.

As shown in Figure 3.2, the P-CSCF must interface to many functions, including access gateways with NAT, network attachments such as CLF and NASS for Fixed Line access, resource and admission policies, and Border Control Gateways. The P-CSCF interacts with the S-CSCF and I-CSCF to process session requests and set up calls.

The P-CSCF is often combined with user-agent facilities to request resources on behalf of the user, acting as an AF that interacts with the resource controller (see Chapter 6). The P-CSCF also plays a part in ensuring security, in combination with access-side security gateways.

3.2.2.2   P-CSCF Functions

The functions performed by the P-CSCF include:

■ Acting as an entry point to the network, as a "well-known" (i.e., published) address

- Forwarding registration requests from the UE toward the I-CSCF, for routing to the appropriate domain, depending on the UE's own domain
- Ascertaining via I-CSCF which S-CSCF to use for the specific UE
- Forwarding SIP messages from the UE to the S-CSCF and responding back to the UE
- Routing Registration requests and INVITE session initiation requests
- Recording and enforcing bearer signaling path
- Detecting and handling emergency sessions
- Remaining stateful throughout the session and generating statistics and CDRs
- Maintaining a security association between itself and each UE over IPSec
- Performing SIP message compression and decompression
- Seeking and communicating allocation of bearer resources

3.2.3 The Serving CSCF

3.2.3.1 Roles of the S-CSCF

The S-CSCF acts as a nerve center of IMS. The primary role of the S-CSCF is to initiate, manage, and tear down the session. It manages the session logic for users who are assigned to it. For this reason it must be involved in all the signaling for its subscribers, both incoming and outgoing. Within an operator's network, different S-CSCF servers can take different roles, and may have I-CSCF combined within.

The S-CSCF supports network admission functions for users, UE clients, as well as ASs. It plays a major role in user-AS authentication, service authorization, registration, and re-registration. For session requests, the S-CSCF supports session initiation, session progression control, session state monitoring, invoking and orchestrating services, and producing charging records. The S-CSCF performs session control services for users as well as application-initiated sessions. It maintains a session state, which is needed to support services for the duration of the session.

The S-CSCF can behave as a User Agent (or a B2BUA) as defined in RFC 3261, which may independently generate SIP transactions. It can do this as either a User Agent Client or a User Agent Server. The S-CSCF also can act as a redirect server that forwards the messages to other servers according to user and service parameters in the SIP message headers and the received SDP information.

The S-CSCF is not an IP version of the real switching function. Although responsible for session initiation and processing, unlike TDM switches the S-CSCF is dedicated to session control and must be devoid of any session features. The features are performed in a number of associated ASs that interface to the session controller via a single standard interface, the ISC. In adhering to this principle, IMS can deliver on the promise of an independent and flexible Service layer.

3.2.3.2 Converged S-CSCF Functions

The S-CSCF server has a wide set of functions, including:

- Registering and re-registering users, as defined in RFC 3261, and fetching their profiles from the HSS
- Terminating session-related signaling, and performing set-up and tear-down of sessions

- Monitoring sessions, getting involved in both session-related and unrelated call signaling flows
- Selecting the appropriate applications according to user subscriptions and service profiles
- Routing to application servers, having established their addresses
- Maintaining user status and reporting on it to Application and Presence Servers
- Updating the Location Servers (HLR/HSS in 3GPP or CLF/UPSF in TISPAN) when registering the user
- Rejecting communication to and from any barred users and according to any other communication restrictions imposed by network policies (e.g., expiry time or bandwidth limitation)
- Providing the endpoints and applications with event-related information, such as tones and announcement notifications and connection to additional media resources
- Generating charging records and providing billing notification to the endpoint

Figure 3.3 shows the S-CSCF with its functional components, supporting registration, routing and media connection, charging, etc. It also shows the main functions that connect to the S-CSCF, namely the data repository (HSS), the network attachment and network admission, the application servers, and the elements that support breakout to CS networks and IP border control.

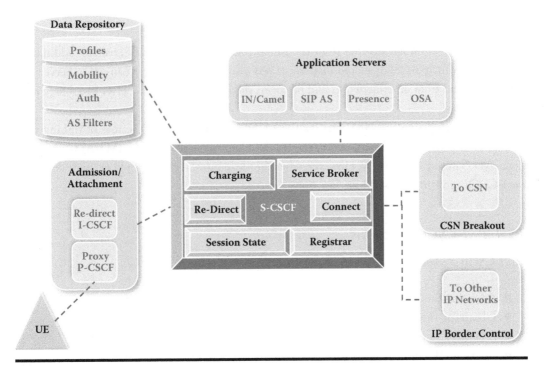

Figure 3.3 S-CSCF components and links.

3.2.3.3 S-CSCF Procedures

The Home S-CSCF receives requests to set up sessions from the subscriber (in originating calls) or to connect incoming calls to the subscriber (in terminating calls). The S-CSCF verifies that it has the user profile for this subscriber and enters session set-up dialog, exchanging session parameters with the other party.

The S-CSCF makes decisions whether to pass on the signaling messages unchanged acting as a proxy or a Redirect Server, and transfer control to other servers. Alternatively, it can act as a User Agent, initiating sessions e.g., connecting to a media server.

After a successful user registration, the S-CSCF downloads from the HSS all the implicitly registered public user identities associated with the registered IMPU (IMS Public User identity). For each identity, the list of subscribed applications and their triggers is also downloaded. These applications are then triggered for originating calls, terminating calls, or mid-call events, according to the predetermined sequence of the Filter Criteria built into the service profile.

The S-CSCF is usually stateful in order to bring in a string of applications. In most cases, the call state is generally maintained by the applications as well. Depending on the type of service, the application may require the S-CSCF to keep state, but in other cases the S-CSCF can act as a proxy only and be stateless (unaware of the call state), which is maintained entirely by the AS. In this way, performance of both the S-CSCF and the AS can be optimized.

3.2.4 The Interrogating CSCF

3.2.4.1 I-CSCF Roles

The I-CSCF function has evolved from the initial role of a SIP Redirect Server. In addition to determining which S-CSCF should be used, it was also intended as the Redirecting Server that connects to and from other networks and performs the functionality of a THIG (Topology Hiding IP Gateway).

While interrogating the HSS remains the I-CSCF main function, the role of linking with other networks has been relinquished to the Border Controllers (IBCF), the IMS-ALG (Application Layer Gateway) and TrGW. However, where border control is not required or is not deployed, the I-CSCF can still perform these functions.

3.2.4.2 I-CSCF Functions

The I-CSCF is responsible for the following functions:

- Accessing the SLF to find the user's record on HSS in a distributed HSS network
- Obtaining the address of the S-CSCF from the HSS
- Assigning an S-CSCF to an unregistered user
- Routing SIP requests received from another network toward the S-CSCF
- In the absence of other servers, performing transit routing functions

3.2.4.3 I-CSCF Procedures

First, the I-CSCF must ascertain whether the user is already registered, and if so, where the S-CSCF can be found. In IMS, the S-CSCF server can be any server within the appropriate network, i.e., the user is not preassigned to a particular server. In distributed large networks, with multiple HSS databases, the I-CSCF must inquire on the SLF to establish which HSS to interrogate. If the user is not yet assigned a S-CSCF, the I-CSCF has the task of dynamically allocating an available S-CSCF to the registering UE. This process is performed within the Home Network, whether or not the user is roaming in a Visited Network.

At the terminating end of a call, the destination user can have an assigned S-CSCF or can be currently unregistered. The terminating I-CSCF needs to find the S-CSCF for the called party or assign a temporary one for such terminating services that may be performed for unregistered users, e.g., forwarding to a voicemail system.

Figure 3.4 shows the I-CSCF accessing the SLF (at the terminating network in this example) to find the appropriate HSS for the user. When the user data is retrieved from the HSS, it may show that the user is currently registered to the network, and the address of that particular S-CSCF is returned, to be used in the subsequent dialog with the calling P-CSCF. If the user is not currently registered, the I-CSCF needs to assign an S-CSCF server to that user.

The assignment may depend on the following aspects:

■ Special needs or capabilities required by the user, as downloaded from the HSS to the I-CSCF
■ Network configuration, setting preference to allocate certain groups of users to certain servers
■ Particular characteristics of the S-CSCF that might be needed for the user type
■ The relative location of the S-CSCF and the UE to optimize network access

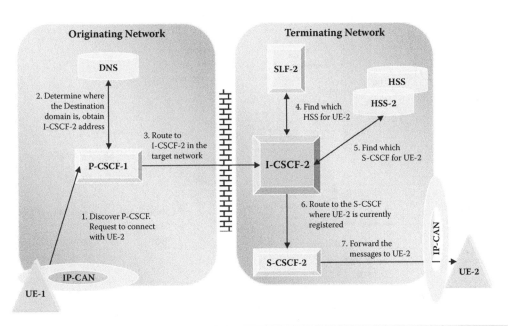

Figure 3.4   I-CSCF role in interconnecting to other IP networks for terminating calls.

■ Load-balancing on the pool of S-CSCF servers

A roaming user contacts a local P-CSCF in the Visited Network. The P-CSCF ascertains the Home Network for the roaming UE and obtains the address of the entry point to it via a DNS query. This entry point can be an I-CSCF or a Border Gateway that forwards messages to the I-CSCF in the Home Network.

The Home I-CSCF interrogates the SLF/HSS to find the Home S-CSCF that serves the roaming UE. The Visited Network's P-CSCF will then forward signaling from the UE to the S-CSCF in the Home Network.

3.2.5 Summary of the Session Control Interfaces

The session control functions as a whole (i.e., the combination of the Proxy, Interrogating, and Serving servers) link to all the IMS functions.

Figure 3.5 shows the various CSCF links:

■ *The ISC reference point.* This reference point is the standard interface between the S-CSCF and all types of ASs. It is SIP based with extensions to control services.
■ *The ISC/Pi reference point.* This interface is between the S-CSCF and the Presence Server. Like all ASs, the interface is ISC, which is extended SIP.
■ *The Mg reference point.* This interface is between the S-CSCF and the BGCF, for SIP messages bound for the border.

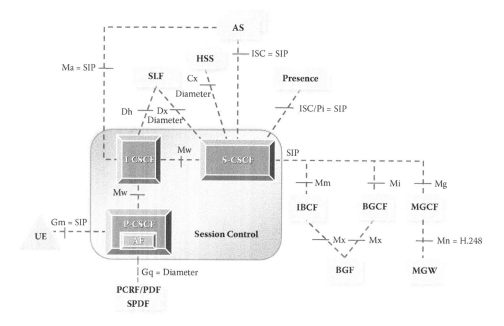

Figure 3.5   The S-CSCF interfaces and reference points.

- *The Mw reference point.* This interface is the internal routing interface between two S-CSCFs, or between an S-CSCF and an I-CSCF or a P-CSCF. This interface is based on SIP.
- *The Mm reference point.* This reference point is between the IMS servers and other IP networks. It is based on SIP.
- *The Gm reference point.* This reference point carries communication between UE and IMS session control elements. It contains messages related to registration, session initiation, and modification requests. This interface is accomplished using standard SIP, as defined by RFC 3261, with other relevant RFCs and additional enhancements introduced to support 3GPP needs.
- *The Ma reference point.* This interface is between the I-CSCF and Application Servers (i.e., SIP Application Server, OSA Service Capability Server, or CAMEL via IM-SSF). It is used to forward SIP requests destined for an application server addressed by the Public Service Identity (PSI), without involving the S-CSCF.
- *The Cx reference point.* This reference point is between the HSS and S-CSCF. It is a Diameter-based interface to User Profiles and Service Filter Criteria on the HSS.
- *The Dx reference point.* This reference point is between the SLF and S-CSCF. It is a Diameter-based interface to SLF, to find the user's HSS.
- *The Dh reference point.* This reference point is between the SLF and the AS. It is based on Diameter. It is used to retrieve the address of the HSS that holds the subscription for a given user, where several instances of HSS are deployed.
- *The Gq/Gq' reference point.* This reference point is between the P-CSCF (as an Application Function) and the policy decision controller (SPDF/PCRF or the PDF). It is a Diameter-based interface to User Profiles and Service Filter Criteria on the HSS. Gq is the 3GPP reference while Gq' is the TISPAN reference, denoting minor differences in parameters.

3.3   The IMS Session Initiation

3.3.1   *Registration*

Circuit Fixed Networks have no concept of registration except for particular services for first-time use. For IP networks, registration and re-registration is an essential and repeated process that establishes the connection path and refreshes the status.

Registration (or log-on/sign-on) is the process of getting attached to a network, with the ability to activate all the capabilities to which the user subscribes. To do that, several processes must take place, including:

- Establishing user identity (i.e., authentication of user credentials)
- Establishing user's location
- Detecting terminal identity and capabilities
- Authorizing access to certain services

Registration is a key interaction between the UE and the network. IP registration is not a one-off process, but a regular action where each registration has a lifetime established in advance. This is not exactly the same as the process for mobile handsets that move between mobile cells and need to re-register in a new cell. Re-registration is a periodical refresh of state, reaffirming the network admission and the live connection.

Chapter 4 provides a full description of registration and authentication processes.

3.3.2 Session Setup Procedures

The process of session initiation involves several aspects:

- User identification
- Service authorization and filters
- Session parameters, QoS setting
- Media type negotiation
- Location and routing management

Once the user is registered, the S-CSCF is ready to receive session initiation requests, either outgoing calls generated by the user UE (originating request) or incoming calls received from the network (terminating request).

First, the S-CSCF checks that the administration has not barred the user. Next, the S-CSCF examines the set of triggers in the Initial Filter Criteria (iFC), in the priority order that has been assigned to the applications. The AS of highest priority is then sent an INVITE to initiate the service.

Incoming calls can be bound toward an unregistered user. In such cases, although connecting the session is not possible, there may still be applications to invoke, for example, voicemail or call forwarding. If not already cached, the S-CSCF must obtain the HSS profile with the relevant iFC information and trigger the terminating services in the indicated priority sequence.

3.3.3 User Identification

The user ID is instrumental in session initiation and affects how registration is performed and the way the session is routed. There are several types of user identities used in IMS:

- *Private User Identities.* The Private User Identity identifies the subscription within the Home Network. This private identity is in the form of a Network Access Identifier (NAI) as defined in RFC 2486 (i.e., user@network format). Users of IMS can have one or more private IDs. It is a network-unique identifier, assigned by the Home Network operator and used for internal processes such as Registration, Authorization, Administration, and Accounting. It is not used for routing of SIP messages.
- *Public User Identities.* IMS users must have one or more public IDs (i.e., contact numbers and addresses). The Public User ID is published so that anyone can use it to contact the user. The Public User ID points to a service profile that includes one or more services, together with their trigger data (iFC). Several public IDs can belong to a single user (e.g., mobile phones, fixed phones at home and at work, plus soft client IDs). Both telecom numbering and Internet naming schemes can be used to identify users. The Public User ID takes the form of a SIP URI (Uniform Resource Identifier) as defined in RFC 3261 and 2396, or the Tel URI ("tel:" format, RFC 3966). An ISIM application in a Mobile handset stores at least one Public User ID (which the user cannot change), but it is not mandatory for all other

Public User IDs to be stored there. The network data can enable linking Public User IDs as "aliases," and if so desired, to share a service profile.

■ *GRUU*. IMS users can choose a unique combination of Public User ID and a UE instance that can be globally addressed. This provides a predictable result when reaching a particular device consistently is necessary. The UE indicates support for GRUU and the association with the specific Public User ID at the time of registration. The UE should assign a single instance ID, regardless of the number of IP-CANs interfacing with that UE. This enables stable addressing while roaming through different ANs.

■ *E.164 Format*. Routing of SIP signaling within the IMS is based on SIP URIs, not numbers as in CSN. Routing to external networks (e.g., PSTN or PLMN) can use non-SIP "AbsoluteURIs" as defined in RFC 2396. E.164 format cannot be used for Public User IDs without a conversion to URI, as telephone numbers are not used for routing within the IMS. This means that session requests based on E.164 format Public User IDs need conversion, which can be achieved via an ENUM server.

■ *PSI*. A general-purpose published identity that can be hosted by a particular AS, hosting applications such as Presence, Messaging, Conferencing, and Group Management. The PSI can be assigned as part of a group of Public IDs (e.g., a conferencing group). These identities are different in that they are associated with a particular service on an AS; thus, the PSI messages are routed directly to the hosting AS. An example of a service that uses PSI is a chat-room service URI to which the users establish a session to be able to send and receive messages from other participants.

Chapter 5 discusses detailed user identity formats and ID management.

3.3.4   Capability Negotiation

Using SIP in setting up sessions for IMS, the parties are able to negotiate capabilities to determine what can be mutually sustained. Capability negotiation can take place before the session, at setup time, at the session acceptance point, and during the session, in response to events.

The user, operator, or an application can initiate capability negotiation. The negotiation compares and considers not only the terminal and its attributes, including codecs and preconditions, but also the user's preferences that are retrieved from the HSS/UPSF. This includes the capability to route the IMS session to a specific UE GRUU when multiple UEs share the same IMS subscription identification, which in turn will affect what features the terminal can have.

3.3.5   Record Session Path

The session path established during the session initiation process must be remembered and even enforced in many scenarios. This enables certain nodes to be kept in the session signaling path, to maintain knowledge of the session, keep "state," and monitor events that may require a reaction. Such nodes are typically performing service control or interconnect function (e.g., MGCF and IBCF).

All initial requests to or from the UE always traverse the S-CSCF that is assigned to the UE. This S-CSCF uses the "Record-Route" mechanism, as defined in RFC 3261, to remain in the signaling path for subsequent requests. This is the norm for IMS connection. However, it may not be necessary for all signaling to traverse the S-CSCF on the way to relevant ASs that are deemed trusted, that is, under the operator's full control. The operator, therefore, can choose to "record-route" or not in certain circumstances.

The recorded path data is stored after a successful registration. The P-CSCF stores the S-CSCF name or the IBCF name for routing to external networks. The S-CSCF stores the P-CSCF name as part of the UE-related information for the session. Subsequently, the P-CSCF and the S-CSCF check that the received message headers contain the correct values. If the header values are incorrect, they can reject the messages or insert the correct values instead, depending on operator policy.

3.3.6 Alerting the Called Party

3.3.6.1 Ring Tones and Media

Alerting the called party in IMS can be performed in more ways than merely providing a ringing sound on the UE device. If the UE is capable of multimedia, text or images can be displayed or video clips played. Ringing tones can be user defined by engaging an application that selects which ringing tone to play, perhaps varied according to who is calling and other parameters.

Playing ringing tones can be performed by a network server, controlling a Media Server (MRF). This is required when interfacing to CPE (Customer Premises Equipment) gateways. The Media Server plays the appropriate sound clips according to instructions from the session controller. Some access points and gateways also take this role, and perform this locally.

3.3.6.2 Forking

Using SIP capabilities, alerting the user to incoming calls in IMS can be performed in parallel on several user devices at once, e.g., when several devices behave like phone extensions for one line. When multiple devices are alerted at the same time, as soon as one of these destinations responds, the session gets connected to that device and all other session requests are discontinued.

This facility of "forking" in IMS is based on SIP functions as described in RFC 3261, 3840, and 3841. If forking is used, a session request toward one IMPU is sent to multiple registered IP addresses for the same subscriber, where the capabilities and preferences for the participating devices allow it. These capabilities and preferences are conveyed as SIP Indicators at registration time of the called party. The multiple instances of the same IMPU are logged on different IP addresses and have different Private IDs (IP Multimedia Private user Identity).

The decision to fork is made by the S-CSCF when several registrations of the same IMPU have matching preferences and equal priorities. Otherwise, the different addresses can be called sequentially, according to the given priority sequence. Alternatively, only one instance of the registered IMPU will be suitable for a particular type of service (e.g., a video terminal for an incoming video call), and therefore forking will not be performed.

Forking may be possible toward legacy devices, where they have an IMS ID stored in the User Profile. However, this may not always be feasible, for example, where the MGCF is not able to perform the forking to and from the PSTN. This is made easier when the AGCF manages this facility, because the session control for both POTS and SIP terminals is integrated. The media gateway control in both cases must manage procedures of "early dialog" that occur prior to the line connection in coordination with the other call attempts, and must terminate these early dialogs if the user responds on other devices.

SIP forking is not the same as simultaneous ringing that is performed by an AS and can alert several independent devices and independent IDs that have been grouped together for this purpose. For example, this feature is needed for Enterprise solutions, to alert an interested group of people for calls to a single company contact identifier. This feature is associated with group management, often coupled with a Presence feature, and managed by the end-user administrator.

Although first thought of as one of the great SIP differentiators, forking presents a number of issues that impede widespread implementation. These issues include privacy and security, risks of looping, and questions of charging. To aid reliability, each forking leg must have its own call sequence number and a dialog that is managed until its individual conclusion. Despite these issues, forking remains a feature frequently asked for.

3.4 The Session Call Model

3.4.1 Session Call Model

Session processing is governed by a "call model" that provides the logic for the session steps. The call model retains session state knowledge, to activate facilities and involve servers as and when the session requires it. When the user initiates a call session, the request is forwarded through the Access Network to the currently assigned S-CSCF. The S-CSCF matches the triggering conditions downloaded for this user and selects the appropriate AS to provide the required service.

From the point of view of the CSCF server, each call has two parts:

1. The incoming leg, which is between the calling UE and the CSCF server, governed by the originating services
2. The outgoing leg, which is between the CSCF server (or could be starting from the AS) toward the called party

As illustrated in Figure 3.6, the S-CSCF call model defines the Incoming Leg Control Model (ILCM) and the Outgoing Leg Control Model (OLCM), each handled as a separate dialog. This diagram also shows the combined ILCM/OLCM, where the S-CSCF handles the logic of both although it can process each of them independently. Several incoming call legs or outgoing call legs can be involved in a session, and their relationship constitutes "call state." At the same time, the AS controls its own call legs, possibly multiples of them, depending on the AS call model.

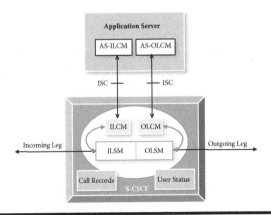

Figure 3.6   Incoming and outgoing call model.

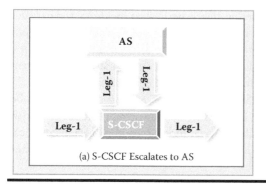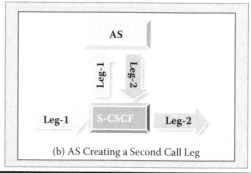

Figure 3.7   S-CSCF connecting to application servers: (a) S-CSCF escalates to AS and (b) AS creating a second call leg.

The S-CSCF selects the application to invoke, as shown in Figure 3.7. Two scenarios are shown:

■ In the scenario in Figure 3.7(a), the AS does not modify the session, although some parameters may have been changed, and the S-CSCF proceeds to forward "Leg-1" to the next hop. Note that Leg-1 is still an outgoing leg in the call model but it is forwarded rather than terminated and another call leg initiated, as in Figure 3.7(b).
■ In the scenario in Figure 3.7(b), the AS creates another session "leg," Leg-2, which could have, for example, a different called party, a different user ID, or a different endpoint. The AS will instruct the S-CSCF to connect the media to the new destination.

The application performs the service and returns the results, and then the S-CSCF proceeds to connect the user. The S-CSCF may find that there are several features or applications to invoke, as shown in Figure 3.8. They should be performed in a predefined order of priority to avoid unwanted feature interaction or unpredictable results. As shown, the call's Leg-1 is forwarded to one AS, and then to another AS before proceeding toward the server that manages the called party. The call leg may or may not be modified when it is processed by any of these applications.

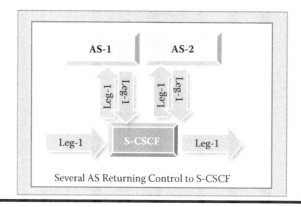

Figure 3.8   Multiple applications in a single session flow.

3.4.2 Activating Services

The Service Subscription Information is the set of all Filter Criteria stored within the HSS for the service profile for a specific user. The S-CSCF downloads user subscription information at registration time and caches it for the duration of the registration period. When an S-CSCF is selected to process a session for an unregistered user (i.e., when an incoming call arrives), it downloads the User Profile. More than one set of iFC can be downloaded during registration if there are any implicitly registered Public User Identities that have different Service Profiles.

Activating services in the wrong sequence can significantly affect the user's experience. It can interfere with the correct performance of session control. For example, offering voicemail to the calling party must occur only after call forwarding and call waiting services have resulted in no connection. This means that each application in a service package for each subscriber must be given a priority level that triggers services in a predetermined order.

For every session initiation, originating or terminating, the S-CSCF examines the Filter Criteria one by one, in the predetermined order of priority, to establish what applications should be invoked and in what sequence they are invoked. When control is returned to the S-CSCF after activating an AS, the S-CSCF will continue to evaluate the next iFC in the sequence of services, taking into account any routing implications resulting from the performed service. If the S-CSCF cannot reach an AS, the S-CSCF follows a default process, either to proceed with examining lower priority triggers or to abandon the attempts to match triggers and release the dialog.

The S-CSCF can share the call state information with the AS, when it is involved, utilizing event management capability, sending notifications whenever the registration status changes. This can be done explicitly (by generating a SIP message containing the information) or implicitly (by passing onward a SIP message that indicates the call state).

Further information about AS triggering is given below and information about service profiles is given in Chapter 5.

3.5   Media Resources

To appreciate how media is controlled in the transport layer, it is recommended to read Appendix A.4.

3.5.1   Media Services

Network media resources can be a number of media processing facilities, such as:

■ DTMF detection
■ Recording and replaying tones and announcements
■ Mid session events that require media interaction (mid-call user input)
■ Unidirectional audio and unidirectional video streaming
■ Conversational video bridging
■ Conferencing: three-way audio bridge, multi-party video bridge, broadcast
■ Transcoding of media codecs

3.5.2   *The IMS MRF*

3.5.2.1   *IMS MRFC and MRFP Functions*

The MRF represents a pool of media resources for IMS that provides support for bearer-related services, such as multi-party sessions, announcements, and conferencing. The MRFC is the signaling controller element that interacts between the media processor and the requestors of media services. The MRFP is the equipment that provides the media resources, that is, it connects, mixes and translates media streams, records, and plays or broadcasts stored media.

The MRFC functions include:

- Controlling the media stream resources in the MRFP
- Interpreting AS and S-CSCF instructions and controlling the MRFP accordingly
- Generating CDRs for charging information

The MRFP performs the following functions:

- Controls bearers on the Mb reference point
- Provides resources to be controlled by the MRFC
- Mixes incoming media streams (e.g., for multiple parties)
- Sources media streams (for multimedia announcements)
- Processes media streams (e.g., audio transcoding, media analysis)
- Grants floor control (i.e., manages access rights to shared resources in a conferencing environment)

3.5.2.2   *IMS MRF Components*

The S-CSCF controls the MRF to play announcements or tones as part of the user interface to the network. The S-CSCF also connects the UE to the Media Server according to instructions from an application (e.g., for a voice response dialog or a conferencing bridge).

Applications are the major users of media resources. The S-CSCF that invoked the application responds to the application's request to involve the MRF by sending a SIP INVITE toward the MRFC on the Mr interface. The MRFC instructs the MRFP, having interpreted the request into a specific media processing instruction that is sent to the MRFP over the Mp interface (H.248).

The S-CSCF connects the MRFP to the UE and allows the flow of media between them. Where the application needs to script the media further, a more complex set of instructions can be sent from the AS to the UE over the Ut interface, which enables the transfer of XML documents.

MRFP media streams can be routed to the CS domain through the IMS MGW, where they are converted from a packet stream to CS network media.

Figure 3.9 illustrates the MRF as two independent elements connecting via the Mp interface. The MRFC negotiates the media resources, and the MRFP delivers the required media functions.

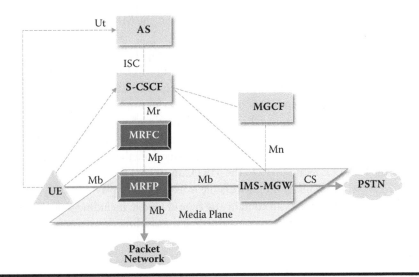

Figure 3.9   Media Server control.

3.5.2.3   MRF Reference Points

The reference points associated with the MRF are as follows:

- *The Mr reference point.* This interface is between the MRFC and the S-CSCF. It carries requests for media resources and the media session requirements. This interface is based on SIP.
- *The Mp reference point.* This interface is between the MRFC and the MRFP. It controls the delivery of the resources. This interface is based on H.248. Where the MRFC and MRFP co-reside, the interface can be internal.
- *The Ut reference point.* This interface is between the UE and the AS. It allows transferring configuration information securely between the UE and the AS for the purpose of performing media manipulation. The Ut interface supplements the session control instructions with more complex service data, usually using XML-derived language.
- *The Mb reference point.* This interface connects to IP media transport services.

3.5.3   Media Resources Management

3.5.3.1   Invoking a Media Server

In IMS, the CSCF interface to the MRF is SIP based, whether the MRFC resides within the CSCF or within the MRF node. The IMS Session Controller (CSCF) always remains responsible for directing the media to the UE address, whether the media was generated by a media server or another UE. Thus, the session controller is fully aware of how the media is used or manipulated. For this reason, some implementations choose to co-locate the MRFC function with the Session Controller, but the MRFP can still be an external sharable network element in a pool of resources.

This enables more than one S-CSCF to utilize resources on the MRFP, and by default, more than one application can use the same Media Server.

The invocation of the MRF utilizes standard SIP messages. The REFER method directs the endpoint to reconnect the media flow to a given address, in this case that of the MRFP. This may complete the call, for example when the user gets connected to leave a voicemail message. Alternatively, it may be a temporary connection of media, for example instructing the MRF to play announcements and collect user input ("Prompt & Collect"), then pass the control to an application, which can complete the second leg of the call depending on the collected information.

Alternatively, the MRFC can accept an INVITE message from the S-CSCF. For simple announcements or tones, the INVITE request sent to the MRFC must contain sufficient information to identify which pre-recorded audio file or multimedia file is requested for playing to the connected party. The file stored on the Media Server can contain any type of media, including tones, announcements, music clips, video clips, voicemail messages, or any digital recording.

3.5.3.2 AS Invoking the MRF

Many applications make extensive use of media manipulation and features, and must be able to control a range of capabilities according to the service logic. The AS may require the MRFP for:

■ *User interaction:* basic user interaction, including tones, fixed announcements, collecting DTMF-based input, and advanced user interaction, including support of IVR and messaging capabilities
■ *Transcoding:* basic transcoding with support for transcoding between voice codecs, and Advanced transcoding with support for multimedia transcoding
■ *Conferencing:* basic conferencing, including three-party calls and multi-party large voice conferences
■ *Broadcast service* (multiparty one-way)

Controlling the MRF is always performed via the Session Controller, the S-CSCF. However, the AS and the UE can communicate directly over an XCAP (Configuration Access Protocol) connection to exchange more elaborate content formatting and scripted controls. This communication is performed over the Ut interface, which is a secure interface, using HTTPS (HTTP over TLS).

An invocation of MRF services by an application can be performed using SIP REFER or RE-INVITE messages, or using the method of Third-Party Call Control (3PCC), through the S-CSCF:

■ The AS uses the SIP REFER method to divert a session toward the MRF. The SIP message must contain parameters indicating the specific UE capabilities that are essential for receiving the media services and the relevant indicators of which functions are to be deployed. For example, when a multi-party conferencing bridge is requested, the MRFC allocates a Conference ID, which will be quoted in all subsequent dialog messages, connecting other endpoints.
■ The AS can use 3PCC to INVITE the MRF to an AS imitated session, then INVITE the UE, and finally link the media between them. The 3PCC method involves the AS creating two separate call legs, one to each party, and then rejoining the media between them. One

such party can be the MRFP, in which case one or both users can be receiving content. The 3PCC method can be used for inviting parties to a conference (giving a Conference ID and the party's URI) or "inviting" announcements to be played to a caller (giving Announcement ID). 3PCC also can be used to invite announcement and transcoding services without an originating UE, i.e., triggered by application logic. The AS (via CSCF) sends an INVITE to the MRFC with details of the type of media and identification of the service, e.g., an announcement number, clip identifier, etc.

The MRF becomes a party in the session, either a terminating party or an additional one in a three-way call. In other scenarios, the MRF can "listen" to mid-call in-band user input, such as DTMF tones, or provide music while a user is waiting to be transferred.

3.5.3.3   The Early Media Facility

The procedure for connecting media between two (or more) parties involves completing a dialog that ascertains the media capabilities of both parties and authorization of a media stream with a certain QoS profile. However, some media services occur before the final OK, for example playing a ringing tone or a "busy" tone, a network announcement, or ring-back preselected media.

Such services are network initiated and therefore can be safely authorized even before the User is fully authenticated and the session bearer resources are fully allocated. The SIP dialog has been enhanced (using PRACK) to cater for Early Media, to enable these temporary media connections to the MRF before the called party has accepted the incoming call.

3.5.3.4   MRF Called by UE

Some traditional media services accompany the generic session with no AS involved. These can be, for example, announcements triggered by certain network conditions. In this case, the operator advertises the name of one or more MRFCs and the UE invites an MRFC to the session.

The INVITE will contain further parameters to define the type of media service required. These parameters may have been communicated directly by the AS to the UE via the Ut interface to define further MRF parameters, transferred via an XML document.

3.5.3.5   Sharing Media Resources

To save costs the use of the Media Server resources must be optimized. In addition, such Media Servers may be equipped with specialty capabilities, such as language and dialect-based announcements, special codecs, or voice recognition engines. Therefore, a particular Media Server can be best selected for each invocation according to the type of connection, location of users, and the application requirements.

Where ASs have integrated Media Servers, they are usually only accessed by applications on that AS, and the allocation of media resources is handled on that server. To share networked media resources, the Media Server MRFP must be external and independent of both the Application Servers and the Session Controllers.

For efficient use of the expensive Media Server resources, a flexible resource allocation system must be in place to respond to the various requesting functions, which could be multiple Session

Controllers and Application Servers. This is a mechanism of balancing resource requests across a pool of MRFPs, addressing not only multiple Media Server nodes, but even blades within each MRFP.

The specific media resource can be reserved in advance or dynamically allocating at the time of the request, according to the type of the application. For example, a large prebooked conference or a broadcast can reserve suitable capacity on a Media Server to accommodate bridging multiple conferencing legs.

More advanced media resource controller function is not yet specified in the standards. The need for this functionality is likely to grow with the advent of large VoIP and multimedia implementations in the future.

3.5.4 Configuring Media Resource Services

3.5.4.1 Configuration Methods

To perform media management for many modern services, an elaborate set of service configuration parameters must be communicated to the Media Server. Typically, session controllers have relatively simple requirements but applications often need methods of conveying service instructions and configurations of complex media resources.

Previously, applications that are media resource intensive (e.g., voicemail, voice response, and conferencing) were often executed on servers with built-in media servers. Such applications often passed service configuration by internal proprietary protocols once the session was established. However, to accommodate the principles of IMS, standard methods of configuring media applications are sought. There are several competing protocols currently in use in the marketplace, including:

- The NETANN (Network Announcement) provides simple SIP-based means of invoking announcements and media mixing.
- VoiceXML is XML based and enables a transfer of instructions and content to the Media Server. It has been popular with IVR system developers due to its programmability that enables downloading a Web page to the Media Server.
- CCXML (Call Control eXtensible Markup Language) enables VoiceXML applications to apply call control over the Media Server activities that are largely lacking in VoiceXML.
- MSCML (Media Server Control Markup Language) is a proposed protocol that is SIP based and is said to be in line with the principles of the Internet and peer-to-peer communications. MSCML delivers enhanced conferencing capabilities and flexible media management. It runs on the Media Server itself, not on the AS.
- MGCF and H.248 (Megaco) are similar syntax protocols and are popular for addressing non-SIP environments, typically used by softswitches.
- MOML (Media Object Markup Language) and MSML (Media Sessions Markup Language) are proposed associated protocols that retain the slave-master model of legacy systems, using MGCP-like syntax within SIP.

3.5.4.2 Conferencing

To implement conferencing, the network publishes the name of the MRFC so that any subscriber UE can address it. The UE will invite an MRFC to a session, asking for a particular media connection, within the capabilities of the MRF. The MRF then returns a Conference ID allocated internally. The UE subsequently uses this Conference ID to create a bridge and conference in other parties. The UE can create two types of conferences:

1. The UE indicates to the invited party to join the MRF session, using a SIP REFER message. This can become a session invitation with consultation.
2. The UE sends the MRF a list of participants. The MRFC uses this information to initiate multiple sessions, inviting the participants to join, and then bridge the media together. This type of multiparty session is without consultation.

Note that some terminals can manage to interconnect three media streams without the aid of the MRF. However, mixing a great number of parties requires the use of a Media Server.

3.5.4.3 Transcoding

An AS can control a transcoding session using MRFP capabilities. The AS sends an INVITE via the S-CSCF to the MRF. The INVITE message contains the session parameters, including what should be transcoded. For dynamic transcoding, the INVITE to the MRF must contain the codecs required for the two legs of the session to be joined.

The MRFC and the AS proceed to negotiate via the process of offer/answer (as defined in IETF RFC 3264) considering any preconditions models for SDP negotiation with the AS. Unless there are no available resources or there is a technical problem, the MRFC will accept the requests, reserve the required resource, and confirm it to the requesting AS and UE.

3.6 Control of Application Servers

3.6.1 Application Server Types

IMS standards neither describe the AS itself nor specify applications, but they do define the interfaces between the AS and the CSCF session controller and the HSS data repository.

IMS supports three types of ASs, in terms of their interfaces to the session control components: (1) the SIP AS, (2) the OSA AS, and (3) the CAMEL IM-SSF. Any of these ASs can offer multimedia services via Core IMS. For each type, there is a set of interfaces (i.e., reference points) by which they communicate with the S-CSCF, I-CSCF, SLF, and HSS.

Figure 3.10 shows the interfaces of the AS to the data repository, the HSS. All types of ASs can access the User Profile data on the HSS using the Sh interface. In addition, access to the HLR and VLR by the CAMEL AS is still available, using the Si interface; that is, CAP (CAMEL Application Part) and MAP are still available for their traditional roles.

While all three types connect to the S-CSCF via the ISC, the OSA AS and IN (Intelligent Network) or CAMEL servers connect through an intermediate interface:

■ The OSA SCS (Service Capability Server) connects the OSA AS to the IMS CSCF.

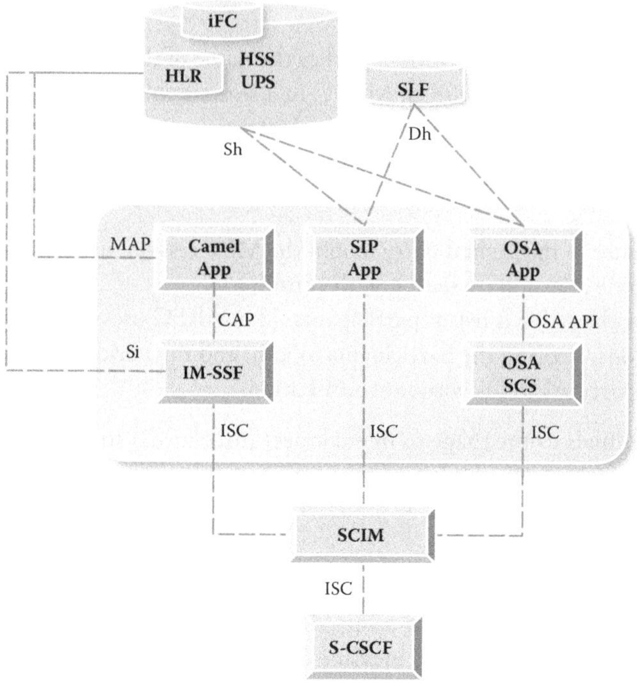

Figure 3.10   CSCF interfaces to AS.

■ The IM-SSF (IMS Service Switching Function) connects the IN or CAMEL platform to IMS.

The OSA SCS and the IM-SSF functions provide conversion between SIP and legacy protocols and APIs.

3.6.1.1   SIP AS

The SIP AS utilizes SIP for service control. SIP ASs can take the role of delivering generic session features to all users, such as call barring, diversions to voicemail, or IP Centrex.

SIP ASs can use different scripts or APIs. The most widely used APIs in SIP AS are:

■ SIP CPL — XML-based scripts
■ SIP servlets — Java-compatible APIs

3.6.1.2   IMS SSF

IN platforms have been providing additional features to CSN switches for many years. They were particularly successful in Mobile networks with a range of services based on CAMEL, e.g., Pre-pay and location-based services. Existing IN platforms are being extended to interwork with IMS and provide not just voice-based services, but also full IP multimedia-based services and integration with Web services.

The interface between the S-CSCF and the IN server is the IM-SSF, which can reside at either end of the connection, added to the IN platform or the CSCF. The interface maps operations in CAP to SIP and Diameter. The IM-SSF main advantage is the ability to address both users on the legacy network and on the packet network by a single service or combination of services, and deliver the sought-after "service mobility."

Note that the Si interface to the HSS is used for CAMEL services only. For accessing a SIP AS controlled by IMS services from legacy SSP on PSTN or ISDN local exchanges, another type of gateway function (e.g., SPIRITS) is necessary. For NGN, the equivalent of IM-SSF is an inter-working module that enables IN service logic programs hosted in legacy SCPs to work for IMS subscribers. The NGN IM-SSF (not the same as CAMEL), emulates the IN Call Model and INAP operations, interpreting SIP signaling, IN triggering, and feature behavior.

3.6.1.3   OSA

Parlay/OSA has been installed for some time, allowing CS networks to accept third-party application access to network subscribers, and interwork with Web applications. OSA can be extended to work with SIP and IMS.

The procedures in the basic S-CSCF do not cover untrusted applications, where external applications must be authenticated and monitored. To do that, third-party service providers run their applications on a carrier's network, according to agreed SLAs, which the OSA gateway monitors.

The OSA AS does not directly interact with the IMS network entities but through the OSA SCS, where the OSA "framework" operates as a filter. The OSA framework provides a standardized way for third-party secure access to IMS facilities.

3.6.2   The ISC Interface to Application Servers

The single interface between all types of IMS AS and the session controller, the S-CSCF, is the ISC. The ISC is based on SIP with certain SIP extensions and SIP profiles that allow flexible application interfaces to the session control.

For an incoming SIP request on the ISC interface, the S-CSCF performs any required filtering, based on the iFC before performing other routing procedures toward the terminating user, e.g., forking, caller preferences, etc. Having performed the service, the AS informs the S-CSCF of the results, if any, via the same ISC interface. The next action may depend on these results; for example, the destination may be modified.

The ISC interface allows the AS to act upon events and changes in the user status and session status. The ISC provides a facility to subscribe to user registration changes, including implicit registrations. When such an event occurs, the subscribing AS is notified by the S-CSCF and can take suitable action. At this point, the AS can also be notified of the registering UE capabilities and make a decision as to which application logic to apply.

ASs can initiate calls to the user and forward them toward the subscriber's S-CSCF over the ISC interface, as part of the service logic. The S-CSCF receiving them will treat the AS as an originator. The request is treated as a terminating call request toward the user that the AS specified. The S-CSCF will then provide the terminating request functionality and activate terminating services before completing the session connection.

The AS also can use the ISC to generate requests to connect the user, acting as a user agent on behalf of the user. Such requests are forwarded to the S-CSCF currently serving that user, and the S-CSCF treats them as originating calls, applying originating services according to the Filter Criteria.

3.6.3 The AS Triggering

3.6.3.1 Initial and Subsequent Filter Criteria

In IMS, even basic session features can be provided by an AS. For originating calls, such services can be, for example, withholding ID, representation ID, or outgoing call barring. For terminating calls, such services can be diversions, media conversion, voicemail, automatic response (IVR menu), etc. Control passes back to the S-CSCF after performing these features.

The S-CSCF fetches information about services and application servers from the HSS during registration. The downloaded data contains a Service Profile per Public ID (IMPU), including the iFC for each service. It may contain further service-specific data ("transparent data") that the AS will make use of, although the HSS may not and need not understand it. If the Service Profile is shared with other subscribers, the S-CSCF needs to reconstruct the full information, retrieving details according to a Service Profile indicator.

The application itself can provide Subsequent Filter Criteria (sFC). Once the AS is engaged, it can download further Filter Criteria that are necessary for the performance of its tasks. The session controller can cache the iFC, but sFC are downloaded from the AS only when it is engaged with the S-CSCF.

Features that are parameter driven within the SIP are still performed by the session controller, depending on implementations. For example, call barring, forking, and simple diversions (especially to voicemail) are usually executed in the S-CSCF. More often than not, services that involve user input, perhaps by self-care, are located on the AS, where the AS holds the service data. In IMS, such service data can still reside on the shared data repository (HSS/UPS), where it can be made available to other applications.

Figure 3.11 shows the iFC retrieved from the HSS at registration time and the downloading of sFC from the application when it is engaged. The S-CSCF has immediate access to SPTs (Service Point Triggers) from the analysis of the received SIP message. The diagram also shows the Service (Platform) Trigger Points that are stored within the AS and affect the service logic.

3.6.3.2 Service Point Triggers (SPTs)

When the S-CSCF receives a request to initiate a session, it must determine which services will be triggered — that is, if and when to trigger them. The SIP INVITE message is parsed to identify the SPTs. These SPTs are compared with the iFC to find a match.

The SPTs are the points in the SIP signaling on which Filter Criteria can be set (i.e., the points where conditions are tested for a match). The S-CSCF session processing logic invokes applications at the points where the conditions in the Filter Criteria are matched with the SPTs.

The following SPTs are defined:

- Any initial SIP method (e.g., INVITE)
- Registration type that indicates an initial registration, re-registration, or deregistration
- Presence or absence of certain header fields

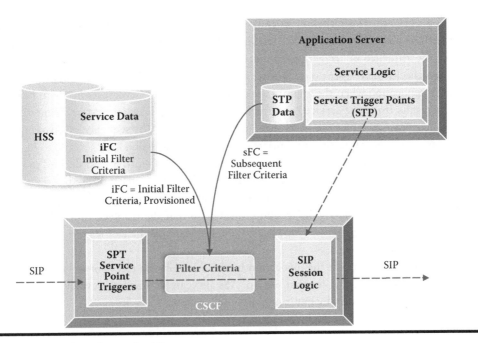

Figure 3.11 Service triggers in AS.

- Content of any header field or Request-URI
- The direction of the request, UE-originating or UE-terminating, to a registered or unregistered user
- Session description information

The S-CSCF session logic recognizes the SPT and acts upon the trigger information. It verifies first that the end user is not barred, before checking if any trigger applies to that end user, according to the Service Profile. Filter Criteria can trigger more than one SPT to get the full information for the AS activation.

3.6.3.3 Mid-Session AS Triggering

The process of matching SPTs with iFC starts from the highest-priority filter criteria — that is, the first service to be activated. If a match is found, the S-CSCF forwards the message to the AS that has been identified for this trigger. If a match is not found, the Filter Criteria must indicate to the S-CSCF whether to abort the attempt or ignore it and go on to process the lower-priority services. The S-CSCF uses the received message and any input data when sending the message to the next AS that matches the next Filter Criteria.

The S-CSCF marks the request message when it is sent to the first AS via the ISC interface, so that the S-CSCF will subsequently identify the reentry point in the dialog, even when certain parameters are changed by the AS during logic execution. For example, the AS might change the destination ID or the disclosed originator's ID, but the session identity and the point in the dialog remain unaltered. The dialog indicator is updated after each hop to an AS. This enables the S-CSCF to examine the dialog indication at any point and determines the next step accordingly.

The applications perform their tasks and, when completed, return control back to the S-CSCF, with or without a resulting output. The results from executing an application in the call path can also alter the sequence of triggering applications, via downloaded subsequent triggers. The S-CSCF then checks the trigger points for the next highest priority, both iFC and sFC, and routes to the next application when the appropriate conditions are met. Several services can be invoked, where the results from one service can affect which service is selected next and how the session logic can progress. The process repeats throughout the session. The sequence of invoking services is, therefore, fundamental to the appropriate execution of the session logic.

The AS service can generate a request to terminate the session. If an AS decides to terminate a request locally and sends back a final response for that request via the ISC interface to the S-CSCF, the S-CSCF abandons the verification and matching of lower priority triggers in the list. The final response still includes the session indicator, so that the involved S-CSCF can correlate the messages. This dialog marker helps the S-CSCF release *all* the call legs that may have been created for this session.

An AS can decide to reject the invocation by the S-CSCF and elect not to be involved in any further filter criteria matching for that session. ASs also can get involved several times in one session if they are invoked again by lower-priority filter criteria.

3.6.3.4 Event Triggering of Applications

Applications can be automatically invoked for a user by detecting particular events and fulfilled conditions, with or without an ongoing call. When the S-CSCF performs the subscriber's registration, the set of applications that could be activated is evaluated. It also identifies applications that register their interest in events relating to the subscriber's status. These applications can be triggered via such events. In such cases, the AS needs to SUBSCRIBE to the user status. This is performed via a "third-party registration" process. This means that the S-CSCF registers the AS as a "watcher," without a user-initiated action.

Once registered for event monitoring, the AS is subscribed to the UE status and will be informed of any registration changes or other significant events for all the associated user identities. Applications can then be triggered by events that fulfill pre-set conditions. Such a mechanism can be used, for example, for voice continuity, triggered by registration to the new domain.

The AS, acting as a user agent, can request initiating new sessions from the S-CSCF, and the S-CSCF will then process the iFC and sFC according to SPTs, as normal, for each leg of the session.

3.6.4 3PCC (Third-Party Call Control)

Applications using Third-Party Call Control (3PCC) often have a profound impact on the session course. 3PCC is performed by the AS to initiate sessions driven by application logic and not by user actions. The AS is said to be acting as a B2BUA, that is, terminating requests and creating new ones, which may have different settings.

Two methods are used to create the 3PCC sessions:

1. The AS initiates two separate connections and arranges for the media flow to go between the two parties.

2. The AS receives a request from the S-CSCF but due to some service logic, creates another connection and then arranges for the media to flow to the new connection.

The AS remains in the signaling path of the 3PCC session, mapping header details according to the logical binding between the two sessions it had created. It allows the AS to intervene and take other actions in addition to merely forwarding messages between the parties.

When a session is initiated by an AS, the charging must be allocated to the parties by that application logic because the session controller cannot decide who the originating caller is, who the recipient is, or who should be the charged party. The fact that the session is initiated by an application is noted within the charging record, to provide an audit trail.

This is a powerful tool to control a session but it is also risky. It is often restricted to a trusted AS only. When the AS maintains call state without the help of the S-CSCF, the S-CSCF is not able to control the session progress and keep track of executed services in the correct sequence. This may result in complications, with messages looping endlessly, call legs not terminated, or other unpredictable results. A 3PCC application must therefore be carefully proven in an extensive number of scenarios.

3.6.5   Service Capability Interaction Manager (SCIM)

IMS allows Web services and IT applications to interact with telephony-style services. This is creating a rich environment for novel services but is also adding great complexity. As already emphasized above, services must be activated in the right sequence and interact properly to avoid unpredictable behavior. Furthermore, user-centric services need to manage service mobility, despite differences across multiple technologies and devices.

For all these reasons, it is perceived that a more advanced "service orchestration" or a "service broker" function may be required to control how services behave and interact, using a predefined set of rules and configuration parameters for each user.

The SCIM function has been identified in the standards documentation and a detailed service architecture is now specified by OMA. The SCIM can be integrated into the S-CSCF. However, to provide a common view in a multi-CSCF network, it may be required to reside on an external server. This can create an issue of introducing yet another possible point of failure and adding another hop to the session processing.

The question of feature interaction is considered complicated and prone to failure. It is notoriously difficult to formulate the rules that must take into account all the permutations of service options and results. For this reason, the definition of a comprehensive generic SCIM may be fraught with difficulties, and implementations may prove imperfect. Resolving feature interaction remains in the scope of service packaging or service brokerage at the application layer.

3.6.6   Representative AS

When scaling up, the network is deployed with several replicas of the same AS, running the same applications so that an AS can be allocated dynamically to a user at registration time. This enables efficient utilization of the AS resources between active users. However, some applications need to retain user-specific data between invocations, and some applications are activated for users who are not registered. In such cases, there are the following options:

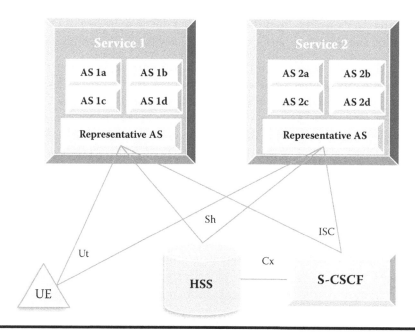

Figure 3.12 Example representative AS.

- Allocating users to a specific AS instance where user-specific data will be kept. This is a rigid and unscalable solution that necessitates some redundancy methods for the AS on 1:1 basis.
- Store the service data together with the user profile as "transparent" data that is not understood by the HSS but is transferred to the AS on request. This enables any AS replica to pick up any user, but burdens the HSS with more database actions.
- Apply the concept of "representative AS," which is described below.

The concept of "Representative AS" associates a particular AS with a user. This AS coordinates application involvement, retains persistent data, and allocates the AS server instance dynamically. To do that, the Representative AS must be named as the contact AS in the iFC. It gets involved in all initial dialog messages. This includes signaling on the ISC interface, the Ut interface, and also perhaps non-SIP functions.

Figure 3.12 demonstrates how the Representative AS is the first port of call for the S-CSCF. It allocates an instance AS for the first service to be triggered, and then proceeds to do the same with subsequent applications.

3.7 Session Routing

3.7.1 Routing to the Destination

3.7.1.1 Packet Routing

Packet routing is based not only on data in the packet header that identifies the origin and the target addresses, but also the attributes of the connection (such as channels, ports, etc.). Routing data is interpreted at different network layers, as shown in the table below:

Type	Description
Layer 1 Physical layer	Routing of packets onto physical ports, optical wavelengths, frequency channels, or timeslots
Layer 2 Data-link layer	Forwarding of user data packets based on information in the packets data link layer headers
Layer 3 Network layer	Forwarding of user data packets based on information in the Layer 3 header (e.g., IPv4 or IPv6 destination address)

While physical handling is performed at the lower layers, routing for IMS primarily utilizes the IP address, at Layer 3.

3.7.1.2   Routing within the Home IP Network

To connect the call, the IMS session signaling needs to identify the destination address in full. For that, it may need to resolve the address via DNS, if necessary, and identify the terminating network. Even within the same network, the routing facility must ascertain which S-CSCF is serving the destination UE.

To initiate a session, the two parties engage in a dialog to agree on session parameters. This is performed via the process of exchanging the SDP Offer and Response. The SDP contains UE capability information that must be negotiated to find compatibility. The called party can respond by another offer of SDP and the two parties continue to exchange information until they agree on the type and level of service.

If the destination is within the same network, the message will be routed to an internal I-CSCF server, which will interrogate the HSS location server, to establish the serving S-CSCF for the destination. Before alerting the called party, the terminating S-CSCF will invoke any appropriate terminating services (such as call forwarding, incoming call screening) according to prestored triggers.

In Figure 3.13, the offer of initial SDP is made in the SIP signaling bound for the destination. The routing path can traverse separate Access Networks to reach the originating caller's Home Network, then send over to the terminating network.

3.7.2   Crossing the Border

3.7.2.1   Network-to-Network Interconnect Functions

Calls connecting to external networks need further functions to be involved in managing the crossing. External networks can be IP networks or CS networks, Fixed or Mobile.

Figure 3.14 shows the CSCF in Core IMS, passing control to the BGCF (Border Gateway Control Function) to decide to which border exit point to route the session. The messages are sent to the MGCF (Media Gateway Control Function) for PSTN breakout or to the IBCF (Interconnecting Border Control Function) for IP interconnection.

The following functions occur at a crossing point:

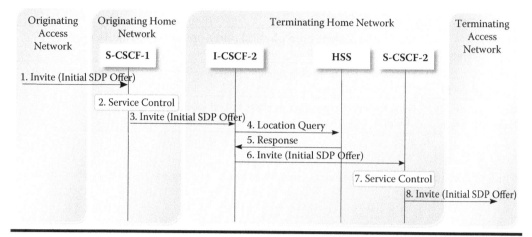

Figure 3.13   Routing between originating and terminating networks.

■ Signaling interworking, either by packet or circuit conversion or protocol conversion, may be needed (even between SIP networks, some SIP profile differences may need to be accommodated)

■ Media transcoding PS to CS conversion or translation between codecs

■ Inter-operator charging information exchange to enable costs and revenues to be apportioned between all the involved operators

■ Service interworking, especially supplementary services such as delivering user IDs.

The BGCF function, due to its nature of redirecting sessions, is often co-located with the CSCF in Core IMS nodes, while the MGCF and IBCF controllers can be located at strategic points for exiting the network. However, all three functions are signaling only and can still be located at the center, in contrast with bearer media gateways that carry large volumes of media packets and are best distributed to the exit points.

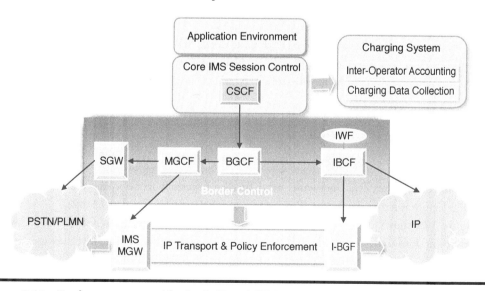

Figure 3.14   Border management for both IP and CSN.

3.7.2.2 Routing to the IP Border

For calls to another packet network, the S-CSCF must obtain the entry point address for this network, from which the call will be routed to its final destination. This address is ascertained by resolving the domain name of the destination URI via a DNS server or a private database look-up, or by interpreting the dialed number dial-code, using ENUM. Such databases contain the Public ID for known entry points, which may be an I-CSCF, an IBCF (Interconnection Border Control Function), or another type of a border gateway controllers. The S-CSCF routes the call via an internal I-CSCF or IBCF before exiting the network.

Chapter 7 provides further details about the IP border.

3.7.2.3 Routing to PSTN Breakout

Routing to a CSN occurs when the destination address contains either the "user=phone" parameter or is formatted in the "tel:" URI format. When encountering the "user=phone" parameter, a routing server — S-CSCF or I-CSCF — must reformat the SIP URI into the "tel:" URI format, as recommended in RFC 3966.

The CSCF will first attempt to establish a routable URI that is outside the Home Network by extracting the E.164 address contained in the SIP message URI and translating it. This may involve a query to an ENUM server, to translate the Tel URI, followed by a query to a DNS server to resolve the domain name. The ENUM/DNS server can be a public server or a private network database. The resulting address, if found, will point to a server across the border, at the destination network.

Alternatively, the path toward a CSN destination might involve carrying it on an IMS Transit facility which is a more cost-effective way of carrying the signaling over IP networks toward a remote CSN breakout point. The conversion to CSN numbers will then occur at the other end.

If the network address translation process fails, the CSCF/BGCF will route to the MGCF to allow exiting to the CSN locally via signaling gateway and media gateway, and route to the destination telephone in the PSTN network, utilizing the embedded E.164 number.

3.7.3 Routing to a Public Service (PSI AS)

The PSI is a published identifier for a service hosted on a PSI AS. This identifier can be quoted as the destination for users wishing to take advantage of the service. Routing to a PSI can be performed even when there is no IMS subscriber involved and no IMS session, e.g., non-session-based chat messaging service. The PSI enables users on both IMS and non-IMS devices to reach an application server in the packet-switched domain.

The example in Figure 3.15 shows the scenario that involves an S-CSCF in triggering the AS hosting the PSI. The I-CSCF retrieves the PSI profile from the HSS and the address of an assigned S-CSCF. The S-CSCF registers the PSI and obtains the identity of the AS hosting this PSI, e.g., the actual AS instance for a particular Conference ID. If the given destination address of the PSI server is not a Fully Qualified Domain Name (FQDN), the I-CSCF will need to resolve the domain name by interrogating a DNS server. The second scenario is shown in path 3a (dashed line). If an FQDN address is given, the I-CSCF will route to the PSI directly.

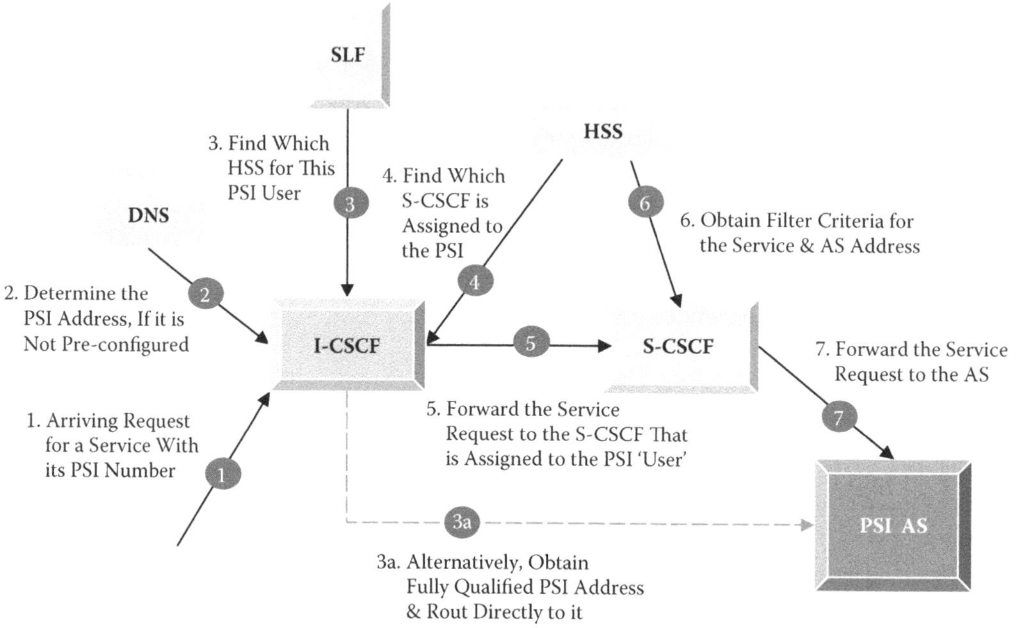

Figure 3.15 I-CSCF routing to a PSI AS via the S-CSCF.

The PSI instance itself is treated as a user, with a user profile on the HSS. The user profile is associated with a particular application and may contain the iFC and other service-related data. This service profile is then shared with all users who access the same PSI service.

When routing to a PSI server within the same network, where there are several HSS databases, an SLF is consulted by the I-CSCF. Having found the HSS containing the "PSI user" data for the requested PSI service, the I-CSCF retrieves the data (filter criteria) from the HSS, along with the identity of the serving session controller assigned to process this PSI sessions. That S-CSCF will then initiate a session toward the PSI AS. In this way, an S-CSCF is assigned and a session is initiated for non-IMS users.

An alternative way is to retrieve the final address of the AS hosting the PSI directly from the HSS, without engaging an S-CSCF, and then forward the session directly to that AS. This may be a preferred option for non-IMS users, if the S-CSCF control and reporting (including CDRs) are not required.

Furthermore, at the operator's option, the I-CSCF can fetch the address directly from a DNS server when resolving the domain name. In this method there is no IMS session, and no IMS network elements involved. This method is preferable for external users seeking to use the PSI service from another network. The DNS approach avoids opening up the HSS to inquiries from external entities.

3.7.4   Delivering Services to Roaming Users

3.7.4.1   Home Services

Roaming users can register to their Home Network via the Visited Network. The Visited P-CSCF forwards their messages to the I-CSCF, via border gateways if implemented, to the I-CSCF or IBCF, and from there to the S-CSCF. An S-CSCF is assigned at the time of registration. The

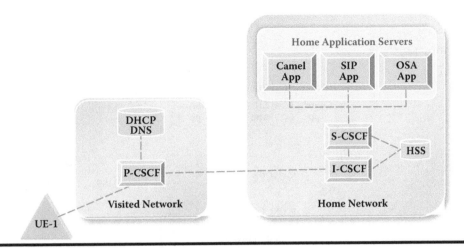

Figure 3.16 Services served from Home Network.

S-CSCF obtains the user profile with the Filter Criteria that invoke applications for the user. These applications are in the same network as the S-CSCF.

Figure 3.16 shows the model of providing services from the Home Network. Operators prefer to keep subscribers on their network and provide them with all their services from the Home Network, even remotely. From the point of view of subscribers, they receive the same services wherever they are. Such service ubiquity is convenient and is welcomed by users.

3.7.4.2 Serving from the Visited Network

When users roam into a Visited Network, the local operator can expect to receive payment for the roaming connection. However, they may see an opportunity to offer local services. Such services can provide information based on the current location, and therefore do not need to know the full user details. Often, these services are based on advertising that draws revenues from the other parties, call centers and local businesses, rather than the subscriber. This avoids the potential problem of obtaining enough information to charge the user or cross-charge through the Home Network operator.

Figure 3.17 shows the user, who is registered to the Home Network, receiving services from the Visited Network. The user can be treated as "Anonymous" for the purpose of invoking the local services for which the user is not charged (e.g., local advertising).

This model exists today in Mobile 2G, and may well exist in 3G and NGN, although this is not covered by the standards as such.

3.8 Mid-Session Management

3.8.1 Mid-Session Modifications

3.8.1.1 S-CSCF Session Monitoring

The S-CSCF is required to monitor the session, that is, to remain in the signaling path, in order to perform special in-session tasks:

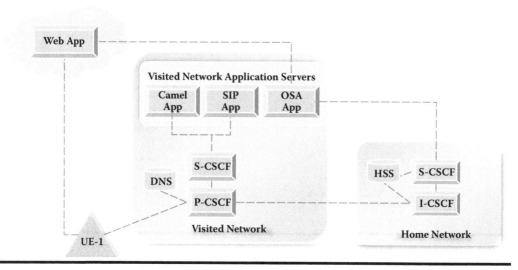

Figure 3.17   Services served from Visited Network.

■ *Media control.* The S-CSCF may be asked, perhaps via an application, to change the session and the media parameters against network policies or as an attempt to defraud the operator. If the S-CSCF is using the SIP "Record-Route" facility and the UPDATE message, it can refuse such a request, and inform the originator.

■ *CDR generation.* The S-CSCF is required to generate CDRs, which are used for offline charging and for statistical purposes. To do that, the S-CSCF must be aware of any session changes, especially session termination. The S-CSCF generates continuation CDRs throughout a long-duration session.

■ *Network-initiated session release.* The S-CSCF can generate a network-initiated session release (i.e., forced termination) for administrative reasons. This means that the S-CSCF must be aware of "hard state" dialogs that need to be terminated by sending an explicit "BYE" message.

■ *Interpreted network identifiers.* The S-CSCF must remain in the path of signaling for special types of identifiers, such as GRUU. The S-CSCF is required to interpret the generic contact into one that is associated with a particular terminal according to the predetermined association for this GRUU (see Chapter 5).

There are still some occasions when the S-CSCF is not kept in the signaling path:

■ Presence status notifications need not go through the S-CSCF if they are generated elsewhere, e.g., by an aggregated Presence server. This is possible because the SUBSCRIBE message ("soft" state dialog) does not need a network-initiated release.

■ Sessions between the UE and an AS, for example for video streaming services or PSI sessions, can be set up by the S-CSCF but may not require the S-CSCF to keep "state." In such cases, the AS must provide the charging information and is responsible for correct termination.

For such services, there is stateless "no-record-routing" parameter in the iFC to inform the S-CSCF that it is not required to remain in the call path.

3.8.2 Handover and Voice Call Continuity (VCC)

3.8.2.1 The VCC Architecture

VCC (Voice Call Continuity) is the ability to hand over session control between a CS network and a PS network, or other Access Network types, without an interruption or degradation of quality or causing other perceived impact on the service. VCC handover can occur within the two domains of a single operator or across business borders.

VCC function logic is controlled by a service that may reside on an AS. This function contains:

- Domain Selection Function
- Domain Transfer Function
- CS Adaptation Function
- CAMEL Service Function

The Domain Selection and Domain Transfer modules are responsible for transfering the session. The decision to transfer is based on various network policies and current status of the UE, that is, where it is registered and whether it has the appropriate capability. Once the decision is taken, the Domain Transfer module performs the session handover. The VCC application in the Home IMS generates charging information for all Domain Transfers for the VCC subscriber.

Figure 3.18 shows the VCC service modules in the VCC AS: the Domain Selection and Domain Transfer modules that manage the handover, and the CS Adaptation, together with the CAMEL Service, that interface to existing mobile GSM SCPs. The service is delivered from an IMS AS, interfacing to IMS core elements. The interface to the S-CSCF is the standard SIP ISC.

The CS Adaptation Function performs the signaling associated with:

- CS originated calls
- CS legs established for Domain Transfer to CS

This module communicates with CS Domain nodes and provides necessary information about the session. It also acts as a SIP user agent for incoming calls from the CS Domain. It can use the dynamic IMRN (IP Multimedia Routing Number) for routing CS calls to IMS.

The CAMEL Service Function operates either independently or in collaboration with the CS Adaptation Function. It enforces CS redirection policy from within the CS Domain. It conveys session details from the CS Domain to IMS. Most importantly, it reroutes the CS calls to the IMS, using the IMRN, and delivers it to the VCC AS, using PSI.

3.8.2.2 VCC Service Principles

VCC service logic operates on an AS in the user's Home Network. It requires dual-mode (or multi-mode) user equipment that can connect to GSM, WLANs, and UMTS. A call originated in the CSN must be anchored in IMS to allow the domain transfer to occur. The session control that performs the transfer is within IMS. Standard CS Domain techniques are used for rerouting call initiations to IMS.

The VCC application utilizes the 3PCC technique to enable inter-domain mobility and domain transfers. This involves two separate call legs initiated from the AS (not the UE), and subsequently

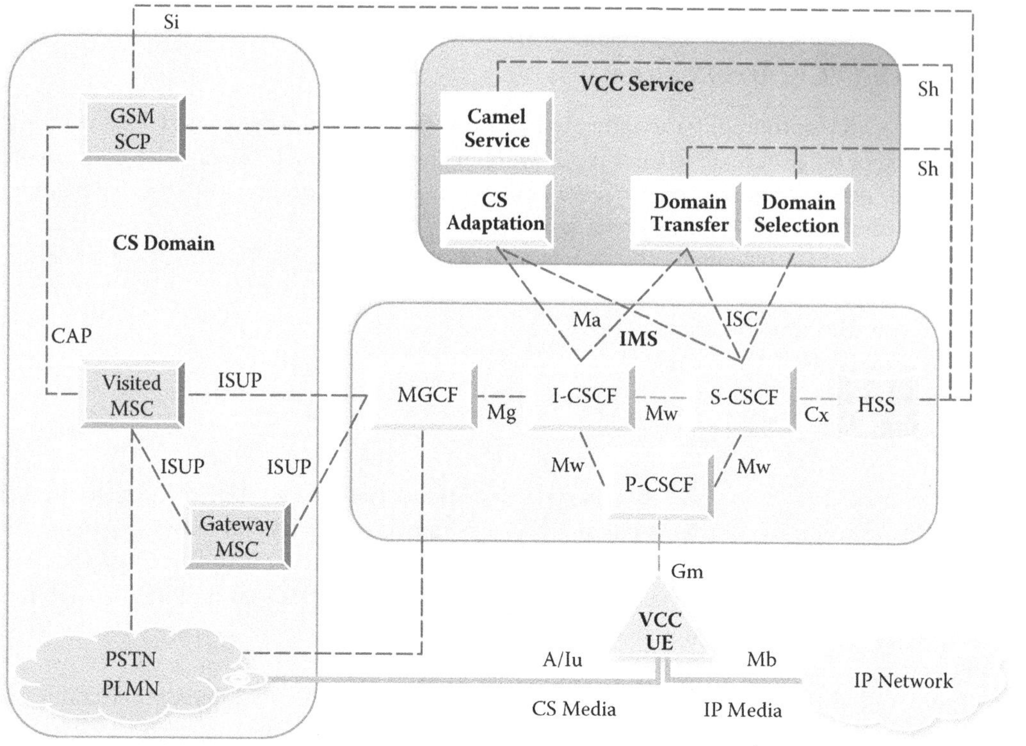

Figure 3.18 The VCC service with dual-mode UE.

joined together. Using a single MSISDN (Mobile Subscriber ISDN) number is clearly very desirable to simplify the transfer and avoid any break in the service that might be detected by the VCC user.

When a call handover is to take place, the UE must be registered in both transfer-out and transfer-in domains. A new call context is established by the VCC UE toward the VCC application in the Home IMS. A new media session is created with bearer resources being allocated in the transferring-in domain. The VCC application in the Home IMS executes domain transfer. When the session has been transferred, the resources in the transferring-out domain are subsequently released.

3.8.2.3 Supplementary Services in VCC

It is not necessarily expected that all supplementary services will work perfectly in VCC. Supplementary services in the CS Domain do not translate exactly to those in the PS Domain. In general, it is assumed that CS-based supplementary services only apply while a VCC subscriber is in the CS Domain, and equivalent services over IMS only apply while a VCC subscriber is in the IMS Domain. However, some supplementary services must be synchronized to retain unchanged and keep the user experience uninterrupted. Broadly, this applies to line identification services that must work in the same way. This includes:

- *CLIP* is presenting the line identity in CS and calling party user ID in IMS. The identity presentation is not changed for the duration of the call, regardless of whether the call undergoes VCC.
- *CLIR* is identity restriction, when it is applicable to the call. It is unchanged by VCC.
- *COLP* is an IMS service presenting called party ID, which is applicable to a call regardless of VCC, where the calling subscriber receives the connected line identity at call setup time and there is no change during handover.
- *COLR* is IMS called party identity restriction. It indicates to the calling party that the ID is not available. This is not affected by call handover.
- *Call forwarding* can operate in the same way with VCC. If the call has been forwarded, the information (dialed number, last forwarded number, etc.) is passed together with the session control.
- *Call waiting* is effective where the waiting call can be reconnected after the handover. The status and call details of the waiting call must be delivered to the new session controller.
- *Call hold* is effective where after handover the call can still be reconnected.
- *Call barring* can be as effective after a handover, where the barred list and options are transferred. However, the new network may have its own barring rules. If a call goes through VCC, which results in the call being barred in the target domain, it is up to the Home operator policy to decide whether the call continues in the target domain/system, or the call terminates, or VCC is not executed for the call.
- *Multiparty call* is effective after the handover, with all parties still connected, and the rules of disconnection apply in the same way.

Information about the charged party must also be transferred to the new domain. Without such information, the service may not be acceptable to users.

3.9 Session Termination

3.9.1 Terminations by P-CSCF

The P-CSCF can initiate session terminations. It can do this on behalf of the calling party or the called party. The P-CSCF may have to refuse establishment of a session and terminate the attempt to initiate the call. This is performed by the CANCEL SIP message. This situation occurs when the resource negotiation shows that there are no sufficient resources for the session, according to its QoS requirements, or the service is not allowed for other policy reasons. The P-CSCF returns a reason code of 503 (Service Unavailable). The same reason is given when the Access Network cannot support the connection (e.g., the handset is out of range or there is no radio signal).

Existing sessions can also be terminated by the network, rather than by the user. This is performed by sending the BYE message and returning a code of 503 (Service Unavailable) or 488 (Not Acceptable Here).

The P-CSCF can terminate an existing session under the following circumstances:

- The user roams out of range or access connection is severed.
- A mid-call policy condition dictates termination of the session.
- Some received SDP parameters are not acceptable or not available.
- A mid-call modification request is not acceptable.

Abnormal terminations can conceivably occur, where the media flow continues after the security association has been deleted or the registration has expired. If there are still such active dialogs, the P-CSCF instructs the Bearer PCEF (Policy Control Enforcement Function) to sever them (via the Gq, Go, Rx, or Gx interfaces), and discard all associated information.

3.9.2 Terminations by the S-CSCF

The S-CSCF can receive a release request from any of the entities involved in the dialog, such as application servers or user agents (AGCF or P-CSCF). In this case, it sends the release request to the destination according to route information in that release request. In this case:

UE (BYE) → S-CSCF → AS (BYE) + Other Party (BYE)

A decision to release the session can be generated within the network, for example, for an administrative reason. In this case, the S-CSCF sends a release request to all the entities involved in this dialog, e.g., to both the AS and the UE that it is serving:

UE (BYE) ← S-CSCF → AS (BYE)

When a session has terminated, it is important to end all sessions on any of the resources and nodes engaged in that session. The S-CSCF should have knowledge of them and it makes sure that they all receive a BYE message.

3.10 Session Charging Procedures

During a session, the P-CSCF, I-CSCF, and S-CSCF generate a CDR (Charging Data Record) for charging purposes. Other IMS elements, such as the media server and the MGCF, also produce CDRs. All these records must be collected in the same data store and correlated for each session, therefore need to carry session identifiers.

During session setup, the P-CSCF generates an ICID (IMS Charging ID) for the originating user, which is a unique charging identifier for this session. This ICID is forwarded to the S-CSCF with the session initiation requests (INVITE). In processing these requests, the S-CSCF includes this ICID in the outgoing message. If the ICID is not present, for example when sessions are initiated by 3PCC on an application server, the S-CSCF can generate one.

The S-CSCF also needs to identify the charging entities to receive session data; that is, the addresses of the OCS and the offline charging data functions (CCF or CDF + CGF) in the Home Network. The S-CSCF obtains these charging function addresses from the HSS and includes them in the outgoing messages. The Local (Visited) P-CSCF can obtain local CCF addresses from pre-configured data.

Interoperability with external network requires that the Home Network is identified so that cross-charging can be achieved. The IOI (IMS Operator Identifier) is a globally unique identifier used for inter-operator accounting purposes. The IOI is carried within the session initiation signaling for the originating user when connecting to an external network, and another IOI can be carried in the called party's response messages. The IOI is always present in messages to and from the MGCF and BGCF that manage border control signaling.

Chapter 8 provides more information about IMS charging.

3.11 Special Sessions

3.11.1 Emergency Session Control

3.11.1.1 Emergency Session Requirements

Emergency sessions constitute calls to any of the emergency services agencies, such as police, fire brigade, or ambulance. Any public communication network must provide facilities to allow users to make such calls and deliver them to the PSAP (Public Safety Answering Point). Emergency call routing must comply with regulations applicable to where the caller is currently physically located.

The special requirements for emergency sessions can be summarized as follows:

- Users must be able to make the call without any restrictions (e.g., barring).
- Calls must be delivered reliably.
- Calls must be routed to nearest PSAP.
- Emergency sessions must be given priority over non-emergency sessions.
- The location of the caller must be indicated to the PSAP.

Location information is needed therefore for two reasons:

1. To route to the nearest PSAP
2. To inform the PSAP of the user's location where an emergency service is needed

When the session destination is recognized as an "Emergency Agency," no further user checks or authentication procedures will be undertaken, and the call is connected forthwith. When the S-CSCF receives an emergency indication, it performs the connection as an anonymous call if the user is not registered.

The call is routed to the emergency services based on location information generated by the network, according to regional-specific regulations. If the call is bound for a PSTN number, the S-CSCF translates the address and routes the call to the BGCF for network breakout.

3.11.1.2 Emergency Sessions in IMS Architecture

In IMS, the P-CSCF must recognize "Emergency" calls and route them appropriately. The P-CSCF role is to:

- Detect an emergency session establishment request.
- Allow or reject unmarked emergency requests.
- Allow or reject anonymous emergency requests.
- Query IP-CAN for location identifier, if necessary.
- Select an Emergency CSCF (E-CSCF) in the same network.
- Allocate high priority level to the emergency session.
- Check the validity of the caller ,Tel URI if provided by the UE, and assert identity or path if needed.
- May respond to the UE indicating the need to initiate an emergency call in the Visited Network.

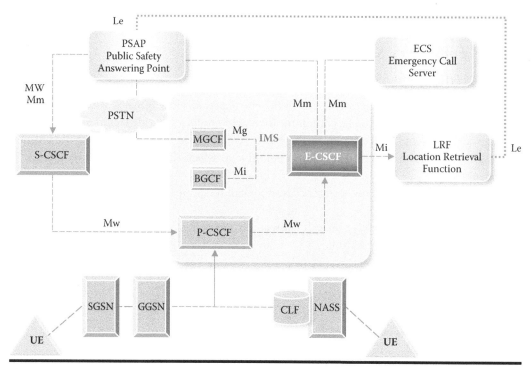

Figure 3.19   Emergency session architecture.

3.11.1.3   Emergency Session Components

The E-CSCF is an S-CSCF with a special task to handle emergency calls. This function can reside in every S-CSCF or in a dedicated server, depending on the implementation. Where the E-CSCF is a dedicated server, emergency calls that reach the caller's S-CSCF are rerouted via the P-CSCF to the E-CSCF. The E-CSCF routes the emergency calls via IMS border control (BGCF and MGCF) to a PSAP in the PSTN Domain. Alternatively, they are routed to the ECS (Emergency Call Server) or a PSAP in a packet network.

The LRF (Location Retrieval Function) is the function that extracts the physical location and returns this information to the requesting PSAP. It receives information from the E-CSCF, which in turn extracts it from the Access Network location server.

The ECS is a special server that manages routing calls in PS networks. The existing concepts of call distribution to the PSAP may be changing as legacy network constraints are removed. In particular, calls can be distributed to any geographic location and need not be routed through a local PSTN switch.

Figure 3.19 shows the emergency session elements: the P-CSCF, the E-CSCF, the LRF, and the PSAP/ECS. It shows the path of the emergency call request from the UE, in any compliant IP-CAN, to the E-CSCF. Location information is conveyed by the IP-CAN, e.g., the CLF for DSL-based Access Network and the GPRS nodes in 3G.

3.11.1.4   E-CSCF Functions

The E-CSCF function, within an S-CSCF or a dedicated server, can be summarized as follows:

- Receive and process emergency session establishment request
- Request location information from the LRF if not included
- Request additional information from the LRF if needed
- Request the LRF to validate the location information if it has been obtained from the UE
- Determine (or retrieve from the LRF) the routing information for the PSAP destination
- Route emergency session requests to an appropriate destination, including anonymous sessions
- Send the contents of the P-asserted ID or UE identification to the LRF, subject to national requirements
- Route the emergency IMS call to the ECS for further call processing, based on local policy

3.11.1.5 Emergency Numbers

Emergency numbers are short codes for emergency services (police, fire, etc.) that are relevant to the locality where the terminal is located. These emergency numbers should be known to the terminal in order to accept them as valid destinations and to initiate a call without authentication or registration. In roaming, the Visited Network can download local emergency numbers to the handset. However, emergency calls in UE Mobile handsets without a SIM/USIM/ISIM must also be possible, even when no other type of calls can be accepted.

IMS should enable facilities such as menu-dial, use of a "red button" or "panic button," or a linkage to a car's air bag control. These facilities can help establish an emergency call without the need to dial a dedicated number or an alternative local number, thereby avoiding wrong connections and delays.

If the UE does not recognize the dialed numbers but the network is able to determine them as local emergency numbers, the session goes ahead as an emergency call anyway. Emergency number identification takes place before any number analysis and takes precedence over any other processing, e.g., supplementary service related.

3.11.1.6 Emergency Registration

Users may be registered or not registered when they launch an emergency call. If the user is not registered, the terminal must be recognized by the IP-CAN but the IMS user registration is not executed in full. If the terminal client (SIM card or laptop client) can provide basic credentials to access the network, the user will be registered temporarily for emergency calls only.

When the UE is already registered, it still needs an emergency registration to enable the emergency service functions. This is performed so that the emergency registration can establish the current location plus a specific access point for further communications, if needed. This Emergency APN (Access Point Name) is kept on the UE separately from the normal APN.

The emergency registration takes place only when the user dials the emergency number, that is, the emergency registration is not on by default. The UE must be capable of managing an emergency registration alongside a pre-existing normal registration.

3.11.1.7   Routing Emergency Calls

An Emergency Service is not a subscription service and therefore is supported automatically in the Visited Network. The roaming UE can initiate an emergency registration with the Visited Network, which still requires the involvement of the Home Network, but the session must be redirected via the Local E-CSCF. The Home Network should be able to detect that the session is bound for an emergency service, whether indicated as such or not, through the analysis of the destination numbers. The Home Network responds to the UE, indicating that the UE should initiate an emergency session in the Visited Network, or may be via the CS domain.

Emergency calls can be routed in either the CS or PS network. Using the existing CS mechanism in legacy networks is an acceptable interim solution. This means that calls can be initiated into IMS session controllers but passed on to the legacy PSTN emergency routing system. A dual-mode UE capable of launching emergency calls in either CS or IMS should give priority to the CS Domain. If the call attempt fails, the UE will automatically make a second attempt on the other domain.

Detection and routing of emergency sessions must be independent of the IP-CAN and are performed by Core IMS components. Therefore, emergency services can operate over cellular access networks, fixed broadband access, WLAN access, and a nomadic NGN access in the same manner.

In IMS, the P-CSCF, in the Home or Visited Networks, detects the request for an emergency session and forwards the request to the E-CSCF in the same network. This P-CSCF must retrieve the location identifier from the IP-CAN and pass it to the E-CSCF. The E-CSCF retrieves geographical location information from the LRF when the geographical location information is not available from the P-CSCF. In fact, the E-CSCF can also use this information to validate location information received from the IP-CAN.

The E-CSCF is responsible for routing the request to the appropriate emergency center, the PSAP — that is, the correct agency and the closest one to the caller — based on various regional rules. The E-CSCF will route to the BGCF/MGCF if the PSAP is PSTN based.

3.11.1.8   Delivering Emergency Caller's Location Data

The determined physical location of the caller is essential to the emergency service. It is used to select which PSAP station should get involved, and it gives the PSAP valuable information about the location of the emergency, independently of the caller's spoken word, to enable them to perform their tasks efficiently.

National regulations decree that the network must provide the caller's location, even when this information is private and normally would not be divulged. This overriding of the caller's right to privacy remains in effect only during the call, or as long as the authorities need it (according to regulations) and, where not entirely clear, subject to the operator's interpretation.

Emergency sessions must be speech calls, not text, but they do involve sending emergency-related data also. Typically, this data enables pinpointing the precise geographic location of the emergency. The data can be sent before the call, at the start of the session, or in parallel. As may be required by a particular PSAP, additional data can also be transferred during the established emergency call, but in all cases, data is only sent to the same PSAP that is receiving the call.

3.11.2   Presence

3.11.2.1   The Basic Principles of Presence

Presence information is exchanged over the network in special sessions. Presence information can be kept for users, resources, or entities that are called "presentities," that is, entities for which Presence status is maintained. The endpoint equipment status of login is the basis of Presence status. Both the endpoint and the registering network server (CSCF) can report on status changes. Applications and users who wish to be informed of the Presence status of any such presentity need to subscribe to notifications of status changes, and become "watchers."

Watchers can maintain a group of users or a list of resources they wish to monitor for Presence, so that they can subscribe to the status of the entire list. This "resource list" is maintained for each user on a server but also can be kept locally on the endpoint. At registration time, the S-CSCF records the user status and sends notifications to the Presence network agent, which forwards information to a Presence server.

3.11.2.2   Presence Elements

A Presence User Agent (PUA) is an entity that provides presence information to a Presence Server. The PUA can be in the UE or a network-based server. User registration is treated as an event, and notifications are sent through to the PS, which in turn alerts the authorized "watchers."

To provide its own Presence information, the PUA acts as the Event Publication Agent) that generates a PUBLISH request. This results in event notifications to be sent to all watchers. The PUA must update the Presence information before the publication expiration time, according to the response from the PS to the PUBLISH request. To prevent excessively frequent publications of Presence status, local configuration can restrict the rate at which the PUBLISH requests are generated.

The PUA can maintain various levels of privacy. This means that it is capable of showing different values of the same Presence attribute to different watchers. It therefore needs to generate a different dataset (a tuple) for every value it intends to show. The published content is produced in the Presence Information Data Format (PIDF).

A watcher is an entity that requests, or subscribes to, Presence information about a presentity from the PS. The watcher can include filters in the body of the SUBSCRIBE request that define what the watcher wishes to know. The watcher can indicate its support for partial notification using the Accept header field. A watcher can subscribe to:

■ Presence information state changes of presentity collections
■ Watcher information event
■ Notification of state changes in an XML document

A Presence Server is an entity that accepts, stores, and distributes Presence information. When the Presence Server receives a SUBSCRIBE request or a PUBLISH request for the information event package, it must verify the identity of the source of the request. If this is a valid subscription, the PS generates a response to the SUBSCRIBE or PUBLISH request.

The Resource List Server (RLS) is the function that maintains Presence for a given list of presentities or resources. The RLS application accepts subscriptions to maintain such lists and sends notifications to update subscribers of the state of the resources in a resource list. When the RLS receives a SUBSCRIBE request for the Presence information event package of a presentity collection, it must

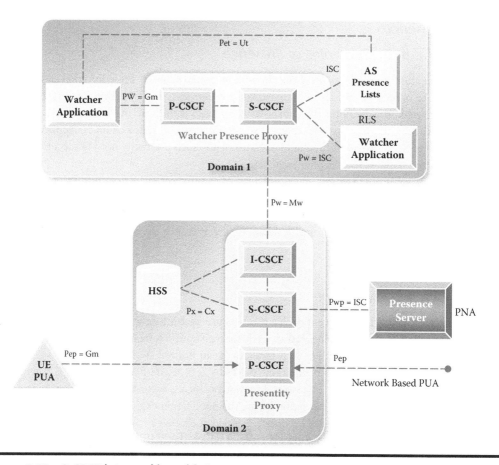

Figure 3.20   S-CSCF interworking with Presence.

verify the identity of the source of the SUBSCRIBE request and, if successfully authenticated, the RLS generates a response by adding a Require header field with the value of "eventlist."

A Presence Network Agent (PNA) provides Presence status from a network-based server rather than the endpoint terminal. The PNA can collect Presence information about the presentity from a number of Core network entities and combine the information to form an aggregated view for the user. Among these Core network entities, the S-CSCF uses SIP to deliver Presence information to the PNA. On receiving a third-party REGISTER request, the PNA will subscribe to the registration event package for a particular user at the S-CSCF. As a result, the S-CSCF will provide the Presence-related information as registration event packages in NOTIFY requests to the PNA.

Figure 3.20 shows the User home network (Domain 2) and the Watcher's domain (Domain 1) exchanging presence messages. The Presence components interact with IMS in the same way as other ASs using the SIP ISC interface. The figure also shows the Resource List Server that can be maintained on an AS.

3.11.2.3   Presence Using the Ut Interface

Presence data is stored using the XCAP format. XCAP is based on the HTTP framework and uses the HTTP methods of PUT, GET, and DELETE for communication over the Ut reference point.

XCAP is used to manage data related to Presence, including user groups, subscription authorization policy, resource lists, etc.

To be compatible with the Ut interface but ensure greater security, the UE should support HTTP Digest AKA (Authentication and Key Agreement), and should be able to acquire the subscriber's certificate from a PKI portal using the authentication bootstrapping procedure. This interface uses TLS to secure the transferred information.

3.11.3   Number Portability (NP)

Number Portability (NP) has been implemented for PSTN in some countries for many years. The concept is to enable customers to move to competitive operators without having to change their telephone numbers. Because the network can be recognized by certain number blocks that have been allocated to it, this imposes a requirement on the carrier to support redirection of calls for subscribers who have moved on to other networks. NP solutions have been implemented primarily as an IN application that fetches the ported number information from a central or operator-based database. Alternatively, the local exchange switches maintained lists that allow recognizing ported-out and ported-in numbers.

In IMS, numbers can be recognized but they are handled by the Tel URI and SIP URI with user–phone parameters. IMS session control must be able to route ported numbers as well as be compliant with regional NP regulations.

In an IP network, the ENUM/DNS server provides a simple solution. ENUM provides mapping of telephony numbers (E.164 scheme) to IP addresses and can supply NP information at the same time. The IMS routing entities need to support NP parameters that are retrieved as part of ENUM/DNS procedures and pass them onward, to be interpreted in the PSTN.

Alternatively, the IMS routing entities can access the NP database in the PSTN via IN/TCAP protocol. This can be achieved by involving the MGCF, using legacy interfaces.

3.11.4   Messaging

This term refers to a messaging function similar to the popular IM but compliant with the 3GPP standards. The size and type of media payload defines how the message contents are carried through. The contents can be text, file, video clips, images, music, and sound clips.

The equivalent to Internet-based IM is called "Immediate Messaging," using the SIP message type of MESSAGE. A further mechanism of messaging has been defined using IMS sessions and their capabilities, referred to as "session-based messaging."

3.11.4.1   Immediate Messaging

The SIP MESSAGE method uses the SIP message type of "MESSAGE" according to RFC 3428. This is designed for quick near-real-time interaction, carried over signaling with *no* media path established. As a signaling message, it is limited in size. This means that a different mechanism is required for transporting large files. It necessitates that the MESSAGE method must be filtered to prevent sending large messages embedded in signaling, which would create congestion in the signaling layer. If the content is not acceptable, due to length or content type, the S-CSCF will reject the forwarding request.

At the receiving end, similar restrictions may be in force. Depending on the operator's policy, when the recipient party is not registered, the MESSAGE function can still operate by forwarding to an application that may be active for unregistered users. Such applications can store voicemail, play an announcement, or divert the call to another predefined destination. Similarly, the immediate messaging text can be stored and delivered to the user later, when the user is "present." The delayed delivery is triggered by the recipient's registration event.

Similar to the popular Instant Messaging, Immediate Messaging is also associated with group management. Users can create and maintain lists of potential recipients and can subscribe to their Presence status. Immediate Messaging allows distributing a single message to multiple parties on such a list. This is performed by creating a group under a PSI name, using an application on the PSI AS. Members are added to the group via the secure Ut interface, in an XCAP document transferred directly from the UE to the AS, or by using the Operations and Maintenance (O&M) operator's provisioning. The UE can then generate messages to send to the PSI destination. The AS hosting this PSI receives the message and creates duplicate messages that will be sent to all group members.

Multiple message distribution also can be facilitated at the recipient end. The application for the destination party can be invoked as a terminating service that sends the received message to a group of public IDs stored on it for such purposes. Such a service is used in multi-party chatrooms, for example, or in multimedia conferencing.

3.11.4.2 Session-Based Messaging

Session-based messaging is based on the same mechanism as any IMS session. The signaling part creates the session between the two parties; but unlike immediate messaging, the media payload flows over a media bearer. In addition, the messages are part of a session dialog that can be controlled by an application, while intermediate messaging is only aware of one message at a time.

The session-based messaging media flow characteristics are defined in the SDP offer allows response between the parties. The IP-CAN bearer can be pre-existing general purpose or a new one can be created. It only needs a "best-effort" level of QoS (i.e., the lowest class of service) because it is intended as a non-real-time service, but may be using a reliable connection-oriented mechanism to enhance data integrity (e.g., using TCP instead of UDP). This session should make use of the security capability of the network to keep the session signaling and the media flow confidential.

Session-based messaging also can be used for a "chat" facility that enables one-to-many or many-to-many message distribution. The group will be associated with a PSI hosted on an AS, where the application on that AS will manage the chat session.

Session-based messaging can provide further facilities, including:

- Charging for the service, per content type or per usage
- Applying policy-determining size, contents, etc.
- Negotiation of maximum message size between the parties
- Messaging sessions with more than one type of media component

Chapter 4

Network Admission

4.1 This Chapter

This chapter deals with the various ways that convergence IMS enables users to be admitted into the network, in either mobile networks or an NGN. The process of admission into the network is described here to put into context the specific functions: (1) Attachment: binding equipment and users with IP addresses and network paths; (2) Authentication: verifying the user or device with the appropriate security according to different methods; and (3) Registration: authorizing services while making the network aware of user profiles, services profiles, and the relationships between multiple user identifiers.

For network Attachment and Authentication processes, the general principles common to all access methods are described first, and then specifications for 3G Mobile and NGN are detailed separately, to explain the different treatment of these functions according to the particular technologies and their own terminology. Also discussed are the procedures for nomadic and roaming users and early implementation methods of "Early IMS" and "NASS bundled authentication."

Finally, the Registration procedures are described. Registration at the IMS level is access-agnostic, and therefore applies to both 3G Mobile and NGN solutions. Explicit and implicit registration, group registration, and AS third-party registration are explained here, along with registration, re-registration, and deregistration procedures.

Call session admission also involves setting up the bearer according to policy and charging rules and session resource allocation for QoS. These functions are described in Chapter 8 and Chapter 6, respectively.

4.2 Network Admission Overview

4.2.1 Multi-Access Admission Control

Admission control is the process that allows requested sessions to connect to the network, ensuring adequate resources. The resources can be bandwidth that supports a certain level of QoS (e.g., latency and jitter), or particular types of ports on processing Transport layer nodes.

Admission control is the set of functions that allows all types of users, both UE and applications, to gain entry and get "attached" to a network, whether they are in their Home Network or a Visited Network. The general network admission activities include:

- IP address allocation and verification, and equipment identification
- Potential network address translation (IPv4 or IPv6)
- Service discovery
- Mobility support, for nomadism (NGN) and full mobility (3GPP)
- Mutual authentication (i.e., authenticating the network to the UE, and vice versa)
- User authentication, implicit and explicit
- Encryption and security for the UNI

Admission control is an essential process that allows users of disparate access networks to receive shared IMS services. While Core IMS elements such as CSCF, MGCP, HSS, and MRF are considered access-agnostic, the binding of UE onto a network, the authentication parameters, and the assigning of network resources are not entirely divorced from the underlying transport and access networks, or the type of end-user equipment.

In 3G Mobile Networks, the admission control is performed in association with the Radio Resource Management functions, which estimate the radio resource requirements within each cell. The process of admission to the network is handled by the GPRS nodes in the UMTS.

TISPAN has defined network admission vision for NGN, which is a Fixed/Mobile converged network model particularly suitable for nomadic users who can move not only between locations, but also to any terminal (e.g., PC). In TISPAN NGN, the NASS (Network Attachment Subsystem) is designed to allow a variety of NGN access forms: xDSL, cable, WLAN, Ethernet, and Enterprise VPN. In WLANs, the air interface process of admission is managed by the WLAN access point and gateways. The WAG and the PDG with an AAA server perform the full procedure.

IMS appears to all access networks as a service, alongside direct access to the Internet in peer-to-peer mode, and alongside flow-based services such as IPTV, video-on-demand, and file transfer. The IMS service is authenticated in a unified process, managed by the IMS Proxy (P-CSCF) or access gateway controller (such as the AGCF). Other services have their own session admission systems, but there is a growing demand to combine them and unify the data.

4.2.1.1 Two-Stage Admission

As the IMS scope extends to multiple access methods, there must be an independent authorization of IMS services. Therefore, there are two stages for network admission:

1. At the IP-CAN level, for initial equipment identification and handshake procedures, and the attachment to the network with a valid IP address

2. At the Core Network level, for authentication of the IMS service, identification of the user, and authorization of subscribed service

In the first stage, the UE gains access to the IP-CAN. It finds and connects to a local access point whose address is published. The UE may need to decide which entry point to select out of a list of available servers. The UE provides details of its equipment identifier to the access point, which then returns an IP address. This IP address can be obtained from a local DHCP server. Once the UE has an address for its first network node to contact, it requests admission to the network through a validation process.

In the second stage, the IMS user identification is performed and usage of IMS services is authorized. This process is managed by the P-CSCF, which is contacted by the access point. The P-CSCF receives registration requests from both NGN and Mobile, to interface to Core IMS. The authentication can be mutual, where the network node must be recognized by the UE and the UE must be identified by the network.

Figure 4.1 depicts three types of bearer IP-CAN networks: UMTS, xDSL, and WLAN. Other types of IP-CANs, including WiMAX, cable, and Ethernet, are not shown here. In this diagram, each Access Network provides its own admission control as first-stage network admission, and then all proceed to IMS service authorization via the P-CSCF. Also shown are the policy decision controllers for all types of access, which are now converging into the PCRF.

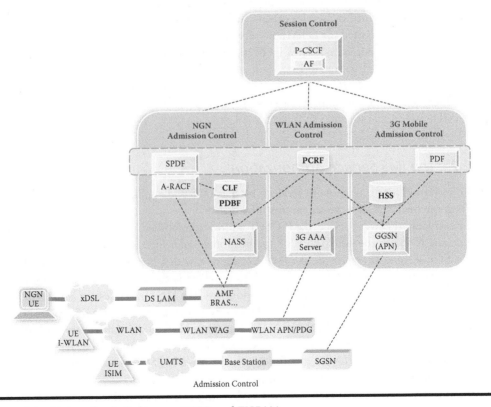

Figure 4.1   Network admission per 3GPP and TISPAN.

4.2.2 Network Admission Generic Processes

4.2.2.1 The Attachment to the Network in Brief

The network attachment processes (i.e., binding user ID and an allocated IP address) are IP-CAN specific and are performed in the NGN or GPRS/UMTS access nodes. Similarly, the network admission processes that authorize usage of resources for the particular user and the type of session are performed in the respective bearer servers.

The 3GPP network admission function relies on information stored securely within the UE firmware. This includes not only the equipment unique identifier, but also a security client, a list of contactable entry nodes to partnering networks, and procedures of authentication. The SGSN and GGSN interact with the Mobile Station (handset) functions, using the USIM or ISIM identities. Network admission is performed at the UMTS nodes. The PDP Context (Packet Data Protocol Context) provides a bearer for the initial dialog, after the UMTS determines the identity of the UE and performs location management.

Initial admission in NGN (according to TISPAN) is initiated by the Access Manager Function (AMF) on behalf of the UE through the NASS. The AMF in xDSL equates to a BRAS (Broadband Remote Access Server). The UE in NGN connects via a static node or CPE, such as residential modems or gateways or IADs (Integrated Access Devices). The addresses of these nodes are generally known to the network by preconfiguration or are registered at installation time.

Nomadic users become attached dynamically to these gateways, obtaining their temporary IP address from preconfigured lists or via a network DHCP server. The Access Point server allows only restricted communication for the purpose of initial signaling. This is a "pinhole" that will not permit any media flow until the service is fully authorized and the session is established, except for "Early Media" when required for DTMF or network announcements.

First-stage session admission for a WLAN is performed by the attaching WLAN access point, as defined by the IEEE (Institute of Electronic and Electrical Engineers) and IETF. User authentication for the WLAN Access Network can be performed utilizing 3GPP facilities or NGN servers. Then, service authorization for access to IMS is performed at Core IMS servers, S-CSCF, and HSS/UPS.

4.2.2.2 The IMS Service Authorization

The authorization for using network services, such as IMS and applications that are controlled by IMS, is granted when users register successfully to IMS.

For entry into the IMS Service, the P-CSCF address is published as the entry point. The Access Node, in the NASS or the GSN (GPRS Support Node), holds internally configured lists of available P-CSCF servers. They provide the P-CSCF address, with a port number or a list of FQDNs for alternative instances of P-CSCF servers. Also provided is the Local DNS server address that can resolve the domain names into unique IP addresses. The UE can request an address of a DNS server IPv4 as per RFCs 2132 and 3361 or a DNS server IPv6 as per RFCs 3315 and 3319.

In NGN, as defined by TISPAN, the NASS is responsible for service discovery, with the assistance of DHCP and DNS servers acting as local and network configuration servers, respectively. In 3G Mobile, the UMTS is involved in finding the available P-CSCF, through preconfigured lists that may include IP addresses or logical names.

The P-CSCF acts as a doorway to the network from the user access side, for both Home users and Visiting users, and for Mobile and Fixed IMS users. When the discovery process results in several P-CSCF servers being offered to the UE, the P-CSCF server is selected through built-in network policies or preconfiguration mechanisms. For roaming users, the P-CSCF forwards the call to the Home Network border entry point. This entry point can be the Home I-CSCF or an IBCF before reaching the Home session controller (S-CSCF).

4.2.2.3   Requesting Resources at Session Admission

Network admission for a connection is not complete until network resources are allocated or reserved for the user session. Resources can be allocated at Layer 2 within the IP-CAN for connections that do not utilize session control, e.g., video streaming and IPTV, although there are arguments that IMS can also benefit these types of connections. Such flow-based connections can be governed by policy that affects the bearer directly, and charged via flow-based charging procedures.

For IMS-based sessions, the AF is the function that negotiates resource allocation. This is a generic function that enables a common interface to Core IMS and performs the appropriate tasks through negotiation with the different IP Access Networks. The AF represents any user agents that require binding of resources, and should not be confused with applications on application servers.

The AF formulates the request of resources according to user information and the requested session type and media type. The request is forwarded to the resource controller that negotiates with the IP-CAN to reserve bearer resources for the session. The AF negotiates resources not only at session initiation time, but also at mid-session when changes are required.

In setting up the resources for the session media flow, the level of QoS must be determined. This affects the level of delay, jitter, and packet loss, as well as throughput or packet rate. Each session is allocated a priority level and resources according to network policies, based on session details, type of service, etc.

The AF can reside within the P-CSCF (for resources toward the IP-CAN requested by the UE), the AGCF (for resources connecting to access gateways and POTS), or the IBCF (for resources toward external IP networks). When UE initiates a session request, the AF in IMS sets up the request for resources and QoS level. It consults the PDF for UMTS IP-CAN or the SPDF for NGN, where service-aware policy decision is required.

The policy decision is conveyed to the network elements that can enforce the policy. A PEF (Policy Enforcement Function) is contained within the IP-CAN nodes so that they have the ability to police packet flow according to class of service or QoS markers (i.e., to "open" or "close" the gate).

Figure 4.2 shows the full scope of converged network admission, including network attachment in UMTS and NASS, and the resource allocation for the bearer, consulting the PDFs in both NGN and 3G. The AF, which is common to all Access Networks, connects to the PDF/PCRF that controls policy in the UMTS IP-CAN, and to the SPDF within the RACS in an NGN IP-CAN.

Note that the GPRS nodes carry both signaling and media, and perform both network attachment and resource allocation for both GPRS and UMTS. In the NGN architecture, the NASS and the RACS are signaling subsystems, and the RCEF (Resource Control Enforcement Function) is a separate layer for the media. The different approach between TISPAN and 3GPP will even out with the developing ideas of LTE, which is currently being specified by 3GPP.

Note that Chapter 6 provides a detailed description of QoS management.

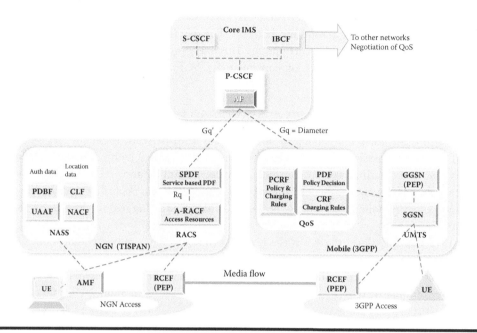

Figure 4.2 Admission with QoS setting in both GPRS and NGN.

4.3 The Network Attachment Process

4.3.1 The Converging Attachment Process

The fixed nature of NGN is evolving with advanced device and access technologies. The device enhancements affect the ways they are attached to the network. For example, as all laptops acquire WiFi capabilities, it becomes very easy to use them as nomadic terminals that are temporarily tethered. Another example of this is the possibility of embedding firmware, like SIM cards, in all Fixed IP telephones and laptops, which would make the binding of equipment and user the same as in Mobile Networks.

As Mobile handsets and laptops, soft clients can connect via one type of access technology — the WLAN or WiMAX — the attachment procedures begin to converge. In particular, when WiMAX becomes available, the constraints of distance and reach of WiFi are removed and nomadism becomes ubiquitous, as in 3G Mobile. Hence, network admission for NGN no longer seems so radically different from that needed for 3G Mobile, and eventually convergence of admission control concepts will also simplify this part of communications.

4.3.2 Network Attachment in 3G Mobile Networks

Currently, the GPRS nodes perform call admission and QoS management in 3GPP networks, together with the CSCF and HSS. As 3G Mobile evolves, the new System Architecture Evolution (SAE) in the LTE design will change that.

The GGSN node in the GPRS network can have DHCP on board or as a linked server, or can use another mechanism for the discovery of the address of the Local P-CSCF, such as local configuration.

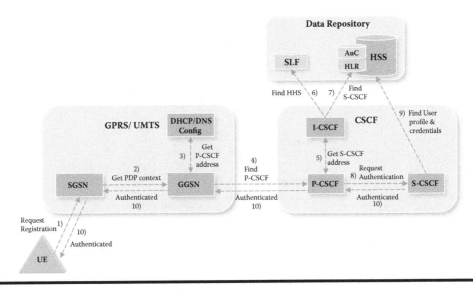

Figure 4.3   Elements involved in 3GPP network attachment.

The process of UMTS network admission in the Home Network involves UMTS binding the UE identifier with an IP address before requesting authorization of resources. As Figure 4.3 shows, the following steps are taken:

1. UE finds the SGSN and requests registration.
2. The SGSN requests the GGSN to admit the UE and activates signaling PDP Context.
3. The GGSN finds the P-CSCF and queries the DHCP/DNS for IP address.
4. The GGSN routes to the local P-CSCF.
5. The P-CSCF forwards to the I-CSCF to obtain the S-CSCF address.
6. The I-CSCF queries the SLF to find which HSS to use.
7. The I-CSCF queries the HSS for which S-CSCF to use.
8. If the S-CSCF is not assigned (user not registered), the I-CSCF selects one and requests registration.
9. The S-CSCF obtains user information from the HSS and can now authenticate and register the user.

4.3.3   Network Attachment in NGN

Admission control in TISPAN is a generic subsystem aimed at describing NGN nodes as well as other access nodes that provide network admission. The NASS provides the mechanism for attaching users to the network. In NGN, user association with an ID does not necessarily rely on firmware, such as ISIM in the IC card. In Fixed line access, the user's CPE can be a fixed node that allows multiple devices to connect to the network (e.g., a multi-line modem or router). The terminals attached to it are therefore known by their lines or ports on that CPE and, once installed, can be "implicitly" authenticated. In addition, NGN needs to provide for nomadic users who not only move with their equipment (e.g., laptops), but also might log in on different devices, yet receive their own network services.

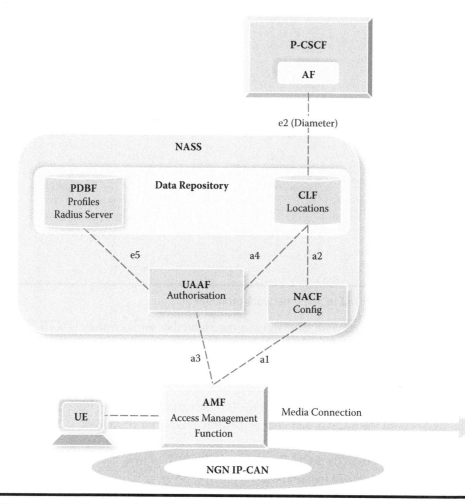

Figure 4.4 NASS: network attachment in NGN.

Figure 4.4 shows the access node with its AMF and the NASS elements: the UAAF, the PDBF, the CLF, the NACF (Network Access Configuration Function), and the reference points between them.

Attaching to the network occurs at the access point where the NGN terminal currently connects. The address of this access point is stored at registration time as the session path to this UE, where the implementation makes no assumption of permanently wired connections. Therefore, at the second stage, the user subscription account is linked to the attached UE.

The NASS provides a number of functionalities. It is responsible for the dynamic allocation of IP addresses, and port and terminal configuration parameters. It coordinates the authentication at the IP layer prior to service authorization by IMS. The NASS authentication is based on locally stored user profiles that contain access equipment details and local user credentials. The NASS also stores the current access path and provides physical as well as IP network location details to IMS servers and applications that need that information.

The NASS has a number of functions that may already exist in some form in current implementations, such as DHCP or RADIUS authentication. Therefore, the NASS functions must be flexible to utilize existing facilities. For example, the NACF can be an existing DHCP server, the

CLF can co-reside with an IMS version of the HLR, and the PDBF can be an existing RADIUS server or joined with an AAA server.

4.3.3.1 The Access Managment Function (AMF)

AMF is an access point in the network, although the UE can reach it via relay nodes (ARF Access Relay Function). The UE discovers the access node address via a Local DHCP and DNS, and makes a request to connect. The AMF is the function that seeks authentication and network attachment from the NASS. At the first step in network admission, the AMF contacts the NACF to request allocation of an IP address dynamically or provide preconfigured information.

4.3.3.2 Network Access Configuration Function (NACF)

The NACF is a server that allocates IP addresses when requested by the UE. The allocation of IP addresses must uniquely identify the UE. The NACF also can be used to retrieve other network configuration parameters such as the address of DNS servers and the address of signaling proxies such as P-CSCF.

The allocation of IP addresses can be performed dynamically, usually using the DHCP mechanism. The allocation of addresses is carried out for either IPv4 or IPv6. This is determined by the capabilities of the UE, the access network and network policies, as well as the NACF capabilities.

4.3.3.3 Connectivity Session Location & Repository Function (CLF)

The allocated IP address is stored on the CLF, which is the NGN equivalent of a location manager. It binds the geographic location information that can be derived from the identity of the APN with the network IP address and username. The NGN location manager supports nomadism (i.e., logging in from different locations). It holds information about the location and path of the session.

4.3.3.4 Admission for PSTN Emulation via the AGCF

The softswitch-based solution for the PES is not linked to IMS, and therefore is not described here. However, the AGCF provides a hybrid access method to IMS, with the AGCF translating H.248 commands to SIP messages toward the S-CSCF. The AGCF appears to IMS as another P-CSCF. It appears to analog phones as a residential gateway.

In the AGCF, the MGC module receives H.248 requests from users behind the local gateway, translates them and sends, via the IMS agent module, SIP INVITE requests toward the S-CSCF or the I-CSCF.

This is a hybrid system where session admission is similar to an H.248-based softswitch but the service authorization is based on IMS. Network admission is therefore based on Fixed-line permanent connection that is preconfigured and requires only basic access-level authentication.

Figure 4.5 shows the AGCF connecting to legacy terminals through a Local Media Gateway (MGW) or a residential gateway. The local MGW acts as an access node communicating to the AGCF MGC module via H.248. The AGCF contains an IMS agent module that interfaces to IMS like a P-CSCF.

Chapter 9 provides additional details about the AGCF.

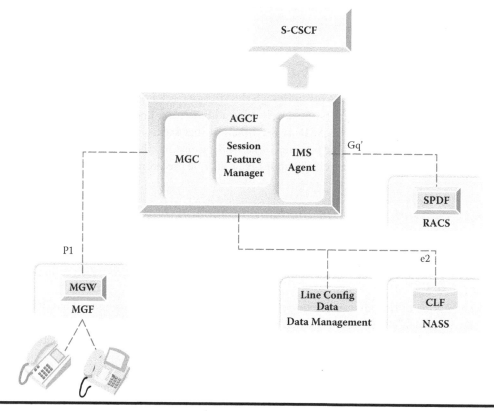

Figure 4.5 AGCF call admission for H.248-based terminals.

4.4 The Authentication Procedures

4.4.1 Authentication Principles

When referring to authentication in general, it is often meant to include the full AAA functionality. AAA mean:

- *Authentication:* verification that a user requesting a service has presented the prestored credentials to prove the user's identity.
- *Authorization*: verification that the requested service is within the SLA or subscription for a particular user, and within the request circumstantial parameters.
- *Accounting:* recording of service usage in the context of an authorized service session so that charging information can be passed to the billing system.

In 3G Mobile, an IMS user has a profile stored on the central repository, the HSS. The generic reference to the repository of authentication data in the HSS is referred to as AuC — the Authentication Center — although some data also resides in an internal database in the GPRS nodes.

In NGN, data for access-based authorization, which is performed by the UAAF in the NASS, is stored in the PDBF and not in the HSS equivalent, the UPS, which contains only IMS-level profiles. This is designed to separate data that may be owned by different business entities to allow business borders between the access network and the user's Home Network. This scenario is more prevalent in fixed networks where opening the local loop process has been going on for some time, while mobile operators generally own both the air interface (the RAN) and the GPRS access network.

After performing the initial bearer authentication for the UE according to the requirement of the particular IP-CAN, the IMS user should be authenticated to the IMS network, to allow activation of services. This process is regarded as service authorization.

4.4.1.1   Authentication Elements

Authentication is performed for several network elements:

- *Device (UE) authentication.* The device needs to prove that it is genuine by authentication of a USIM/ISIM through a process of "challenge and response." In NGN, a similar concept is under consideration for nomadic equipment. Otherwise, tethered equipment is known to the network via the physical connection of the IAD or residential gateway and line or port number.
- *User authentication.* The user, independently of the terminal used, also must be recognized by the network. This process associates a user profile and a user account with a specific terminal and a current IP address.
- *Mutual network-UE authentication.* To prevent signaling interception and "replay" attacks, the UE can confirm that the network connection is genuine via the MAC method.
- *Application authentication.* It is often necessary to check the authenticity of an application requesting network resources. It also may be required to monitor the level of authorization for specific software operations. This is usually part of an application gateway, such as defined by OSA, and is a separate process from network/UE authentication. Some applications require user authentication and authorization, but in most cases they utilize "single sign-on" or federated authentication and accept the network-based service authorization
- *Transaction authentication and non-repudiation.* For some business transactions that are carried out using communication devices, it is necessary to digitally sign the transaction with a user private key, specifically where there is a need for non-repudiation to prevent users from denying having made the transaction.

4.4.1.2   User Authentication Data

IMS authentication of users is based on data stored securely on the user equipment in the ISIM. The subscriber is allocated a private user identity by the Home Network operator plus a number of contacts or numbers associated with devices or soft clients. These identifiers form an essential part of the data for user authentication, which is stored in the ISIM. This preconfigured data also contains security parameters and keys to secure the process of registration to IMS and allow users to get authenticated independently of the Access Network and the access technology. This process should therefore be the same for both 3G Mobile and NGN.

The main user authentication parameters on the ISIM include:

- *IMPI.* The IMPI identifies the user account within the Home Network.
- *One or more IMPUs.* The IMPUs are various contact details that are quoted by incoming callers.

■ *The Home Network domain name.* The domain name is used to build the destination address for signaling messages (i.e., REGISTER and INVITE).

For early implementations, when the UE does not contain the ISIM, there are mechanisms in 3G Mobile ("Early IMS") and in NGN ("NASS bundled authentication") to provide an interim alternative. These solutions are based on ways of generating a private user identity, a temporary public user identity, and a Home Network domain name. See details below.

4.4.1.3 The Authentication Architecture

User authentication must be a secure process. In particular, the transfer of keys for authentication and the process of registration must be protected. To that end, the 3GPP defines a generic authentication architecture. It is divided into two strands, following two main methods of authentication:

1. *Shared secret.* The shared secret method requires a bootstrapping mechanism to establish the shared secret that will be used later for authenticating the user.
2. *Certificates.* The certificate method needs a Certificate Authority that securely generates the certificates and delivers them to the user equipment, to be cited in later communication.

Figure 4.6 shows the generic architecture and its relationship with the HSS and the UE. It shows the shared secret bootstrapping mechanism that generates the shared secret if it is not already stored on the HSS. Alternatively, it shows the certificates method, backed by the SSC (Support Subscriber Certificate) that uses the PKI (Public Key Infrastructure) portal to generate certificates.

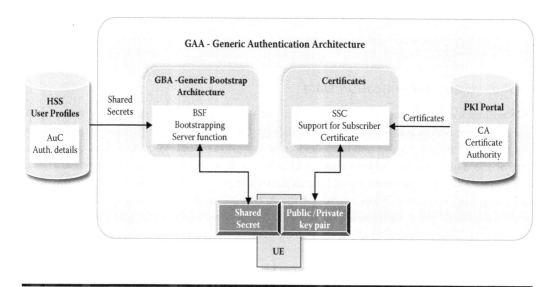

Figure 4.6   Authentication architecture.

4.4.1.4   The Generic Bootstrapping Architecture (GBA)

Before communication between the UE and any NAF (Network Application Function) can start, the UE and the NAF must first agree on whether to use shared keys obtained by means of the GBA. If the UE does not know whether to use the GBA with this NAF (which can be a P-CSCF), it uses the bootstrapping procedure via the Bootstrap Server Function (BSF).

The set of all GBA User Security Settings is stored on the HSS, or in the AuC module of the HSS. If the subscriber has multiple subscriptions, the HSS can contain one or more such security profiles mapped to one or more private identities, i.e., IMPI for IMS and IMSI (International Mobile Station Identity) for 2G and GPRS.

4.4.1.5   The Bootstrapping Server Function (BSF)

The BSF facilitates the initial negotiation of security keys with the UE. The BSF can establish bootstrapped security association with the UE by running the bootstrapping procedure based on HTTP Digest AKA when nothing else is available. The generic BSF and the UE need to be mutually authenticated using the AKA Protocol, and agree on session keys that are applied afterward between the UE and a NAF.

A bootstrapping session on the BSF also provides security-related information about the subscriber (e.g., user's private identity) that is used in subsequent communications. After the bootstrapping has completed, the UE and the NAF can run application-specific protocols where the authentication of messages will be based on those session keys generated during the mutual authentication between the UE and the BSF.

Figure 4.7 shows the relationship between the BSF and the UE, over the Ub interface. It also shows the NAF interface to the user (Ua) that utilizes the security keys produced by the BSF. The interface between the NAF and the BSF (Zn) enables the NAF to obtain the matching keys for the communication over Ua. Note that the NAF can reside in another operator's network, and therefore may be interworking through a proxy, passing through border controls. During the bootstrapping procedure, the BSF also uses the Zh interface to request authentication vectors from HSS and obtain the user security settings.

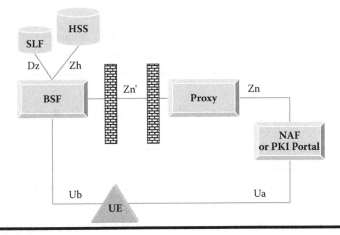

Figure 4.7   Bootstrapping model.

The interfaces to the user (Ua and Ub) can be further secured by running them over TLS. During the TLS handshake, the NAF can indicate to the UE that bootstrapped security association is required by policy. It also can indicate that the bootstrapped security association has expired and needs to be renegotiated with the BSF. The UE can also initiate the process without a prompt from the NAF, and will contact the BSF to provide bootstrapping authentication whenever it detects that there is no existing security association between the UE and the NAF.

The triggers for utilizing the BSF include:

■ The UE has received a message that bootstrapping initiation is required.
■ The NAF indicated to the UE that bootstrapping renegotiation is needed, due to an inadequate security level or other local policies.
■ The lifetime of the key in UE has expired or nearly expired.

The end result of the bootstrapping procedure is that both the BSF and the UE have a security association in the form of a bootstrapping transaction identifier and security key material. This bootstrapping procedure is based on HTTP Digest AKA, as described in RFC 3310.

If the UE and the NAF already share the key required to protect the connection, they can start immediately to communicate securely. The bootstrapping session is kept "alive" for a certain time period, and is deleted in the BSF when the session expires.

4.4.1.6 Generic Bootstrapping Reference Points

■ Ub is the reference point between the UE and the BSF. It provides mutual authentication between the UE and the BSF. It allows the UE to bootstrap the session keys based on 3GPP AKA infrastructure. This allows using HTTP Digest AKA (RFC 3310) with USIM and ISIM.
■ Ua is the reference point between the NAF and the UE. It carries the secured application protocol using the keys that have previously agreed with the BSF. The PKI portal also can be regarded as an NAF, and this interface then carries the subscriber's certificate produced by the PKI portal.
■ Zh is the reference point between the BSF and the HSS. It allows the BSF to fetch the required authentication credentials.
■ Zn is the reference point between the NAF and the BSF. It is used by the NAF to fetch the keys agreed upon during a previous HTTP Digest AKA protocol run over the Ub interface from the UE to the BSF. It is also used to fetch application-specific user security settings from the BSF, if requested by the NAF.
■ Dz is the reference point between the BSF and the SLF. It allows the BSF to get the name of the HSS containing the required subscriber data.

4.4.2 Authentication Techniques

4.4.2.1 Authentication Using Shared Secret

There are several authentication protocols that rely on a prestored shared secret between the two communicating entities. Popular examples include HTTP Digest and IKE (Internet Key Exchange) with a pre-shared secret and a mechanism based on username and password. The main issue with these mechanisms is the weakness of the procedure to agree on this preshared secret. To overcome this for IMS, an AKA-based mechanism can be used to provide both communicating entities with a preshared secret.

Typically, the shared secret can be used for:

- Distribution of symmetric ciphering key (CK) and integrity key (IK) for securing applications running between the UE and a server in the network
- Distribution of passwords and PINs for third-party applications
- Protecting the distribution of certificates between the UE and the certificate authority

Shared secret-based authentication utilizes the following credentials and produces the listed "key materials" listed in the following table:

Parameter	Function	Description
K	Shared secret	Preconfigured secret
RAND	Random authentication	Random challenge generated by network elements toward the UE
AUTN	Authentication token	Included in the RAND challenge sent to the UE
RES	Response	Generated by the UE, applying a secret key to RAND
SQN	Sequence number	Used for tracking the dialog and avoiding replay-back attack
AUTS	Authentication token sync	Resynchronization initiated by UE ISIM to ensure correct sequence
CK	Cipher key	Generated during RAND procedure and used for subsequent encryption
IK	Integrity key	Generated during RAND procedure and used for mutual authentication

The AKA process is started by the ISIM application, using the stored shared secret and the AKA algorithm to generate the authentication request message, masking the shared secret. The authenticating server can send a RAND (Random) challenge at this point or at a periodic reregistration time. This challenge requires the UE to respond in the right way, following a sequence of messages. The Sequence Number (SQN) is remembered at both ends and is checked during this authentication dialog. The Response (RES) to the RAND is calculated by the UE, using the received authentication token (AUTN) in the RAND message. This constitutes a mutual authentication because both ends must agree on the exchanged information, and produce a pair of keys (the CK and the IK) that are used later for communication between the UE and IMS servers.

4.4.2.2 Certificates and Public–Private Key Pairs

An alternative to using shared secrets for authentication is to rely on asymmetric cryptography. This assumes that the entity to be authenticated already possesses a key pair (public–private) and a corresponding digital certificate. The certificate validates the key pair and binds the key pair to its rightful owner. An example of a protocol that uses authentication based on a key pair (public–private) is HTTP over TLS (commonly called HTTPS), as specified in RFC 2818. The main disadvantages of this type of authentication are that a PKI portal server is needed and that asymmetric key cryptographic operations often require substantially more computational effort than symmetric key operations.

Certificates can be used for authentication in the following scenarios:

- When it is necessary to check the identity of the end user.
- When the application security protocol (e.g., normal TLS) works best with key-pair (public–private) authentication and subscriber certificates.
- Where there is a need for non-repudiation and where the user is required to digitally sign the transaction with a user's private key. The fact that the private key in the key pair has been sent by the user acts as non-repudiation proof, unlike the matching of a shared single key in AKA.

To use the asymmetric encryption technology, a client must have a digital certificate that is created by a Certification Authority (CA). It also needs a (public–private) key pair. The certificate and key pair can be prestored on the client terminal, secure from user tampering. If they are not found, they must be generated or obtained dynamically, including the corresponding digital certificate.

Mobile UE can obtain digital certificates from the PKI portal. To do that, it sends a request to the Certificate Authority (CA) in the Home Network and receives the issued certificate in the reply. This process of certificate enrollment and the communication between the UE and the PKI portal is in itself a Mobile application that requires security protection. As in all cases, there is a choice of preshared-secret-based or asymmetric cryptography with certificate-based security that may be used for this purpose.

Protection of the request for a certificate by the certificate-based security method means that when requesting a new certificate, there is one still valid. More common is the scenario of protecting the certificate enrollment by a preloaded shared secret on the PKI portal and the UE. If there is no such shared key, the bootstrapping mechanism is used to generate it.

Having requested a certificate in a secure session, the UE will receive a certificate corresponding to its key pair (public–private). This certificate, together with the key pair, can then be used to authenticate the user for any new session.

4.4.2.3 Public Key Cryptography

Public key cryptography (also called asymmetric cryptography) is based on using a pair of two different keys, a public key and a private key. A public key is "public" because it is generally available to everybody and can be used either to encrypt messages intended for the owner of the corresponding private key or to verify the digital signature of that owner.

Corresponding to the public key is a private key, typically known only to one principal. The private key is used to decrypt the message. Because it is uniquely bound to an individual, a private key can also be used for a digital signature on a message; however, for improved security, different keys and different algorithms should be used for decryption and digital signatures.

To use a public key, the entity using it must know which principal is bound to the public key. This binding is usually accomplished by a certificate, typically a record asserting such binding, containing an indicator of timeliness and signed by a well-known and trusted third party.

4.4.2.4   HTTPS Authentication Proxy

HTTP over TLS can be used for authentication between UE and AS in certain circumstances. This can be performed through an "authentication proxy," a server that supports authentication for several ASs behind it.

The authentication proxy becomes a secure end node for TLS. The servers behind it need not use TLS within their trusted area, but only toward users for authentication. This proxy often operates as a "reverse proxy." This means that messages toward the AS are intercepted and redirected to this proxy so that they can be authenticated before sending on to their original destination.

4.4.2.5   "Replay" Attack and Mutual Authentication

One method used by intruders is to "listen" and record an exchange of registration messages between the UE and the authenticating server. Later, the intruder plays back these messages, pretending to be the server. The UE can then proceed to reveal more private information that may be utilized for fraud or nuisance. The intruder need not understand what is conveyed in the protocol, but merely persuade the UE to respond as if connected to the proper network server. This method of attack is referred to as a "replay back" attack.

To protect users against this, the remote network element must be authenticated to the UE, and messages must be proven genuine and not old, or extracts from different dialogs. The protection mechanism must ensure that the protocol contains values, such as message sequencing and marking, that can prove that the messages were generated in the current session. These values also make sure that the messages fall within a certain time period and that they have not been already acknowledged and accepted.

4.4.2.6   Authentication of AS Users

The AS application may need to validate users. To do so, the application can use previously received information about the user in the REGISTER request from the S-CSCF. The AS performs user identity verification, for initial or stand-alone AS service requests, using user credentials that have been provided via P-Asserted-Identity header (the Authorization header) or other mechanisms.

For ID verification, the user identity received in the SIP message requesting AS action is compared with information obtained at registration time, passed on to the AS in the P-Asserted-Identity header. If the user is meant to be anonymous (i.e., the Privacy header value is set to "id" or "user"), the AS will go no further. When the user is not marked as anonymous, the P-Asserted-Identity header supplies internal verified information.

This information, which is contained in the P-Access-Network-Info header and P-Visited-Network-ID header, also can be used for user verification, even when the credentials have been supplied without the P-Asserted-Identity header. The AS then can check that these credentials are correct and, if not, issue a random challenge, as in a normal authentication process.

External users can still be regarded as anonymous if the FROM field indicates it. External users may be allowed to use some services, even if they are anonymous, and therefore may not need to be fully authenticated. For all other services, the user request will be rejected and the user will not be granted authority.

4.4.3 The 3G Mobile Authentication

4.4.3.1 Authentication Requirements of IMS Elements

The set of all user security settings for GPRS authentication is stored on the AuC as part of the HSS. In the case where the subscriber has multiple subscriptions (i.e., multiple ISIM or USIM applications), the HSS can contain several sets of security settings that can be associated with one or more private identities, IMPIs and IMSIs.

The HSS requirements for supporting 3G Authentication are:

- HSS to provide the only persistent storage for GBA user security settings.
- The user security settings will enable authentication interworking between networks; that is, the use of proprietary security procedures should be avoided to enable free roaming.
- The user security settings must include parameters enabling authentication bootstrapping. For example, there should be an indicator whether a UICC-based procedure is available to participate in the initial issuing of authentication keys.
- The user security settings must specify the key lifetime duration, indicating the time when the security settings were last modified by the HSS.
- It should be possible to assign application-specific user settings to a group of Application Function agents.

UE requirements for supporting 3G Authentication are:

- Support of HTTP Digest AKA Protocol
- Capability to use both a USIM and an ISIM in bootstrapping
- Capability to select either a USIM or an ISIM to be used in bootstrapping, when both of them are present
- Capability of the internal application to indicate the type or name of UICC application to use in bootstrapping
- Capability to derive new key material to use
- Support of an NAF-specific application protocol

4.4.3.2 The IMS Authentication Process

To receive services from the IMS network, the user must be registered. Users are authenticated when they request to register. Every registration that includes a user authentication attempt produces new security associations. If the authentication is successful, then these new security associations replace the previous ones.

To protect the IMS signaling between the UE and the P-CSCF, it is necessary to agree on shared keys that are provided by IMS AKA, and a set of parameters specific to the protection method. The security mode setup is used to negotiate the SA (Security Association) parameters required for the IPSec ESP tunnel, which provides a secure bearer. The UE needs to authenticate the network server to make sure it is connected to the right server. It does this by calculating a MAC (Message Authentication Code) and comparing it with the received MAC that the server has sent.

Two pairs of (unilateral) SAs are established between the UE and the P-CSCF based on the IDs. The subscriber may have several IMPUs associated with one IMPI, which can belong to the same or to different service profiles. But currently there is a limitation that permits only *two* pairs of SAs to be active between the UE and the P-CSCF at any one time. These two pairs are updated when a new authentication of the subscriber has been successfully completed.

For IMS services, a new SA is required between the UE and the IMS before access is granted. This is created using IMS AKA, which is a challenge–response protocol, with parameters derived from data in the AuC in the Home HSS. The identity used for authenticating a subscriber is *the private identity*, which has the form of an NAI (user.name@domain.name). The AKA Protocol is a secure protocol that was developed first for UMTS. IMS AKA uses the user's IMPI in the process. The HSS and the ISIM in the UE share a "long-term" key, which is associated with this private ID. The methods of calculating the parameters in IMS AKA are the same as in UMTS AKA, except that the IMS RES is not sent in cleartext as in UMTS, but it is encrypted with a combination of other parameters for added security.

4.4.3.3   IMS Authentication When Roaming

The IMS subscriber profile located in the HSS in the Home Network contains private information and security credentials that may not be revealed to external parties. At registration, an S-CSCF is assigned to the subscriber by the I-CSCF. The subscriber profile is then downloaded to the assigned S-CSCF over the Cx reference point from the Home HSS (Cx-Pull). When a subscriber requests access to IMS services, this S-CSCF will check by matching the request with the subscriber profile information. If there is a match, the subscriber is allowed to continue with the request.

When the user is in a Visited Network, the Local P-CSCF finds the S-CSCF in the Home Network and the authentication is processed during registration to the Home IMS service. Roaming users are given a provisional permission to access the Visited Network for the purpose of authentication. The Visited Network verifies that the user's Home domain operator is genuine and is an existing connectivity partner before forwarding the messages to the Home Network. If the user is authenticated, the Visited Network will proceed to authorize bearer resources for the media flow with its associated QoS.

In the case of roaming, a Quintet containing the challenge is sent from the Home Network to the Serving Network. The Quintet contains the Expected Response (XRES) and also a MAC for mutual authentication. The Serving Network compares the response from the UE with the XRES and, if they match, the UE is authenticated. The UE calculates an Expected MAC (called XMAC) and compares this with the received MAC. If they match, the UE has authenticated the Serving Network.

4.4.3.4 Early IMS Authentication

Early IMS has been defined to enable quicker and easier introduction of IMS. This should ease migration, reduce the impact on existing networks, and attempt to utilize existing system servers, handsets, and facilities. However, Early IMS Authentication is inferior in several respects and should be used only as a temporary measure for the following reasons:

◼ Any type of early solution, short of full IMS Authentication, has a profound effect on the level of security afforded, to both users and network. In all cases, Early IMS provides inferior service, lower security, and access dependency.
◼ Due to the linking of IMS service authorization and the admission to the PS Network, there is a dependency between SIP and the PS bearer that does not exist with the full IMS security solution. IMS Authorization is therefore not access independent, but is valid only for GPRS and cannot support scenarios where IMS services are offered over WLAN.
◼ Because IMS AKA is not used, there is no key agreement and no IPSec security associations between UE and P-CSCF. To avoid breaches in security, authentication without IPSec should only be conducted within the Home Network; that is, the authenticating GGSN must be in the Home Network, so roaming authentication is not enabled.

For all these reasons, the Early IMS security solution and user authentication must not be used as a long-term replacement for the fully compliant IMS security solution. When both the UE and the Network support full IMS security, there must be a mechanism that prevents a "bidding down" type of attack that persuades the network to choose a weaker security scheme.

There are several methods to provide Early IMS Authentication and Registration, including:

1. Early IMS Authentication can rely on the binding of the HSS public–private user identity with the IP address that has been allocated to the user in the GPRS nodes. This interim solution does *not* authenticate users at the IMS level. Instead, it reutilizes the security procedures of the bearer resourcing function at the GPRS or UMTS PS level.
2. The HTTP Digest scheme potentially can be used but it provides lower protection. In addition, the fact that the username or password must be embedded into the handset is a major disadvantage, while the GGSN binding solution reuses IMSI/MSISDN, which are in the handsets already.

One advantage of Early IMS Authentication is that it can utilize RADIUS rather than Diameter as the authentication protocol. It is also possible to utilize RADIUS-to-Diameter conversion in the interface between GGSN and HSS, where the GGSN has only RADIUS capability. This enables reuse of existing methods and existing systems.

In the absence of ISIM, the GPRS makes use of the IMSI and MSISDN stored on the USIM rather than the IMS identities (IMPU and IMPI). Where only 2G SIM is available, it is still possible to authenticate using GSM numbers, but with even weaker security.

In Early IMS, user authentication is performed by linking the IMS registration procedures, which normally uses an IMPI (an NAI format of "user@domain") to a PDP Context that has been authenticated at the GPRS level, using an IMSI. Where IMSI is available, it is parsed to provide Country Code (first three digits), Network Code (next two or three digits), and Subscriber ID (last nine or ten digits). The Country Code and Network Code make up the Home domain name, and the remainder is used as a temporary IMPU.

The Early IMS authentication mechanism assumes that only one contact IP address is associated with one IMPI. Although the user may have several IMPUs, for Early IMS there can be only one IMPU associated with only one IMPI. This is due to the one-to-one linking of an IMS registration (IMPI) to a PDP Context that is based on an authenticated IMSI. This also means that the same public user identity cannot be registered simultaneously from multiple terminals, as it could in full IMS.

The GGSN provides the user's IP address to the HSS with the IMSI and MSISDN that were utilized to verify the user's identity. The HSS should already have the binding between the IMSI/MSISDN and the IMPI/IMPU, and therefore can bind the IP address to the user. The GPRS nodes interface to the RADIUS server in the AuC part of the HSS, over the Gi interface, when a PDP Context is activated toward the IMS system. The GGSN informs the HSS when the PDP Context is deactivated or modified so that the stored IP address (or IPv6 prefix) can be updated in the HSS.

For Early IMS, IPv4 rather than IPv6 is used to make use of servers and handsets that have not migrated to IPv6. When using IPv6 with Early IMS, the primary PDP Context is bound only to the 64-bit prefix of the 128-bit IPv6 address, not the full address. This means that, in Early IMS, the "IP address" is contained in the first 64-bit prefix of the IPv6 full address. The IP address (or IPv6 prefix) binding is checked in later communications. The GGSN would not allow a UE to transmit an IP packet with a source IP address different from the one assigned during PDP Context activation, to prevent "source IP spoofing." There is also a check that the SIP header address is the same as the IP source address noted by the GGSN, because there is no NAT between the EU and the GGSN.

4.4.4   NGN Authentication

4.4.4.1   Authentication Levels

In NGN, the user and the equipment are not exclusively bound if there is no user identifier built into the firmware. The terminal equipment facilitates multiple combinations of users and service sets. There are two stages for the authentication: (1) in the access layer, recognizing the equipment and access path; and (2) at the IMS service level, with the user account and user public contacts involved.

In NGN, these two levels of authentication are:

1. *At network attachment level between UE and NASS.* The authentication considers access point information, public user identities, and pre-stored user credentials before allowing access to a basic network service. This is a process that is performed between the UE and the local network nodes, with or without contacting the Home Network in the case of nomadic users.
2. *At the service layer level, for IMS.* The authorization process checks the user identity and retrieves the user's subscription information and service authorization credentials. This process is performed between the UE and the sub-system application.

The two levels of procedures for user identification and service authorization are shown in Figure 4.8 — one at network access for admission (below) and the other at the IMS level for service authorization (above). IP connectivity must be obtained first, with authentication at the NASS, before authorization of IMS services can take place, involving the CSCF and the UPSF. Note that the information stored in the CLF is made available for IMS.

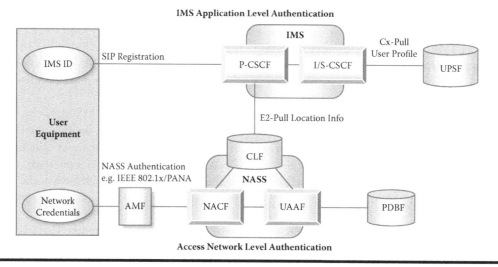

Figure 4.8   Two-level NGN authentication.

4.4.4.2   NASS Authentication Elements

The NASS system provides a generalized view of the process of attaching and authenticating user terminals to a wired, or tethered, access connection. This includes some cooperation between customer premises gateways, clients at the access nodes, and the NASS elements.

The UE connects to the NASS via intermediary nodes that need to participate in the process, either to relay messages or perform some local tasks. The generalized definitions of such customer premises equipment are:

- The CNG (Customer Network Gateway), which can be CPE that connects to the IP Access network local node.
- The CNGCF (CNG Configuration Function), which is the local function that provides network configuration to the UE when it logs on.

The access network, whether it is xDSL, WLAN, or other, can consist of ARF and AMF servers:

- The ARF relays the UE messages from the CNG to the AMF that requests authentication and network admission.
- The AMF is the contact with the NASS. The AMF can reformat the messages and convert protocols to communicate with both the UAAF and the NACF.

The NASS itself contains two main components that are involved in network admission and authentication:

(1) the UAAF performs the authentication and confirms the IP address allocation. It uses the PDBF that contains the authentication data for the user access terminal. The UAAF performs key and token management and holds the authentication dialogue, including the "Challenge and Response."
(2) The NACF that allocates IP addresses and binds the physical link (e.g., line ID, or modem or router) to the logical network address is the element involved in network attachment. This server can be a DHCP server or a local server that maintains fixed IP addresses.

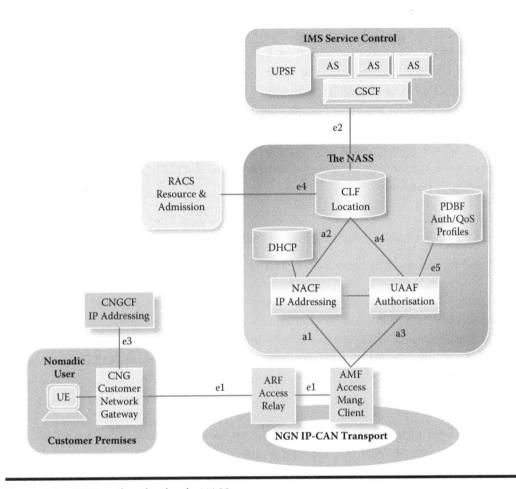

Figure 4.9   NGN authentication in NASS.

The results of the successful authentication procedure are stored on the CLF database, and can be interrogated by the P-CSCF in the IMS subsystem, where service authorization for IMS will take place, using the UPSF user records. The CLF data is also used by the RACS for allocation of QoS levels and resources.

Figure 4.9 shows the elements involved in the authentication process, including their reference points. These functions reside in customer premises gateways, IP-CANs, NASS, and IMS.

4.4.4.3   The UAAF Access Authorization Processing

UAAF performs user authentication based on comparing the identifiers received from the UE with prestored data on the PDBF. The UAAF also can perform various accounting procedures for each user authenticated by the NASS. The UAAF can be part of an xDSL access node or the element in an AAA Server for WiFi or WiMAX access.

The UAAF can act as a proxy in the Visited Network or the Home Network server. As a proxy, it can forward admission requests and accounting messages it receives from the AMF to the UAAF that is acting as the Home UAAF server. The UAAF proxy also relays responses received from the Home UAAF back to the AMF in the Visited Network.

The UAAF uses the PDBF in the authentication process. The PDBF delivers to the UAAF the UE identities; methods of authentication, keys, other credentials material and other network configuration details. In particular, the PDBF contains the QoS Profile, also referred to as a User Network Profile, to distinguish it from the User Service Profile. This QoS Profile provides information required for setting up resources for the user's sessions. Therefore, the PDBF contains complementary data to the User Profile that is kept on the UPSF. The PDBF data is also different from the CLF that stores network attachment details and therefore can indicate the user's location at a given point in time, according to current registration. The PDBF can be co-located with the UPSF or stored separately.

4.4.4.4 NGN Explicit and Implicit Authentication

NGN authentication can be implicit or explicit:

- *Explicit NGN authentication* is a procedure that explicitly examines user credentials and network details in an exchange of information between the UE and the NASS. It requires a signaling procedure to be performed between the UE and the NASS.
- *Implicit authentication* does not require a direct dialog between the NASS and the actual UE. However, the NASS performs a procedure based on identification of the bearer connection for this UE when it is associated with a known access device or a wired gateway.

Explicit authentication is performed during the processing of registration requests of the UE by the NGN subsystem of the NASS. The type of explicit authentication mechanisms used depends on the Access Network configuration and on the operator policy. The procedures vary according to the type of terminal and access. For example, in xDSL access, the authentication mechanism is provided by PPP, IEEE 802.1x, or PANA (see below).

A CNG acting as a routing modem connecting to a private IP realm needs to initiate explicit authentication. For an external Access Network (e.g., in an Enterprise), the NASS authenticates by an interaction with the Access Network node. For a multiple-line CNG, explicit authentication can be requested per line attached to it.

Where the UE is tied to an entry point in the Access Network (e.g., a wired broadband terminal connecting to a DSL access node), the user can be registered implicitly without an explicit registration request for each instance of UE used. Where the user terminal links to a local CNG (e.g., a DSL modem or router), the CNG is authenticated when it is connected to an access node within the operator's network. The CNG identity is validated by the NGN authentication process.

Line authentication is, in fact, a form of implicit authentication. The CNG is usually a host for more than one line. These lines can be implicitly registered, without individual authentication, depending on network policy. Each line is identified by its ID generated by the CNG for routing calls to it. Line authentication ensures that an access line is authenticated according to the operator's defined Line ID. It is based on the activation of the Level-2 connection between the CNG and the Access Network.

4.4.4.5 The NGN Authentication Process

The authentication process for NGN requires recognition of the UE identity, whether it is a gateway connecting device (CNG), a SIP phone, or a PC with a soft client. The device must gain admission into the network by getting attached to a network IP address.

First, the UE identifiers are authenticated and then an IP address is allocated to bind the logical and physical locations. Both steps are performed prior to the IMS services authorization. The UE identifier can be the CNG ID or Line ID. At this stage, the user QoS Profile is retrieved, as well as the specific configuration profile. When the authentication has completed successfully, a confirmation of the authentication, together with current location details (the ID of the relevant IP edge node) and QoS Profiles are stored on the CLF.

For accessing the Core IP network, an IP address and IP configuration are required. At this part of the process, the NACF allocates the IP address and provides prestored configuration information. The NACF establishes the mapping between the allocated IP configuration information and the Line ID. The IP address for the UE and the details of the binding with the physical elements are stored on the CLF. This information is retrieved by the RACS for resource provisioning, and ultimately by the CSCF for session establishment.

4.4.4.6 PPP-Based and DHCP Authentication

The main authentication methods currently employed by UE and CNG devices in NGN include:

- Using PPP
- Using DHCP

An alternative to DHCP authentication is the static line authentication that can be implicitly performed in cases where nomadism is not possible.

4.4.4.6.1 PPP-Based Authentication

PPP authentication allows terminals that use this protocol for Internet access to obtain permission to use the operator's network. The PPP must be converted to the prevailing IMS management protocols — RADIUS, Diameter, and SIP.

For a PPP authentication request, the AMF terminates the PPP request and sends a translated request using an AAA protocol (RADIUS or Diameter) that is acceptable to the IMS elements. The AMF acts as a RADIUS or Diameter client, interfacing to the RADIUS or Diameter server. When using PPP, the physical access ID can be provided from the UAAF to the CLF.

Figure 4.10 represents PPP-based authentication:

1. UE/CNG initiates a PPP Request to apply for an IP address. PPP is used for access and line authentication.
2. AMF translates the PPP Request to an Access Request, and forwards it to the UAAF. The UAAF performs authentication according to data retrieved from the PDBF.
3. When a confirmation of successful authentication reaches the AMF (Step 2a), it sends the Configuration Request to NACF to obtain the IP address and other parameters, including the IP address of the IMS entry node (e.g., P-CSCF).

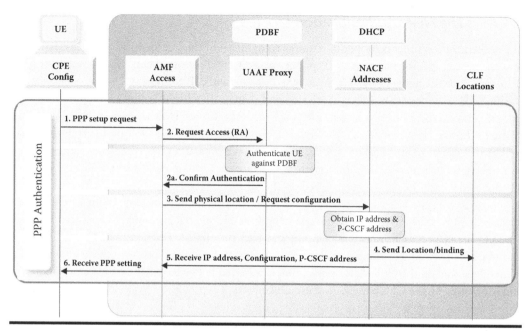

Figure 4.10   Signaling flow for PPP-based authentication options.

4. The NACF generates an IP address or FQDN for the UE and sends it to the CLF, with the binding information including Line ID and IP Edge ID. The CLF pushes the binding information to the RACS via the e4 interface.
5. AMF sends the IP address of the Core IMS server to the UE or CNG within the PPP configuration option extension.

4.4.4.6.2   DHCP-Based Authentication

DHCP-based authentication is better suited to provide mobility to nomadic users in NGN. DHCP is also the main method of attaching IP addresses in 3GPP.

Figure 4.11 depicts an example call flow for a procedure based on DHCP requests from the UE/CNG, with no PPP. The sequence is as follows:

1. The terminal equipment or CNG initiates authentication based on IEEE 802.1x or PANA, for example.
2. The AMF contacts the UAAF for authentication. The UAAF performs authentication against credentials data stored on the PDBF.
3. After successful authentication, the UAAF informs the CLF that a UE/CNG is authenticated and stores the data there, although there is no associated IP address as yet.
4. The UE initiates a DHCP request for an IP address. This request is relayed by the AMF to the NACF (Step 4a), which contains a DHCP server.
5. The NACF allocates an IP address to the given UE identifier and informs the CLF that an IP address is now associated with the UE/CNG. The CLF will push this binding information (between allocated IP address, line ID, and IP edge node ID) to the RACS via the e4 interface.

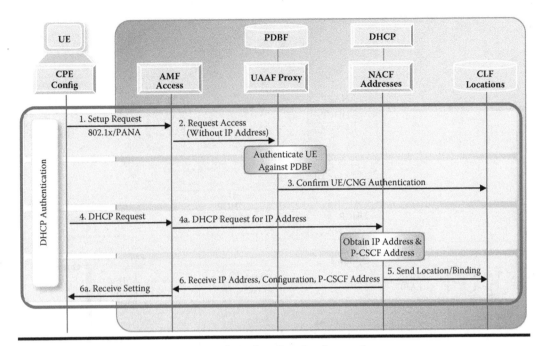

Figure 4.11 Signaling flow for DHCP-based authentication options.

6. The NACF returns the allocated IP address to the AMF, which, in turn, relays it to the UE/CNG (in Step 6a). Also returned is an address or FQDN for the entry point to IMS, which is used to initiate calls.

4.4.4.7 NGN Authentication Methods and Scenarios

There can be many paths for the NGN UE messages to reach Core IMS Network Elements. They can traverse several transit intermediary networks, Visited Networks, and eventually reach the Home Network. Every network in the path gets involved in the process of authentication.

Figure 4.12 describes the two authentication options. Option 1 defers authentication to the AMF in the IMS operator's network and uses DHCP or PANA/EAP. Option 2 (darker servers, dotted lines) performs PPP-based authentication with the UE in the Visited Access Network before passing control to the UAAF in the IMS operator's network.

In Option 1, AMF-1 resides within the same network as IMS. The UE initiates a DHCP request or an 802.1x/PANA request, carrying EAP. The Access Node may have routing agreement and passes these messages to an AMF within the operator network as they are. The AMF formulates the authentication requests and launches the two-step process, one to obtain the IP address and the other to verify user authenticity against stored data.

In Option 2, AMF-2 belongs to the Access Network or an intermediary network. In this example, the AMF-2 receives PPP requests. It must translate them to RADIUS (or Diameter) before forwarding to the UAAF that contains a RADIUS/Diameter server and to DHCP to connect to the NACF.

Brief descriptions of protocols are given in Appendix A.

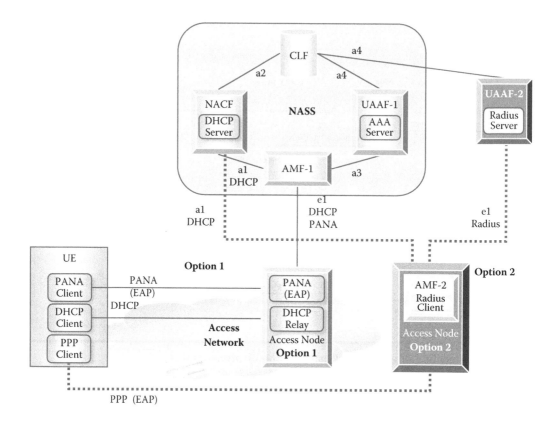

Figure 4.12 Optional implementations of authentication.

4.4.4.8 NASS Bundled Authentication

As in Early IMS Authentication for 3G Mobile, NGN can also provide interim authentication solutions. IMS normal authentication for both 3G Mobile and NGN requires an ISIM with the preconfigured values for the private user identity, one or more public user identities, and the Home Network domain name. These identifiers are used to populate the SIP REGISTER request.

When there is no ISIM application, the UE can generate alternative identifiers, deriving them from the IMSI, when an IMSI is present in a USIM. The UE generates the following values, deriving them from the IMSI:

- A private user identity
- A temporary public user identity
- A Home Network domain name

However, unlike Mobile handsets, in NGN solutions there may be a need to connect equipment that has neither ISIM nor USIM to IMS. In this case, the wired connection (i.e., the CPE or access gateway device) may have a fixed connection identifier that can be associated safely with a particular user. A temporary public user identity is allocated to the user by various means, for example, based on hardware identifiers or preconfigured information delivered from a CNGCF to the UE. Such configuration data can be received by the UE in the P-Associated-URI header in the SIP message.

This process relies on the NGN authentication at the access level, performed by the UAAF against credentials in the PDBF. It can reuse existing methods based on RADIUS servers. The use of HTTP Digest user name or password is commonplace in NGN, but this provides for weaker security due to the lack of IMS AKA and IPSec to secure the registration dialogue.

The UE commences network attachment procedures after it has been successfully authenticated by the NASS. The CLF in the NASS gets updated with the binding of the allocated IP address and the line location information, which consists of a Line Identifier and xDSL node identities.

The next step is to obtain the user profile from the UPSF, via the I-CSCF. The authentication method must indicate that this user can use the NBA (NASS Bundled Authentication), otherwise, full IMS Authentication will be required. The UE sends a SIP REGISTER request to the P-CSCF. The P-CSCF ascertains whether a security association is required for the connection, according to the SIP signaling, local policies, and addresses in Layer 2 and Layer 3.

The registration procedure includes the P-CSCF retrieving the UE location from the CLF. The P-CSCF can find the CLF according to the UE's access network and IP address or the path of received packets. The CLF responds to the "Location Information Query" over the e2 interface, according to the given UE IP address. The location from the CLF then is appended to the registration request, which is sent to the I-CSCF/S-CSCF.

The S-CSCF queries the UPSF over the Cx interface and receives back the prestored location associated with the IMPI and IMPU. The S-CSCF compares the location received in the REGISTER message with the one received from the UPSF. If they match, the UE is successfully authenticated. This means that for NBA, there is a dependency between the PDBF and UPSF that does not exist in conventional NASS or IMS. Thus, the NBA sacrifices the ability to roam by relying on a pre-stored location.

4.4.4.9 Authentication Options for the Nomadic User

When a fixed network user appears on the network that does not hold this user's subscription, the user is deemed nomadic. In this case, the user will need the Visited Network IP-CAN to assist in registration to the user's Home Network and cooperate in the authentication process.

There are three options of authentication to consider for nomadic users:

1. Authenticate via a Proxy UAAF communications with the Home UAAF, performed in the Home Network.
2. Authenticate in Visited Network but forward location information to the Home CLF.
3. Authenticate at IP-CAN level only in Visited Network and provide local services.

Option 1 using V-UAAF to H-UAAF. The V-UAAF recognizes the nomadic user, verifies that the Home Network is on the agreed roaming network list, and sends the authentication request to the H-UAAF. The H-UAAF uses its own PDBF data to authenticate the user.

Option 2 using V-CLF to H-CLF. The Visited Network has sufficient information to authenticate the user locally. This depends on the type of access and the level of risk associated with the requested service. The local authentication data may have been supplied from the Home Network or is based on credentials supplied by the user. The results of the authentication, the allocated IP-address and binding, are then stored on the Visited CLF, and forwarded to the Home CLF. The

V-CLF location is required to facilitate routing of incoming calls, even when no Home Network services are otherwise requested.

Option 3 for Local authentication only. If there is no roaming agreement, users may still be allowed into the network for a limited service. Users can have a separate agreement for accessing the Local IP-CAN, such as agreements for hot spots WiFi service.

The AMF initiates the first step in authentication, on behalf of the NGN UE, toward the NASS. The UE obtains configuration details from a Local DHCP (the NACF) server, in the Visited Network. The request is passed to a proxy UAAF, also in that Visited Network. The Proxy UAAF can belong to another operator's network, so the registration messages can then traverse the access network, intermediary network, and Core IMS network, all belonging to different business entities.

The UAAF-proxy will forward access and authorization requests as well as accounting messages, which it receives over interface a3 from the AMF to the UAAF server over interface e5. Responses received back from the UAAF server over interface e5 will be forwarded to the AMF over interface a3.

Note that the DHCP can reside in the first Access Network the UE encounters, but the IP address must be acceptable also to the Visited Network, i.e., taking care of private IP addressing with NAT and IPv4/IPv6 translation. The example in Figure 4.13 shows the components, according to TISPAN, within the Visited IP-CAN, the Visited Network, and the Home Network. The diagram shows the three options: Option 1 (dashed line) between V-UAAF and H-UAAF, Option 2 above (dash and dot line) between V-CLF and H-CLF, and Option 3 (dotted line) in the Visited Network only.

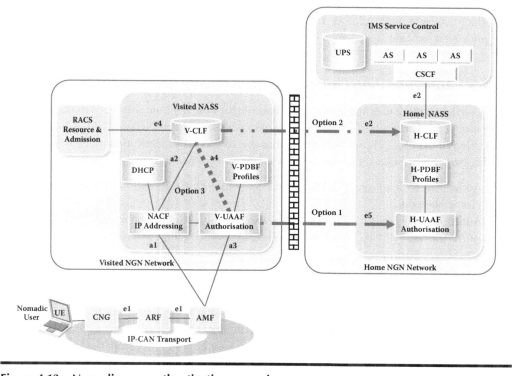

Figure 4.13   Nomadic user authentication scenarios.

4.4.4.10 Storing NGN Location

During the authentication process, the IP location of the user is logged in the CLF. This information is used for several purposes. For example, security functions check the location to ascertain that the media payload packets follow the same route as the signaling and are not diverted for fraudulent purposes. Subsequent signaling messages will reassert the location in the header to ensure that the media flow is delivered to and from the same authenticated UE that has successfully registered.

When sessions are initiated, the P-CSCF retrieves the location and details of the path and the addresses of the access servers that performed the registration, and sets this information as the session connection path. Incoming calls are routed to the user according to the path in the CLF.

The CLF path information is also required in other instances for call session admission, to make policy decisions and allocate resources to satisfy a particular QoS level.

4.4.4.11 NGN Reference Points

The following are reference points used in NASS sub-systems:

- a1 is the reference point between the AMF and the NACF. This reference point allows the AMF to request the NACF for the allocation of an IP address to end-user equipment as well as other network configuration parameters. It does not carry DHCP signaling but transfers the resulting IP address and configuraton in DHCP.
- a2 is the reference point between the NACF and the CLF. This reference point allows the association, or "binding," between the allocated IP address and the user identity that will be notified and registered on the CLF through information received from the NACF. The location information can include details of the IP Edge Device ID or the Line ID. The information flows on this interface between the CLF and NACF include Bind-Indication, Bind-Acknowledgment, and Unbind-Indication, using DHCP.
- a3 is the reference point between the AMF and UAAF. This reference point allows the UE via the AMF to request authentication by the UAAF. The a3 may use either RADIUS or Diameter.
- a4 is the reference point between the UAAF and the CLF. This reference point provides the CLF with indicators of the user's preferences for privacy of location information, as provided by the UAAF. This reference point is also used for the CLF to obtain the user's QoS Profile, which is provided from the UAAF database, the PDBF.
- e1 is the reference point between UE and the AMF (via ARF). This reference point enables the UE to initiate requests for IP address allocation, network admission, and authentication, and receive network configuration parameters to access the network. Through this reference point, the UE provides the NASS with credentials (e.g., password, token, certificate) needed for authentication. It also enables the UE to validate the network, via the mutual authentication process. Based on the authentication result, the AMF authorizes or denies network access to the UE. The e1 interface is also used between UE and ARF and between ARFs for messages relayed to the AMF. When traversing an ARF, it adds its location information to record the path to the UE. The e1 protocol is access network specific and can be PANA (EAP) or DHCP.
- e2 is the reference point between the CLF and the CSCF. This reference point enables various network elements (applications and session controllers) to retrieve network location information from the CLF. The location information is essentially the assigned IP address allocated to the UE and the access route. The e2 uses Diameter.

■ e3 is the reference point between the CNGCF and the CNG. This reference point allows the CNGCF to configure the CNG, which is considered part of the user terminal. The e3 interface is used during initialization to enable local equipment to access the network. This interface uses HTTP, FTP, and TFTP (Trivial File Transfer Protocol) for the information transfer.

■ e4 is the reference point between the CLF and the RACS. This reference point allows the RACS to inquire and receive user IP location information and geophysical location details from the CLF. This data is needed to consider resource allocation. The information that can be transferred over the e4 reference point includes binding of the logical access ID (the Line ID) with the assigned IP address and with the user network profile information necessary to make resource allocation decisions. The e4 uses Diameter.

■ e5 is the reference point between the UAAF proxy and UAAF server. This reference point is used for processing authentication requests between a Visited UAAF and a Home UAAF server in different administrative domains, where the Home Network performs the authentication against stored user information. It also allows the UAAF-proxy to forward accounting data for the particular user session to the UAAF-server. As a roaming interface, the e5 can use Radius or Diameter, depending on what is agreed between the networks.

4.5   Registration Procedures

4.5.1   Principles of Registration

4.5.1.1   Registration at Access Level and at IMS Level

In 3GPP, the ISIM in the handset provides initial authentication and allows for creating a PDP context. In NGN, the NASS provides such authentication, with related entry path and local data stored on the PDBF. In a WLAN, the AAA Server provides the access-related authentication in conjunction with the local Access Point. Once such entry is gained to the Access Network, IMS-level (or Application-level) authentication and registration can take place.

Registration in IP networks is required to bind users with their current location on the network, assuming that they are able to move around, or that they log on to different fixed devices. The registration also provides a means of tracking active users. Registration in IP networks has a limited time assigned, after which the registration expires. This enables the network to ensure that the connection is alive through a series of periodical re-registrations.

4.5.1.2   Explicit and Implicit Registration

IMS registration can involve explicit and implicit user profiles, so related profiles are activated at the same time. A user can have more than one Public User ID. For example, the user has a wireline IP phone number, IMS mobile, and a URI soft client on a laptop. When a user registers a Public User ID, it is an explicit registration of that ID. However, for convenience, when one Public User ID is registered, all the other associated Public IDs (which are stored in the user profile on the HSS) can be registered at the same time as a registration set. Such secondary registration is called "implicit registration."

Implicit registration is characterized as follows:

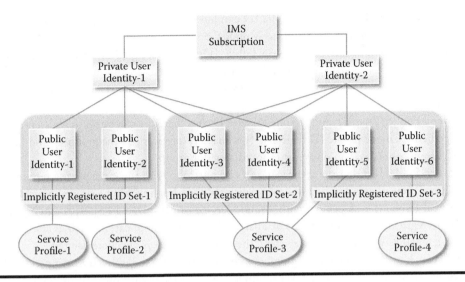

Figure 4.14   Implicit registration of a set of Public User IDs.

- User Public IDs are not in any hierarchical arrangement, with no single ID acting as master.
- All Public IDs in the implicit registration set must be associated with a single Private ID.
- This does not prevent a User Public ID from being associated with another Private ID that is not registered.

Figure 4.14 provides some examples of the relationships between the Private User Identifiers (IMPIs) and the Public User Identifiers (IMPUs), and the implicit registration groups they can form together.

In the case of terminals with no ISIM, a group of Public User IDs can still be implicitly registered using a temporary Public ID. This requires the assignment of a Temporary Public ID for an initial registration process, and then the entire set can be registered at once. The Temporary Public User Identity cannot be used unless an implicit registration is activated. The HSS will contain such an indicator to ensure that the Temporary Public ID is not misused.

Note that more information about the data structure appears in Chapter 5.

4.5.1.3   Group Registration Options

The Group Registration facility allows registration of a group of Public IDs in one registration request. This means that a single registration request and the following dialog results in many Public IDs being registered all at once, avoiding multiple dialogs for individual registration. This is an important tool to reduce the load on IMS nodes when many users try to register at the same time. Such scenarios arise, for example, when recovering from a power failure or a server failure.

There are several ways of grouping user IDs for a group registration, including:

- A user can have several Public IDs for different devices, for example, fixed IP phone, IMS mobile phone, and soft clients on other devices associated with a single Private ID. All of these IDs can be grouped as an implicit registration set. When one ID registers, the rest are automatically registered. Each Public ID can have its own service set or can share services. This is a convenient feature for Fixed Mobile Convergence users in particular.

■ Implicit registration sets can contain Public IDs that belong to different Private IDs (see Figure 4.14). Once registered through implicit registration, an alternative Public ID can be activated for the same Private ID. The new registration does not depend on the lifetime of the first.

■ Several Public IDs can belong to one Private ID, and therefore can be set to register implicitly together when one Public ID registers. This enables, for example, the grouping and handling of enterprise accounts in one registration dialog per group.

■ Several Public IDs that share the same service portfolio with the same parameters are called aliases. These aliases can be implicitly registered together, invoking a single Service Profile.

■ Several Public IDs associated with a single service can be given a PSI. This can be used for group registration of all members of a list used by a single application, such as chat rooms or conferencing. When a user deregisters, the PSI for that users also deregistered.

■ UE can have a temporary or long-term association with a Public ID, using the GRUU format. It is possible to have several Public IDs associated with one UE in one GRUU set registered as a group. The UE can be an instance of a network server that terminates sessions, such as the MGCF.

Chapter 5 provides descriptions of ID types, including PSI and GRUU.

4.5.1.4   AS Third-Party Registration

Applications can be subscribed to registration events for the user, so as to receive NOTIFY messages about new registrations, expirations, re-registrations, and deregistrations. Such events can trigger particular services. This also provides a mechanism for the AS to discover all the implicitly registered public user identities and to obtain the current capabilities of the UE.

When the S-CSCF registers a user, it downloads the user record with the service profile from the HSS. If the registration request matches a particular trigger for a "watcher AS," the S-CSCF performs a "third-party registration" to the relevant AS.

The third-party SIP REGISTER message is so called because it is initiated by network servers, not the UE, but it associates with the other two parties, the AS and the UE. The third-party REGISTER is populated with the public user identity (or one of the implicitly registered IDs) and the S-CSCF address. It can also contain application data associated with the filter criteria obtained from the HSS, such as the IMSI for a CAMEL application. The ICID, IOI, and charging function addresses are also included in the message, where appropriate.

4.5.2   IMS Registration Process

4.5.2.1   Access-Agnostic Registration

The registration process involves establishing association of the allocated IP address with the user subscription, which requires the UE to be authenticated to the network at both the IP-CAN level and the IMS service level. Note that the IMS process is independent of the type of access.

Registration for IMS in NGN is largely the same as for 3G Mobile or WLAN. The functionality and relationships of the P-CSCF, I-CSCF, and S-CSCF are the same. The UPSF is used in

NGN in the same way as the user profile part of the HSS to obtain information for the user identifiers and the associated service profiles.

4.5.2.2   Selecting the Registrar Serving CSCF

When it is determined that the user is not currently registered, the registration process commences. The first step is to assign an S-CSCF to perform this task. The selection and allocation of the S-CSCF is performed by the I-CSCF. It is based on a number of factors. Operators can be influenced by traditional methods to allocate blocks of numbers to reside on specific servers and override this only when that S-CSCF is not available. Other considerations are often topological, considering the relative position of the P-CSCF and the S-CSCF. In some cases, different server capabilities can dictate the selection, e.g., selecting by regional requirements, local regulation, and language.

Where the selection of the S-CSCF process is free from permanent association, a pool of servers can be used for greater efficiency and resilience. Because users are not permanently tied to a specific server, an overflow situation is handled automatically by selecting alternative servers. This provides IMS with one of its greatest advantages.

The S-CSCF performs registration for the user who has been assigned to it, when the user logs on to the network. Once a user is registered to a certain S-CSCF, it becomes the user's Home session control server for all communications to and from that user.

When users are not registered, there are still some terminating services that should be performed, such as call forwarding or voicemail. Another type of call processing without registration is making calls to the Emergency services. A default S-CSCF is therefore selected to handle incoming calls for unregistered users.

4.5.2.3   The Registration Process

The S-CSCF receives a REGISTER SIP message from the user, through the local P-CSCF that the Access Network had discovered. The REGISTER request contains the Public User Identity and the proposed expiration time for the registration "life." According to preconfigured network policies, the S-CSCF can modify the registration time, reject it, or accept it. When registration expires, the UE needs to reregister.

The call flow in Figure 4.15 describes a scenario within a single Mobile Network (i.e., no roaming). The registration procedure initiated by the IMS UE begins with the discovery of the local P-CSCF (Step 1). To validate the user, the storage of the User Profile needs to be located by the I-CSCF. In a network where storage is distributed, the I-CSCF interrogates the SLF to establish which HSS server contains the User Profile (Steps 3 and 4).

The I-CSCF interrogates the appropriate HSS and finds whether or not the user is already registered (Steps 5 and 6). If not, the I-CSCF will assign a new S-CSCF for this user. If several S-CSCF servers are deployed in this network, the I-CSCF has to select which one to assign. The User Profile is read by the S-CSCF (Steps 8 and 9), obtaining credentials from the HSS that are used in the authentication process and can be used again in random reauthentications.

During the registration process, the S-CSCF retrieves the User's Profile from the User Profile storage (the HSS or UPSF), including the filter criteria for subscribed services and the trigger points to activate them. This process first establishes whether or not the user is roaming, and if

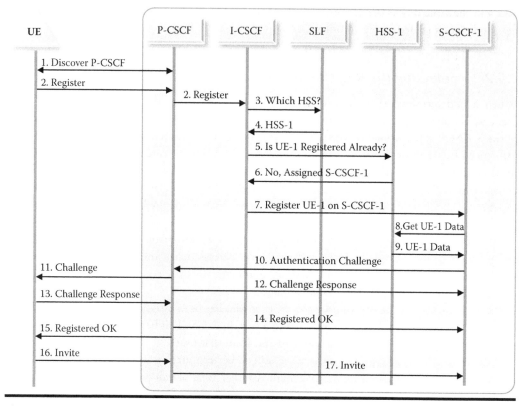

Figure 4.15 Registration in multi-CSCF scenario.

not roaming, whether the user is already registered. If no registration is found, the authentication process will commence (Steps 10 to 13).

If an external Presence server is deployed or there are other applications that subscribe to the user's registration status, the S-CSCF will send notifications of registrations, re-registrations, and deregistrations to the servers that have subscribed to such events.

Once authentication is successfully completed, an S-CSCF is assigned, the user status is updated, and the user is registered. The assigned S-CSCF retrieves details of the subscriber from the HSS and caches them in local memory. The S-CSCF keeps user status (Presence) and data for the subscribed services, including the iFC to manage the sequence of service invocation.

4.5.2.4 Registration Data

Information kept on each of the IMS functions changes during and after registrations. There may be variations, depending on specific implementations, especially when business borders are crossed between the access network, the intermediate network (such as WLAN), and the Core IMS network.

The following table shows what data is stored during the various stages of registration:

Node	Network	Before Registration	During Registration	After Registration
UE	Home or Visited	Authorization credentials, Home Domain name, Proxy name/Address	Authorization credentials, Home domain name, Proxy name/address	Authorization credentials, Home domain name, Proxy name/address + UE GRUU
HSS	Home only	User Service Profile	User Service Profile P-CSCF network ID	User Service Profile S-CSCF address/name
P-CSCF	Home or Visited	Routing Function	Initial network entry point, UE address, Public and private user IDs	Final network entry point, UE address, Public and private user IDs
I-CSCF	Home only	HSS or SLF address	S-CSCF address/ name, P-CSCF network ID, Home network contact details	Note: The I-CSCF data is transient
S-CSCF	Home only	Note: no data until it is assigned to a user	HSS address/name, User profile, Proxy address/ name, P-CSCF network ID, Public/private user ID, UE IP address, UE P-GRUU and T-GRUU	HSS Address/name, User profile, Proxy address/name, P-CSCF network ID, Public/private user ID,UE IP address, UE P-GRUU and T-GRUU + Session state data

Source: Based on 3GPP TS 23228.

4.5.2.5   Registering When Roaming

Roaming UE needs to discover the local Proxy, that is, the P-CSCF that will relay the registration request to the Home Network. The discovery utilizes a DHCP server, local access point configured data, or other methods. After contacting the P-CSCF in the Visited Network, the UE requests registration in its Home Network.

The P-CSCF examines the "home domain name" to discover the entry point to the Home Network. This entry point can be an interconnecting BGF where the communication crosses business borders, or an I-CSCF, where there is a trust relationship between the networks. The Proxy should never use the entry point details that have been cached from prior registrations, but the entry point server address is obtained anew by resolving the domain name given by the UE.

4.5.2.6   Re-Registration

When the registration assigned period expires, the UE client initiates a re-registration process. This is usually transparent to the user. It is performed to reaffirm that the connection is live. To

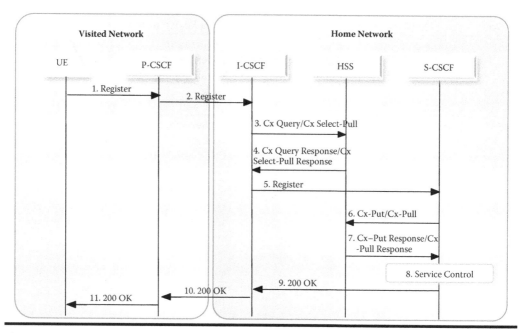

Figure 4.16   Registration while roaming.

lighten up the process and avoid unnecessary processing in frequent re-registrations, the dip into the HSS data is omitted because it is already cached. Re-registration also can omit the requirement for full authentication to be performed afresh.

In Figure 4.16, reregistration is performed without repeating the authentication. To reregister, the UE sends a new REGISTER request. The REGISTER information to the P-CSCF includes Public User Identity, Private User Identity, Home Network domain name, UE IP address, capability information, Instance Identifier, and GRUU support indication. Steps 3 and 4 can be omitted in re-registrations because the data is already cached, but the HSS is updated again in Steps 6 and 7 with the new expiration time obtained by the re-registration.

4.5.2.7   Deregistration

Deregistration can be initiated either by the UE or by a network server. UE-initiated deregistration is accomplished by a registration request that has an expiration time of zero seconds. This may be an ungraceful deregistration, for example, when the battery is flat or the equipment fails to operate properly. A graceful deregistration occurs as a result of:

■ Operator action to suspend service for this user, due to congestion or suspicion of fraud
■ Serious error detected, for which the system must shut down
■ Duplicate registrations or inconsistent data found, e.g., when roaming to another registration area without deregistering the previous one
■ Application decision, perhaps when service delivery is completed or discontinued
■ Subscription management decision (e.g., contract expiration)
■ UE requirements have changed and can be satisfied only by another S-CSCF

To perform deregistration, the HSS needs to find the registered user's record and determine that this Public User ID is actually registered. The I-CSCF finds the address of the currently serving S-CSCF through a name-address resolution mechanism, and sends the DEREGISTER message to it.

When one of the Public User IDs within an implicit ID set is deregistered, all Public User IDs that have been implicitly registered are deregistered at the same time.

Based on the filter criteria, the S-CSCF sends notifications of the deregistration events to the applications that are monitoring user status. This allows such applications to perform any necessary service closure procedures. The service control platform removes all subscription information related to this specific Public User ID when it is deregistered. The HSS then can either clear down or keep the association of the S-CSCF with that Public User ID, according to network policy.

The deregistration procedure can be initiated from certain IMS functions, usually the S-CSCF, depending on the exact reason for initiating it. The HSS, which already knows which S-CSCF is serving this user, can initiate a Cx-based DEREGISTER message towards that S-CSCF. Conversely, the S-CSCF can initiate a Cx-Put to inform the HSS. Other trusted or secured parties also can initiate deregistration to the S-CSCF.

4.6 Network Admission Comparison

Convergence of network admission methods occurs only at the IMS layer, while IP-CANs continue to have different requirements. This brings additional complication to the User Terminal data. Therefore, the ability to admit terminals with existing IDs will be desirable for a long time, even when the level of security is lower. The table below gives a comparison across the range of authentication methods and registration identities.

	IMSA	*UMTS AKA*	*Early IMS*	*NASS Bundled*	*HTTP Digest*
User ID	ISIM (IMPI/IMPU)	USIM (IMS/ MS ISDN)	SIM/USIM (IMSI/ MSISDN)	Wireline CPE ID	Username/ Password
IP-CAN	UMTS	UMTS	GSM/GPRS	Fixed (e.g., xDSL)	Internet
Roaming	Y	Y	N	N	N/A
Mutual Authentication	Y	Y	N	N	N
Security Level	High	High	Medium	Medium	Low

Chapter 5

Profiles and ID Management

5.1 This Chapter

This chapter describes the network data repository, and its architecture and capabilities. This includes not only the 3GPP-defined HSS and its equivalent TISPAN-defined UPS, but also mobility servers interfacing to 2G (HLR, VLR) and to Fixed Network access locations (CLF), plus other authentication servers (PDBF, AuC) and routing servers.

This chapter begins with the principles of a central, sharable repository of data, and the benefits from network data treatment as a Data layer. The approach of combining FMC data through data modeling is discussed against the definition of a generic mechanism of data retrieval via the GUP (Generic User Profiles).

The 3GPP comprehensive definition of data management and the HSS are summarized next, followed by a description of the TISPAN data servers in NGN. In both cases, mobility data management is also described and reference points are detailed.

ID management is described in the subsequent sections, with definitions of the different identifiers, their roles and structures, and the relationships that can be forged between them, such as IMPU and IMPI, implicit sets, GRUU, and PSI. The use of the appropriate identifiers — in particular, procedures such as group registration, emergency calls, or anonymous calls — is also explained.

Also described are server identifiers with their formats and roles, along with service profile data, service trigger points, and invocation priorities. And finally, to complete the data picture, there is a brief discussion of routing database data, such as ENUM, DNS and NP data.

5.2   Data Repository Principles

5.2.1   Scope of Central Data Repository

IMS principles lend themselves to the concept of the central data repository, where data management interfaces are declared and specified, and therefore the data is open for use by disparate network nodes.

Figure 5.1 provides a suggestion of such architecture. This architecture contains the following main data management components:

- Identity management, including various types of user IDs as well as PSIs
- Consolidated user profiles, from both NGN and 3G Mobile
- Authentication center that contains procedures and data for Mobile, NGN, and WLAN
- Mobility data management, dynamically tracking the user location
- Routing addresses, handling domain names (DNS) and telephone/URI conversions (ENUM).

Generic central data repository management — the GUP — may provide central retrieval facilities from various data repositories.

5.2.1.1   Areas of Convergence

The independent data repository is crucial to the IMS to ensure full interworking between multiple network servers and functions. It is also becoming increasingly important to make sure that this data is sharable with disparate applications, regardless of the access mode.

The need to combine information between technology domains is apparent already in any operator's network, where 2G and 3G services are to be offered, or NGN and WLAN. There is also a trend toward consolidation of Fixed Line networks and Mobile networks within a single organization as communication requirements converge.

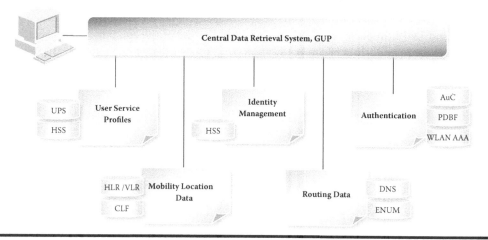

Figure 5.1   The scope of central data management.

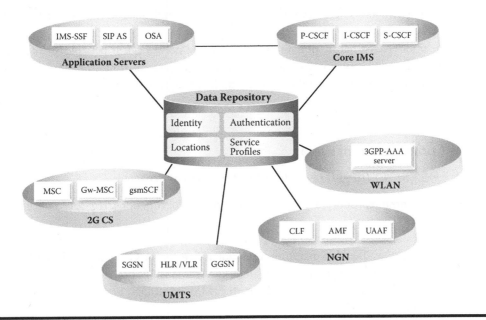

Figure 5.2 Multi-access network data repository.

Figure 5.2 demonstrates the range of networks and functions that may have an interest in the same central data. These networks can belong to one operator or to several business administrations, in particular in the case of VLR location information and exposure applications.

5.2.1.2 Data Repository Features

In a converged network, there are even greater numbers of varied data structures but they can still be rationalized in terms of a single repository structure. Location information, for example, is not expressed in the same way in GPRS/UMTS as it is in DSL Broadband. However, the same principles of maintaining central, secure, unambiguous data remain valid. The central repository must ensure:

- Standardized access to data
- Open interface to applications
- Efficient static data and dynamic data storage
- Data integrity and security
- Privacy
- Fraud protection

There are multiple sets of data that can be integrated into the network's central repository. This repository acts as a master data store, where subsets of data are cached in various network nodes as required for best performance. Data also can reside in a distributed form (i.e., in specialized data servers), yet all of it is still accessible via a standard common interface.

During an active service, data can be modified by local servers, where the changes are first stored. Then the data modifications are uploaded back to the master database to get it fully synchronized and redistributed if necessary. Similarly, customer self-care systems allow users to modify their master data using a Web interface. This data is uploaded to the Master Repository and redistributed to bring into line all relevant servers.

With the multifaceted nature of data in the converged network, it is possible for the user to change a single piece of data by more than one mechanism at the same time. Therefore, there must be a mechanism to resolve such conflicts, for example, giving priority to one method of update over the other.

5.2.2   Convergence of Data

5.2.2.1   HSS as an Integrated Mobile Data Repository

This network's central repository needs to support a myriad of functions, regardless of the access type:

■ *User identification.* The HSS maintains users' identifiers for the CS Domain, the PS Domain, and IMS, and keeps the appropriate hierarchy and relationships between them. This includes IMSI and MSISDN IDs for the CS Domain; IMSI, MSISDN, and private identities and public identities for IMS.

■ *User security information.* The HSS supports the authentication procedures with authentication information. It stores users' credentials and generates authentication keys, integrity, and ciphering data from existing data. It provides this data to various entities, namely MSC (Mobile Switching Center), VLR (Visitor Location Register), GPRS nodes, 3GPP AAA Server, or the IMS CSCF (Call Session Control Function).

■ *Access authorization.* The HSS authorizes the user for Mobile access when requested by the MSC/VLR, SGSN/GGSN, 3GPP AAA Server, or CSCF, by checking that the user is allowed to roam to that Visited Network.

■ *Mobility management.* This function supports user mobility through the CS Domain, PS Domain, and IMS.

■ *Call (2G) and session (3G) establishment support.* The HSS supports initiation and establishment procedures of sessions in the CS Domain, PS Domain, and IMS. For terminating calls, it provides the identity of the serving call session control entity that hosts the called party.

■ *Service authorization support.* The HSS provides basic authorization for session establishment, the identification of services, and their invocation. The HSS furnishes the service control entities with the service data, parameters, and filter criteria to determine triggering and priorities.

■ *Service provisioning support.* The HSS provides access to the service profile data for users that can be updated when service provisioning details are changed. This includes data for SIP applications, for CAMEL-based services and for Public Service ID (PSI ASs) that are treated as special users.

■ *Service exposure.* The HSS communicates with the SIP AS and the OSA-SCS to support applications that provide services for the network subscribers. It can also communicate with the IM-SSF to support CAMEL service operating in the CS space.

■ *Generic user profile data repository.* The HSS supports generic data retrieval systems for IMS user profiles and related data.

5.2.2.2   Achieving Convergence of Data Repository

The HSS was originally conceived as the IMS master data store that contains both static and dynamic data for subscribers, and their service profiles and triggers. However, the scope of the HSS has grown as the benefits from a shared repository of data became clearer. In a transitional phase toward Core IMS, the HSS contains information that is required by both CS and PS networks.

At the same time, users may have similar data records on both Mobile packet and NGN packet networks. Other access data, such as WLANs, create yet more duplicate records, with minor variations that may be unique to the type of access. With the trend to add more types of access (cable, Ethernet, WiMAX) comes the requirement to increase the scope of the central data repository and enable data consolidation where possible.

Location information is a major driver of the consolidation of data. Location is utilized not only by numerous Location-based Services, but also by regulation, such as Emergency Calling and Lawful Interception. This applies to both mobile (3G) and nomadic (NGN) users. Although the logical network location (IP address) is known at the Core IMS servers, the actual physical location (geographic positioning or map grid coordinates) and the routing path (access nodes and proxies) are determined in the Access networks (e.g., GPRS or DSL).

In Mobile networks, the location information is stored at the HLR. Together with the HSS, this data enables consolidation of both 2G and 3G. The equivalent of HSS in NGN is the UPSF, which stores the static user records with application subscriptions and triggering data, but the dynamic information of the access path is kept in a separate database, the CLF in the NASS. The CLF full description (as later amended) is "Communication session Location & Repository Function."

5.2.2.3   Modeling Converged Data

The process of converging data involves modeling common data and data trees. Data can be converged better at the top of the "stack." Static user profiles and associated service profiles can easily share the same schema. Authentication data (key materials, encryption data) vary according to the authentication methods used by the various access networks, but certain credentials can be shared and some methods can be utilized in more than a single domain. The differences in the location data and access information vary more widely and are more specific to each access network.

In principle, it is possible to map data contents between NGN and 3G networks, as shown in Figure 5.3 where data for 3G Mobile networks is shown on the left and data defined by NGN is on the right. The UPS contains the user profiles, such as HSS, but not the HLR. The authentication part of the HSS is similar to the PDBF used by the NASS, and the HLR/VLR mobility data is parallel to the NGN access details identifying the entry path.

Data modeling across NGN and Mobile can be performed for each network. The data modeling can yield a model as suggested in Figure 5.4.

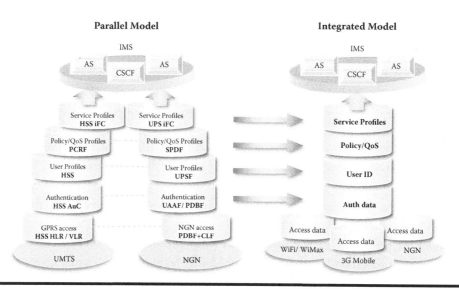

Figure 5.3   Data for both GPRS/UMTS and NGN access.

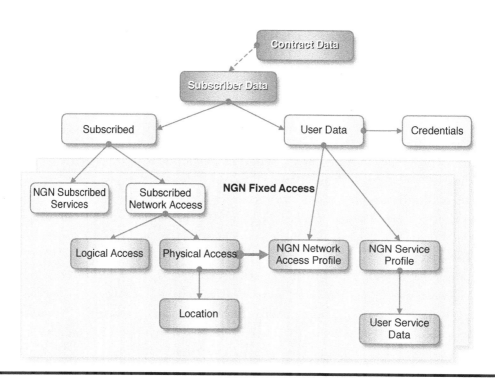

Figure 5.4   Modeling common data.

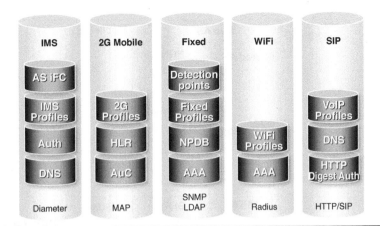

Figure 5.5 Silo-based architecture of data in telecom domains.

5.2.3 The Layered Approach

5.2.3.1 Evolving Data Management

Current data is considered an integral part of every network element in the network. Where multiple instances of the same function are deployed, there is an opportunity to maintain a high-availability disk array with a shared database, as often found in large IT systems where multiple report generators are able to "mine" the data for various purposes.

Figure 5.5 shows the silos created by separate data stored within different domain systems. The examples of data stores shown within each silo are further segmented when different vendors provide platforms.

Naturally, vendors create a database for any of their products, expecting their database to be dedicated to their product. This makes the operational control of the single platform easier and the performance of it can be modeled more accurately. However, managing data in isolation does not help the operation of the network as an interwoven set of services involving many data components. As more domains and more applications are utilized for the same market, data is duplicated within each system.

The same principles of the layered approach in IMS session control should also applied to the data management. In a layered data plane, network data is stored on a unified data center that is open (i.e., provides standard interfaces) to all internal servers that may need to utilize that data.

Figure 5.6 depicts the transition from the current silo architecture to a layered one, where technology-domain data can be organized in a sharable structure with a unified retrieval protocol.

5.2.3.2 Shared Data

There are significant advantages in consolidating data in a central repository. In operations, the following contribute greatly to simplify and reduce costs:

- A single repository will reduce operator costs greatly, as a result of simplifying fragmented data on many database types and hardware.
- Maintenance of the data, when it is integrated via modern tools, is easier and requires less effort from technical staff.

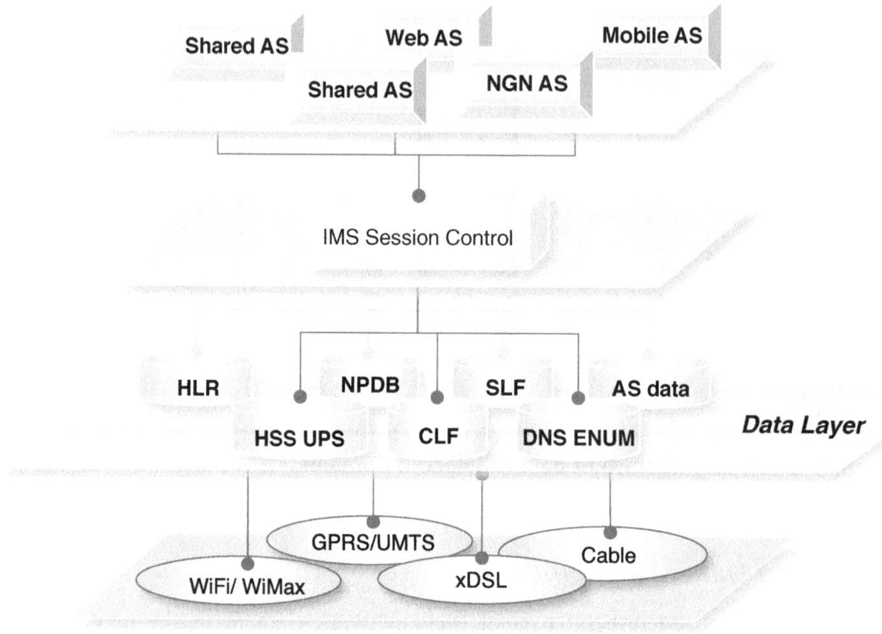

Figure 5.6   From silo data architecture to layered central data store.

- If data is correlated and integrated, duplications of the same data can be avoided. There is lower risk for data corruption to occur. Where common data (such as static user information) is held only once, data conflicts are avoided, thus reducing costs, but more importantly enhancing customer satisfaction.
- Data integrity is enhanced where all user data is brought under the umbrella of a secure and confidential database, while disparate data stores may have variable levels of vulnerability.
- Data retrieval standards for most IMS elements are specified in some detail. This means that a mix of vendors' servers can address the central store with confidence that they will be compatible. This reduces the need for customization and reduces interworking uncertainties.
- Operational staff will find such a central store much easier to manage, needing less training with clear, standard interfaces.
- Staff security levels (data encryption, privilege levels) are easier to maintain where sensitive data is held centrally under the same security facilities.

Sharing stored data is the easiest form of unifying different access modes and different terminals, yet providing a consistent service environment to the user. This is a great enabler of the much-talked-about "user-centric" service because the shared data enables services to collaborate and utilize data from other services.

Making user data independent of any session controller or application is an essential step toward implementing the IMS principle of separating the service layers from the session control layers. It is an important step toward Fixed/Mobile convergence, where the applications still operate in different technology domains but could utilize each other's generated data in their service logic, and enhance the user's perception of convergence and service integration.

Most importantly, great benefits derive from the fact that many applications will be able to share service data, and not merely user static data, but also dynamic status, such as location and

Presence. This means data provided by one application could affect service logic for another application, thus enabling more powerful and more integrated services.

5.2.4   *The Generic User Profile (GUP) System*

The GUP was introduced in 3GPP Release 6 and updated in Release 7. Although developed for mobiles, it seems applicable to NGN as well. The GUP is a server that authenticates access to data and facilitates it. The GUP gives access to multiple data repositories that constitute the primary master copy of user data. The GUP interfaces to the RAF (Repository Access Function), which provides standardized access to the GUP Data Repository.

The storage formats and the interface between the RAF and GUP Data Repository are not specified by the GUP standards and depend on implementation. The RAF, the GUP, and the Data Repository usually reside within the same network.

5.2.4.1   *GUP Functionality*

The following summarizes functionalities that can be incorporated in the GUP:

- Harmonized access interface, providing standard interface and ready templates for the consumers of the data, independently of the profile structure and the data storage.
- Single point of access, simplifying data retrieval and hiding actual data locations. A discovery service, as specified by the Liberty Discovery Service Specification, can be used if the access point is not known.
- Authentication of profile access (validating that the requestor is who it purports to be), adopting generic mechanisms such as used for the OSA framework, in particular authentication mechanisms from the Liberty Alliance Project.
- Authorization of profile access; that is, maintaining access levels based on user-specific rules or common privacy rules as defined by network policies, examining the requestor's identity and the type of requested data. In addition to the generic authorization data, additional service-specific data can be defined, e.g., for LCS (Location Services).
- Authorization of level of access, enabling the GUP to apply different authorization criteria, policy control, and load control to Home Network services applications.
- Privacy control, providing filtering according to a subscriber's specific privacy requirements or external and access rights.
- Synchronization of data storage, where the HSS is the master copy and other data repositories can hold cached data. This means that the changes to the master copy of the data are propagated to other entities.
- Access of profile from Visited Network with a single point of access.
- Location of profile components kept on the GUP.
- Charging for data access, if implemented (e.g., to enable transaction- or event-based charging). This necessitates the use of mechanisms for the correlation of the charging information produced by the GUP Server and GUP Data Repository.

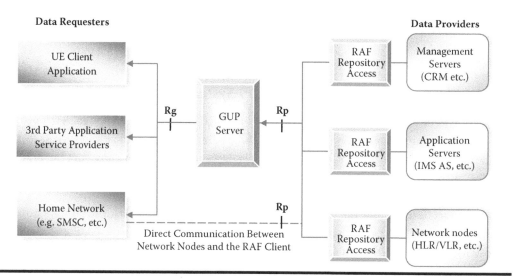

Figure 5.7 AS accessing GUP user profile via RAF.

5.2.4.2 The Repository Access Function (RAF)

The RAF stores standard formats for data retrieval. The RAF is accessed via the GUP, which validates the requestors and permits access to data. After obtaining permission, the requestor (usually an application) can send requests directly to the RAF.

In a single retrieval mode, the GUP facilitates data transfer one by one, as and when the items are requested. Alternatively, the GUP may be asked for several pieces of data, which it retrieves and collates before transferring to the inquiring party. The multi-item retrieval mode is, of course, more efficient.

Figure 5.7 describes the GUP function, which allows multiple functions to access multiple sources of data. The RAF client resides with the source of data. An inquiry from an AS or UE is processed by the GUP. First, the GUP must validate the rights of the inquirer to have the information. When approved, the GUP retrieves the address of the appropriate RAF and enables the information transfer.

5.2.4.3 GUP Reference Points

- Rp is the reference point between the HSS and the GUP client, the RAF. This interface utilizes the RAF to extract data from multiple sources via the GUP server. The GUP therefore can access the HSS as one of the data stores and retrieve user information. Application servers (ASs) can also use the Rp interface for direct access, without the help of the GUP Server. The Rp implementation is based on the Diameter protocol.
- Rg is the reference point between the GUP and the AS (Application Servers) or other external entities. This interface is used by the AS to interrogate the HSS via GUP (instead of the direct interface of Sh). The Rg reference point allows applications to create, read, modify, and delete any HSS user profile data using GUP methods, which is independent of where the data is located.

5.3 Mobile Networks Data Repository

5.3.1 3G/2G Mobile Data Architecture

5.3.1.1 The IMS HSS

Mobile subscriber data is stored in the network for simultaneous use by both 2G and 3G; that is, there are no separate data stores. The HSS contains permanent subscriber data that defines user identities as well as transient subscriber data to support the call control and session management by IMS. The HSS holds a variety of data items for IMS, including:

- User identification, numbering, and addressing information
- User ID structure, with associated public and internal IDs
- User authentication information, including passwords, integrity check, and ciphering
- User location information, i.e., node address where the user is registered (S-CSCF)
- User profile information, including service triggers and preferences

Figure 5.8 illustrates the HSS functions. When the HLR/VLR is considered part of the HSS, Mobility Management is contained within (as shown). However, because IMS is becoming a unifying architecture for multiple access networks, Mobility Management can be split into *access-related mobility* (GPRS/UMTS, WLAN, DSL) and *user-based mobility* (IMS).

The HSS also must support automated provisioning, including the Self-Care user portal and Web-based management portals. Management activities are simplified and enhanced by the central repository approach.

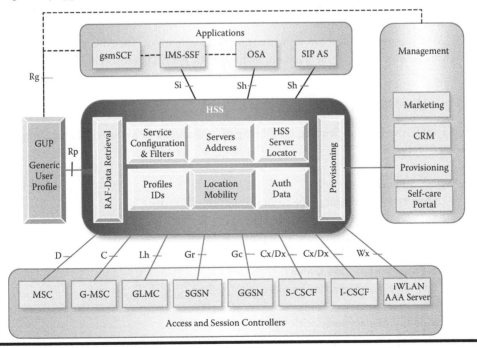

Figure 5.8 HSS components.

Figure 5.8 also shows the interface to the GUP, which allows standardized data retrieval. A client for the GUP system resides within the HSS, interfacing through the Rp interface to the GUP Server. Application servers of all types then can retrieve data using the Rg interface to the GUP.

5.3.2 The 3G Mobile Data Components

5.3.2.1 The HSS Relationship with HLR

The HSS is notionally a single data repository but it may consist of several database structures in multiple instances or distributed data blocks. The organization of the data is not mandated and is dependent on the size and topology of the network and other operational considerations — hence, the lack of clarity regarding the location management.

Although keeping data centrally located is important for data integrity, this does not preclude the option to cache local data within different servers and subsystems, as long as the HSS still serves as the master data and reliable synchronization procedures are maintained.

The HSS, as the master database for IMS, can integrate heterogeneous information and enable enhanced features in the CN to be offered to the Application and Services Domain, in a homogenous way. In particular, both 2G and 3G services can be served by one HSS. Existing GPRS data is stored in the existing HLR, which is responsible for mobility management across both CS and PS Domains. The HLR data contains the 2G identifiers (IMSI/MSISDN), as well as the relevant temporary data for GPRS mobile subscribers who register for either CS or PS (data) GPRS services.

For IMS, HLR is sometimes considered a subset of the HSS functionality, and sometimes as an external, complementary data store for access-based location data. The VLR contains roaming subscribers' data that is required for call handling for mobile subscribers currently located in the capture area of a Visited Network.

The AuC is currently used for authentication of mobile users. The AuC evolved to allow for more advanced methods of authentication, including the AKA Digest specified by IMS.

5.3.2.2 The HSS Data Items

The IMS HSS, the HSS, depending on phasing of migration and modes of access, supports the following functions:

- Session support:
    - Identity management, public and private, cross domain
    - Support for registration and presence
    - Call or session establishment support
- Authentication and admission support:
    - Service authorization support
    - User access authorization support
- Mobility management:
    - Mobility management, with handover policy
- Service support:
    - Support for Application Services

- Maintain service profiles and service triggering
- Application services support, including CAMEL services support
■ Data management:
  - Data distribution, data integrity, and synchronization
  - Service provisioning support
  - Data retrieval functions, including privacy policy and GUP Data Repository facilities

The HSS data contains static user profiles and dynamic registration details. The static user profiles data is stored as a master copy on the HSS and can be cached by the assigned S-CSCF at the user's registration time. This data includes:

■ UE information
■ User profile, with details and options
■ Identities, including Public and Private IDs and their relationships
■ Current registration status with name and address of currently serving S-CSCF
■ Application subscriptions with their triggers, parameters, and user options
■ Administrative user information (e.g., barring, suspension)
■ Charging information and options, including Serving Charging Gateway address
■ The list of authorized Visited Network identifiers for this user
■ Services for this subscription related to Unregistered State, including default S-CSCF
■ The set of identities that make up an Implicitly Registered Public User ID Set

More data items can be added by operators as deemed necessary.

5.3.2.3   *The Subscriber Location Function (SLF)*

The SLF is a database that contains knowledge of a distributed HSS system, including the IP addresses for the HSS servers and other database particulars, such as regional servers, server hierarchy, and distribution parameters.

The SLF is not required in an environment with a single HSS and therefore is considered a scalability function. The SLF is also unnecessary when the Home Network is based on preallocation of users to specific HSS servers, or when there is a mechanism of calculating the HSS server identity from the value of the user ID.

The SLF can be accessed by:

■ CSCF during the registration and session setup
■ S-CSCF during the registration
■ AS to get subscriber-specific service data
■ 3GPP AAA Server during registration and session setup time

All the interfaces (Dx, Dh, and Dw) are based on Diameter dialogs.

The SLF must be kept up-to-date regarding all the HSS servers within the Home Network, with their contents and IP addresses. The synchronization between the SLF and the different HSS servers is handled by the management provisioning system.

Figure 5.9 shows the SLF interfaces to the session control servers. The SLF is primarily queried by the I-CSCF, which is responsible for routing to the correct servers during registration time

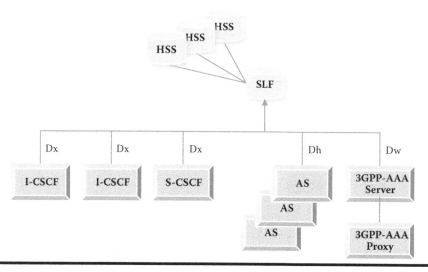

Figure 5.9   SLF interfaces.

and when routing outside the network for session setup. The SLF is also interrogated by ASs and WLAN 3GPP-AAA servers when they need to draw on the user's shared information.

5.3.2.4   The Home Location Register (HLR)

The HLR can be considered part of the HSS data repository or treated as a separate entity. The HLR contains a subset of the network information that relates to IP location and mobility, that is, information required for internal routing of registrations and connection requests.

The HLR helps bridge legacy communication and the new IMS. It is already an integral part of both GSM and GPRS networks. The HLR takes care of the following:

- Providing IP location management for CS Domain entities, the MSC server, and Gateway MSC servers, through the C and D interfaces and for CAMEL SCP servers via the Si interface
- Providing location management for PS Domain entities (the SGSN and GGSN), through the Gr and Gc interfaces, respectively, and interfaces to the G-MSC (GPRS-enabled MSC) server
- Providing location management for 3GPP AAA server in access via I-WLAN through the D'/Gr' interface

5.3.2.5   The Visitor Location Register (VLR)

The VLR, like the HLR, is already an integral part of GSM and GPRS networks. The VLR serves users who visit another network while out of reach of their own Home Network. The VLR facilitates user registration by interacting with the Home Network HLR. To do that, it can use the MS given ID, that is, IMSI or MSISDN, or a generated temporary number for routing to the Home Network.

The VLR function keeps track of visiting handsets in GSM and in packet networks. It records the location of the MS (Mobile Station) with the transmitted ID as it moves in the Visited Network. These details are then used for routing incoming calls for the subscriber.

User mobility in IMS follows the same principles as in GSM. The same handset can be used for both IMS/UMTS and for GSM/GPRS in a dual mode.

The VLR maintains the following identifiers:

- The IMSI
- The MSISDN
- The Mobile Station Roaming Number (MSRN)
- The Temporary Mobile Station Identifier (TMSI), if applicable
- The Local Mobile Station Identity (LMSI), if used
- The location area where the MS is registered
- The identity of the SGSN where the MS is registered, in GPRS
- The last known location and the initial location of the MS

5.3.2.6 The 3G Authentication Center (AuC)

The AuC can be considered a subset of the HSS data that supports authentication of users in both CS and PS Domains. The traditional AuC is associated with an HLR and resides in the same network. Users served by an instance of the AuC must be served by the HLR associated with it. The same rule applies to a VLR. The AuC also supplies the SGSN with authentication information for access-level network admission, unless such information is stored locally on the SGSN itself.

The AuC stores an identity key for each mobile subscriber registered with the associated HLR or VLR. This key is used to generate security data for the mobile subscriber. The interface between the AuC and the HLR or VLR is the "H-Interface." This interface is not standardized and can vary by implementation. Through the H interface, the HLR or VLR or the SGSN requests authentication and ciphering from the AuC and receives authentication details and keys.

5.3.3 The HSS Protocols and Interfaces

The HSS as a combined central data store must, by definition, interface to numerous network elements, as well as provisioning systems and business support systems.

Figure 5.10 shows the combined data store for the GPRS/IMS environment, containing the HSS User and Service Profiles, Authentication (AuC), Location (HLR), and Server Location (SLF).

The GPRS Support Nodes (SGSN and GGSN) maintain copies of subscriber data that are needed for transmission and routing of calls for mobile handsets currently found in their area. The GMLC (Gateway Mobile Location Identifier) contains location information from the HLR that can be provided to external clients for ASs with LCS. Also interacting with the data stores are interworking WLAN servers, the WLAN 3GPP AAA servers and proxies. The term "HSS" is used to refer to all these elements interchangeably.

Figure 5.10 also shows the interaction between the Mobile HSS and those network servers that depend on its data, that is, the IMS session controllers (I-CSCF and S-CSCF) and applications on ASs, including SCP Intelligent Network servers via the IM-SSF.

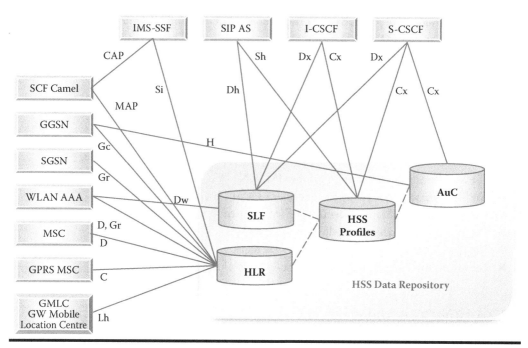

Figure 5.10   3G Mobile data repository architecture.

5.3.3.1   The HSS Reference Points

The reference points associated with the HSS are as follows:

■ *The Cx reference point between the HSS and CSCF.* The Cx reference point supports information transfer between the CSCF and HSS. This interface supports procedures for assigning an S-CSCF to a user and retrieval of user profiles from the HSS. This interface is also used for transferring authorization methods and user credentials from the HSS to the S-CSCF during the registration process and for checking roaming permissions, among other things. This interface is used to obtain the subscriber's service profile with service triggering data, precedence, and other parameters. This information helps the S-CSCF to determine how to process the session, which AS should be involved, and in what order.

■ *The Sh reference point between the HSS and the SIP ASs.* The Sh interface is used for transferring User Profile information such as user service-related information, user location information, or charging function addresses. The same interface is used between the HSS and the OSA AS. This interface is also the tool of transferring data between operators and can carry service data as is, without any interpretation by the HSS. The Sh interface is based on Diameter.

■ *The Dx reference point between the SLF and the I-CSCF or S-CSCF.* The Dx interface allows the CSCF to use the user's Public ID, to find which HSS contains the user profile. The SLF returns the HSS server address. The Dx interface is based on Diameter.

■ *The Dh reference point between the SLF and the AS.* The Dh interface allows the AS to inquire on the HSS Server Location for a user, giving the user's Public ID. The Dh interface is based on Diameter.

- *The Dw reference point between the SLF and the 3GPP AAA Server.* The Dw interface allows the 3GPP AAA Server to inquire on a user, giving the user's Public ID, to find which HSS contains the User Profile. The Dw interface is based on Diameter.
- *The Wx reference point between the HSS and the 3GPP AAA Server.* The Wx interface is used for authentication of users of WLAN access when they register to IMS. The Wx interface is based on Diameter.
- *The Gr reference point between the SGSN and HLR.* The Gr interface provides the basic ability to send and receive packet data within the area of the SGSN. The SGSN informs the HLR of the location of an MS (Mobile Station). The HLR downloads to the SGSN data for this Mobile subscriber. Occasional exchanges of data also occur when the Mobile subscriber activates a particular service or changes preferences, or when some parameters of the subscription are modified by administrative means. Signaling on this interface uses the MAP Protocol.
- *The Gc reference point between the GGSN and HLR.* The Gc interface can be used by the GGSN to retrieve information about the location and supported services for the Mobile subscriber, to be able to activate a packet data network address. This interface uses MAP and TCAP over SS7. It can be used where there is no SS7, but it needs a protocol converter.
- *The Gf reference point between the SGSN and EIR (Equipment Identity Register).* The Gf interface is used by the SGSN to exchange data with the EIR so that the EIR can verify the status of the IMEI (International Mobile-station Equipment Identity) retrieved from the MS. Signaling on this interface uses MAP and TCAP.
- *The H reference point between the HLR and AuC.* The H interface is used to send authentication data from the AuC to the HLR, when the HLR receives a request for authentication and ciphering data for a Mobile subscriber for whom the HLR has no information. The protocol used to transfer the data over this interface is not standardized.
- *The Si reference point between the HSS and the IM-SSF AS.* The Si interface is used for transferring IMS CAMEL-specific information. The protocol on the interface between the HSS and the CAMEL Service Control Function is the MAP Protocol.
- *The Le reference point between the GMLC (Gateway Mobile Location Center) and AS.* Where it is implemented, information relating to the location of the user is not revealed to the AS without authorization and only according to the target user's privacy rules for filtering, which are performed by the GMLC.
- *The Lh reference point between the GMLC and the HLR.* The Lh interface transfers location information from the HLR to the GMLC, where applications can retrieve it as and when it is needed for the service logic.

5.3.4   Mobility in UMTS Networks

5.3.4.1   HLR, VLR, and GPRS Location Data

GPRS nodes serve as a Mobile access network for UMTS, extending their ability to support IP Data Services to IMS communication. As the handset in Mobile networks moves around, the location details change frequently and must be updated constantly. The existing CS Network keeps location data in the HLR and VLR. PS Networks also retain knowledge of the locations in the HLR/VLP.

The GPRS nodes themselves store user data to facilitate call admission. The GGSN records associate routing information with the registered subscription. This is needed to tunnel packet data traffic from the GGSN toward the SGSN where the MS is currently registered. The GGSN keeps dynamic session data for users with at least one active PDP Context Media flow stream.

A number of temporary identifiers, including TMSI, P-TMSI, TLLI, and LMSI, are used in UMTS:

- The TMSI is used to support the need to keep the roaming user's identity confidential. Both VLR and SGSN can use TMSI. They must be able to correlate TMSI and IMSI to identify a particular MS. An MS can be allocated two temporary identifiers for use in CS and PS Networks. The TMSI is provided for services controlled by the MSC, and the P-TMSI (Packet TMSI) is provided for interworking with the SGSN.
- For addressing resources used by GPRS, a Temporary Logical Link Identity (TLLI) is used. The TLLI identifier is constructed from either the P-TMSI (local or foreign TLLI) or is generated directly as Random TLLI.
- In the VLR, a supplementary LMSI is defined for fast subscriber data retrieval. The LMSI can be allocated by the VLR at location updating time, and is sent to the HLR together with the IMSI. The HLR makes no use of it but includes it together with the IMSI in all messages sent to the VLR concerning that MS.

5.3.4.2   The Authentication Data for Roaming Users

An HLR is associated with an instance of the HSS/AuC. For each subscriber on an HLR, data must exist on the AuC/HSS. If this MS is not yet registered in the VLR, the VLR and the HLR exchange information to allow the MS to register.

Roaming authentication data contains:

- An identity key for each Mobile subscriber registered with the associated HLR. This key is used to generate security data.
- Data for mutual authentication of the IMSI
- A key used for checking the integrity of the communication over the radio path between the mobile station and the network
- A key used to cipher communication over the radio path between the MS and the network

In roaming situations, the HLR requests the data items that are needed for authentication and ciphering from the AuC via the H interface. The AuC will respond only to the HLR associated with it. This HLR stores the authentication data and delivers it to the VLR and SGSN that perform the security functions for an MS.

5.3.4.3   Service Restrictions for Roaming

While roaming, the possibility of a mismatch of services supported by the Visited Network elements and the Home Network must be taken into account. The mutual capabilities are discovered and compared when a session is initiated. If there are services that cannot be performed in the Visited Network, they will be barred or restricted, despite existing subscription data for the user.

The service restriction indicators are contained in network-generated temporary service data, which must be kept separate from the permanent service data. This data is stored securely on the HLR or VLR and must not be open to users to tamper with or modify. Where it contradicts subscription data, the service restriction data takes precedence. Finally, when the user roams into another area or onto the Home Network, the service restriction data is changed or erased, and the original service data is restored.

Complete barring of roaming users can also occur. The Barring of Roaming indicator can be set to:

- No barring of roaming
- Barring of roaming outside the Home network
- Barring of roaming outside the Home country

This is permanent data and is stored conditionally in the HLR, both for non-GPRS and GPRS subscriptions.

5.3.4.4   List of Authorized Visited Network Identifiers

Roaming is based on existing roaming agreements between network carriers. Each Public User ID is given a list of approved networks that it can roam. This list is stored on the HSS and is permanent, static data. The list may contain groups of networks described as "wildcard." The format of this list is decided by implementations.

Lists of authorized Home Networks for visiting users are also kept on IP-CAN nodes to verify that a roaming or nomadic user can connect to their Home Network.

5.3.4.5   Location Services (LCS)

Location services are required not only to provide user services, but also for network-based services, some of which are regulations driven. The use cases for LCS are:

- Applications that base their service logic on the user's location, often referred to as LBS (Location-Based Services). Such services can initiate alerts (e.g., road and travel conditions) or local information (e.g., nearby hotels and restaurants).
- Location is required for services that are transparent to the user. For example, handover between networks during roaming to other networks or other access technologies such as WiFi and GSM handover. Some supplementary services and applications rely on location services, such as IN/CAMEL Pre-Pay. O&M facilities may also require location notifications for several reasons.

■ The regulatory Emergency service specifications dictate that LCS will provide reasonably accurate location information to the Emergency agencies in a timely fashion, to enable locating callers automatically. This service is mandatory in some countries, (e.g., the United States). In IMS, location information must be provided to the E-CSCF.

■ The regulatory requirement for state security via Lawful Interception (LI) includes location information. The LI platform, therefore, will subscribe to the LCS.

There are several competing and complementing technologies that provide positioning of the MS. The following are methods to be supported in UTRAN:

■ Cell coverage based
■ Observed TDOA (Time Difference Of Arrival)
■ Assisted GNSS (Global Navigation Satellite System)
■ U-TDOA (Uplink – Time Difference Of Arrival)

The LCS system functions provide the following:

■ *Location system control function* for coordinating location requests
■ *Location system billing function* for charging activities related to location and roaming
■ *Location system operations function* for provisioning of data of positioning capabilities, clients and their validation, fault management, and performance management
■ *Location system broadcast function* for broadcasting LCS information when necessary
■ *Location system coordinate transformation function* for converting an area definition, expressed in a geographic terms, to network identities (Cell Identity)
■ *Location subscriber authorization function* for authorizing UE or AS clients to receive LCS (this function translates pseudonym to identify and validate subscribers who have successfully subscribed to LCS)
■ *Location subscriber privacy function* for performing privacy-related checks and authorizations, to ascertain whether to divulge the location information

The interface to IMS is performed via the LIMS-IWF (Location IMS Interworking Function). This function enables routing of LCS requests based on an IMS Public User Identity (SIP-URI) to the Home Network of the target user. The LIMS-IWF in the Home Network of the target user is responsible for determining the appropriate HSS and for obtaining the MSISDN associated with the IMS Public User Identity from the HSS. This function is optionally combined with the GMLC.

5.3.4.6 Gateway Mobile Location Center (GMLC)

The GMLC is the gateway that supports mobility between GSM, UMTS, WLAN, and applications. UE requesting mobility information contacts a GMLC. The GMLC can request routing information from the HSS and the HLR (via Lh) to find the serving server (MSC or CSCF), and obtain authorization for the request. It can then contact the serving server via the Lg interface to obtain the latest location information.

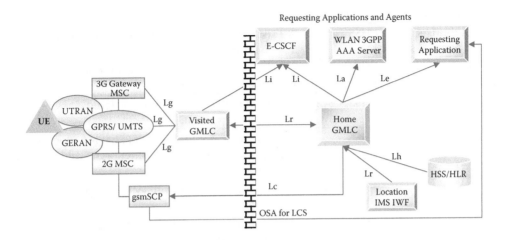

Figure 5.11   GMLC interfaces to provide convergent mobility information.

There may be more than one GMLC within the PLMN and they convey these details between them on the Lr interface. The GMLC also checks on the user's privacy settings, as provided by the Home HSS or MSC profile data, and applies these rules in the authorization process. The Lr interface also carries privacy rules between the Home Network and Visited Network GMLCs.

Figure 5.11 provides a network view of the GMLC, with a user requesting location information about the target user in another network. The request is then processed via the Visited GMLC. Servers that request LSC also are shown here — the Emergency P-CSCF, the 3GPP AAA Server for WLAN, SCP applications (via OSA), and SIP applications.

5.3.5   Summary of 3G IMS Data Items

Most of the data elements defined in 3G for IMS are applicable to any access technology. The following table summarizes the data items, their uses, and where they are stored:

Item Description	Stored on:	Data Uses
IMPI — Private User ID	HSS, CSCF	Auth uses IMPI or (for GPRS) derives one from IMSI
IMPU — Public User ID	HSS, CSCF, BSF	One of several instances of published contacts
PSI — Public Service ID	HSS, CSCF	Published address for a service that can be used by external parties or subscribers.
PSI — Private Service ID	HSS, CSCF	Operator's defined address for a service
Barring Indicator per IMPU	HSS, CSCF	Operator's indication for barring IMS calls. Users can still register.

Item Description	Stored on:	Data Uses
List of authorized Visited Networks	HSS	List per IMPU. If the IMPU is in an implicit set, the list applies to all.
Services for unregistered user	HSS	List of services that are triggered for the user when unregistered.
Implicit set	HSS, CSCF	Set of public IDs that are registered if any one of them requests it.
PSI Activation State	HSS	State indicator for users of PSI. It is temporary for specific users and permanent for a wildcard value (group).
Display Name	HSS, CSCF	A string associated with IMPU, for presentation name.
Registration Status	HSS	Temporary data resulting from registration and de-registration.
Serving CSCF names	HSS	The assigned S-CSCF name is temporarily held for registered users. PSI service instances also have a temporarily assigned S-CSCF name. Default S-CSCF (when unregistered) is permanent data.
PSI AS	HSS	The AS that hosts the PSI service is a permanent item for PSI services "user" records.
Diameter client and server address	HSS	Diameter Client (on S-CSCF) and Diameter Server addresses are used by HSS for Diameter dialog
Security keys	HSS, CSCF	Materials used for authentication, ciphering, encryption, and creating keys.
Server capabilities	HSS	S-CSCF capabilities that allow I-CSCF to select appropriate S-CSCF.
iFC	HSS	Stored per user per application to identify conditions for service invocation. For PSI, iFC is mandatory.
Shared iFC ID	HSS, CSCF	This ID identifies a set of iFC that can be shared by several IMS subscribers and PSI service instances.
AS data	HSS	Service options per user, such as Service Key, Trigger Points, and Service Scripts.
Transparent data	HSS, AS	This data is stored on the HSS for one or more AS and is used transparently by these servers.
Media Profile ID	HSS, CSCF	This identifier points to a list of media parameters that the user is allowed to request. The ID is interpreted by the S-CSCF.
Service Profile list	HSS, CSCF	This is the list of all service IDs that the user subscribes to. Policing the allowable services is performed in the S-CSCF and also in the AS.
Event Charging Function Address	HSS, CSCF	Primary and secondary Event Charging server addresses are stored to enable server redundancy.

Item Description	Stored on:	Data Uses
Charging Gateway Function Address	HSS, CSCF	Primary and secondary Charging gateway server addresses are stored to enable server redundancy.
gsmSCF address	HSS/HLR	The allocated SCF server address is stored for users of Combination services (CSI).
IM-SSF address	HSS/HLR	IM-SSF server address is stored for users of Combination services (CSI) and is used to send Notification on Change of Subscriber Data.
CAMEL data for CSI	HSS/HLR	Service information to communicate to CS for originating or terminating combination services, such as service key and DP (Detection Point).

Source: Based on 3GPP TS 23008.

5.4 NGN Data Repository

5.4.1 NGN Data Architecture

The Data Repository for NGN Fixed networks is defined in TISPAN. The NGN data is organized into three distinct data stores:

1. *UPSF*: This is the store of user identity structure and the associated service profiles. It is said to be the equivalent of HSS without the HLR element.
2. *CLF*: This is the location data store for NGN, containing the association of user ID with an IP address, plus details of the access path. The CLF registers the location and conveys this information to the P-CSCF in order to route calls to the NGN UE and to obtain the QoS profile for interaction with the RACS.
3. *PDBF*: This is the data store that contains the users' credentials and is involved in authentication procedures with the UAAF. This can be an internal database for the UAAF, or part of the converged HSS.

The split of data elements kept on the UPSF, the CLF, or the PDBF may not be exactly the same as the division of work between the HSS, HLR, and AuC in 3G Mobile systems, but the functionality is very similar.

UPSF maintains user identities at the IMS application level, with the user's service profile and application data, but the connectivity details for the user are kept on the CLF, and the authentication information and settings are on the PDBF, where they are available to the access nodes. Authentication data and network access information are used by the NGN IP-CAN (e.g., the xDSL broadband access nodes or the WLAN AAA Server).

As shown in Figure 5.12, the master data repository is the UPSF, supported by data stored within the NASS, in the CLF, and in the PDBF.

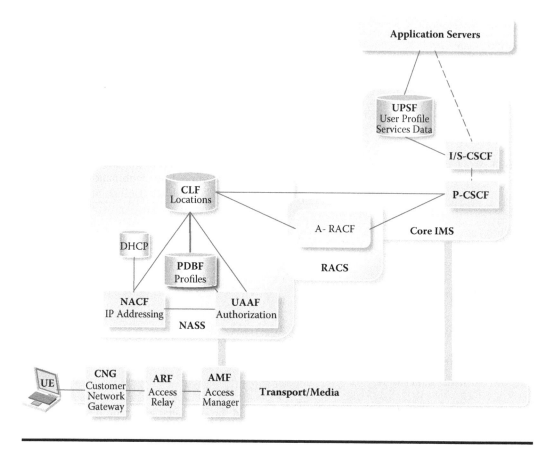

Figure 5.12 NGN data storage elements.

5.4.2 NGN Data Elements Details

5.4.2.1 The User Profile Server Function (UPSF)

The UPSF is responsible for holding service-level user-related information, to distinguish from access-related information, which is stored in the PDBF and CLF. For IMS, the UPSF stores. This data includes:

■ IMS user identification, numbering, and addressing information
■ IMS user security information, including security keys and materials for IMS authentication
■ IMS network location information (i.e., serving CSCF server addresses) supporting user registration
■ User profile information, including service profiles

The UPSF can store user profile information related to one or more service control subsystems and applications. All IMS applications should utilize data from the UPSF and contribute data to it.

As Figure 5.13 reflects, the UPSF interfaces are mostly the same interfaces as for the HSS. The UPSF interface to the application servers is the Sh reference point. The interface to the session controllers (CSCF) is the Cx reference point. Note that the interface to the IM-SSF is somewhat different from 3GPP because Fixed Line also requires interworking with ETSI-defined Core

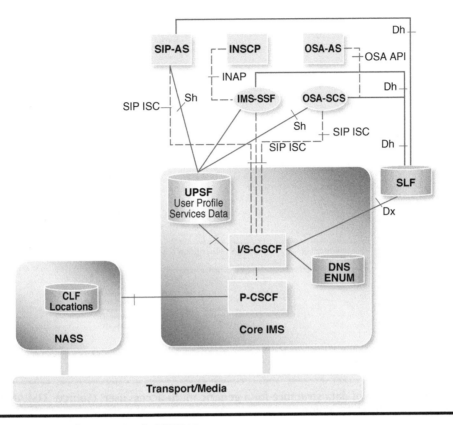

Figure 5.13   UPSF data storage in TISPAN.

INAP rather than purely CAMEL. The SLF interfaces are also the same as in 3GPP, as is the DNS/ENUM server.

5.4.2.2   NGN Data Modeling

The data items for NGN are based on the relationship between:

$$\text{User} \rightarrow \text{Subscriber} \rightarrow \text{Subscription} \rightarrow \text{Services}$$

as shown in Figure 5.14. The data structure allows for multiple access networks and multiple service instances.

The credentials apply at the user level. Network access details, including network access credentials, are defined for each access type, within the subscription. The service profile contains relationships to User Public IDs (implicit sets, shared iFC IDs, default IMPUs, etc.), as well as details characterizing the service package.

5.4.2.3   The Profile Database Function (PDBF)

The PDBF is the data store used in the NGN authentication process. It mainly contains per-user identity: supported methods of authentication (Certificate or Shared Secret), credentials to

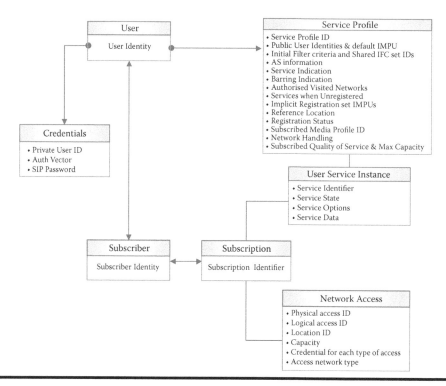

Figure 5.14   Example of data structure for a generic multi-access user. (*Source:* TISPAN Draft TS 188 002-2 and 3GPP TS 32808.)

support challenge-and-response security dialog, keys and certificates, and other parameters and network access configurations.

The PDBF can be co-located with the UPSF in the same way that the AuC is part of the HSS.

5.4.2.4   *The Connectivity Session Location and Repository Function (CLF)*

The CLF is the Connectivity Location Function, or as it has been later named, the Connectivity session (Location and repository) Function. The CLF keeps a record of users' location details as determined when they register to the NGN network. It is involved in the access-level authentication and retains addresses of the access nodes in the connection path for this user. It also can store the association with a User Identity, established during authentication performed by the UAAF, along with the level of QoS and Privacy preferences (i.e., anonymity and location privacy). Where authentication data is stored on a separate database (i.e., the PDBF) accessed by the UAAF, the CLF should be able to fetch the data from that database.

Every active session is recorded on the CLF; therefore, the CLF contains dynamic data, such as the HLR for Mobile handsets. This data can be used by applications for Presence status and for location-based features.

The CLF also communicates with the RACF when a session is negotiated according to network policies and available resources. The CLF data includes QoS settings. This is the subscribed QoS profile with service precedence and priority, reliability, delay, and throughput for the sessions. It is possible for one user to have a configured access profile that contains multiple QoS profiles.

5.4.3   NGN User Mobility

5.4.3.1   Nomadic Users' Addresses and Configuration

In fixed connections, the user access identifier is often defined by the intermediate equipment between the terminals and the network, the CNG. This can be a CPE such as a modem or router, or a network-based access point gateway in the operator network. These network elements, or combinations of them, constitute the physical location of the user. The nomadic user may be linking through such devices when roaming to another location; therefore, the binding with the CNG location is only temporary.

When the NGN terminal is tethered at a certain point in the network, it gets an IP address allocated to it via the local NACF, usually a DHCP server. This IP address is a logical network address, not a geographic location. However, when the user registers, the CLF records the binding of the physical geographic location details of the APN and CNG equipment with their IP network address. Fixed users (e.g., where the CNG is located at the user's Home and connects permanently to a SIP phone) need not generate IP addresses and fresh binding every time.

Figure 5.15 shows four main strands of the procedure running between the involved components:

1. CPE configuration
2. Network configuration and resource allocation
3. Authentication and service authorization between Visited and Home Networks
4. Location management, with dynamic status and physical path recording

The authentication and authorization procedures take place, passing through the intervening networks. The AMF accepts the CPE requests to register and forwards them to the UAAF. If the UAAF is in the Home Network, it performs the authentication. Otherwise, the proxy UAAF connects to the Home UAAF server to perform the authentication, using prestored security data in the PDBF.

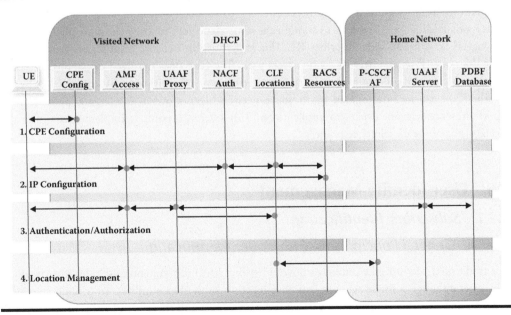

Figure 5.15   NGN registration procedures.

The visited APNs may contain names of domains that the UE could be connected to i.e., domains with interconnection agreements.

The IP configuration involves gaining access to the visited IP-CAN by discovering the access node and starting a dialog via the visited AMF. It also involves the AMF dialog with the local NACF to attach the UE to the network and record the location of the nodes. The information of the access nodes will be used further to allocate resources that can be agreed upon with the Visited Network.

5.4.3.2   NGN Mobility Location Data in the CLF

The CLF is the location register, equivalent to the HLR in Mobile networks. The CLF keeps information of every active session. It contains the binding of the network access profile for the nomadic terminal with the logical IP address, allocated by the NACF. The location information maintains knowledge of the intervening networks, including cross-business borders.

The CLF can be interrogated by service control servers and applications, requiring location information. It will return information in a variety of forms, such as:

- Network location IP address
- Geographical coordinates
- Post mail address
- Access node identity
- Line identity

The CLF can be interrogated by IMS control servers to make control decisions or establish routing. Before divulging this information, the CLF checks what level of privacy applies to this request and acts accordingly. This is governed by operator's policies and in compliance with regional regulations regarding data protection.

It is possible for several records to contain the same identifiers, binding the same physical access ID, logical IP address, and subscriber ID. This is permissible because a subscriber can establish more than one IP session over the same access path. The CLF does not need to establish any link between such records, although it may do it for the purpose of optimizing its storage capacity.

The CLF can be interrogated by Presence network agents over the reference point (Pn) to report Presence-relevant events to applications. This reference point is implemented using the mechanisms of the e2 reference point.

5.5   User Identity Management

5.5.1   Subscriber Identification

5.5.1.1   General Principles for Subscriber Identification

In a traditional network, the publicly known identifier (the telephone number) also identified the subscriber although a single account can have multiple subscriber numbers. In Mobile networks, it is assumed that the handset is always carried by the user, and therefore identifies the user. Any serving network that the UE is roaming to must recognize the user according to the embedded

ID within the UE. The ID is used to establish which realm the user belongs to, and how to route to the Home Network.

By contrast, in a Fixed IP network, the IP user can log on to any device, and the user identifier is not necessarily embedded. Therefore, user identification must rely on a registration and association of the user account with the current connection point IP address.

5.5.1.2 The SIM (Subscriber Identity Module) and USIM/ISIM

In Mobile networks following 3GPP specifications, the identity of a subscriber is encoded in an identity module within the handset firmware, the UICC. This is a removable component of the UE that can be taken out and inserted into another compatible handset, thereby allowing certain user service mobility. The UICC makes the IMEI unreliable as a unique user identifier, but is still used to identify equipment capabilities and software compatibilities. The IDs in the USIM (UMTS SIM) and ISIM (IMS SIM) are associated with the subscription, not the equipment, and therefore they represent the user identity.

The UICC contains a user ID application module, for use in GSM, UMTS, and IMS:

- SIM in GSM
- USIM in UMTS/GPRS
- ISIM in 3GPP IMS.

These modules keep the UE identifiers and associated procedures that enable the UE to access various networks. It is possible for UE without ISIM to register to the network and obtain IMS services, utilizing USIM procedures.

5.5.1.3 The ISIM Data

The IMS SIM card (ISIM) contains not only the IMS identifiers (Public and Private IDs), but also Home domain name, security keys, various access rules and preconfigured data that enables network admission procedures, authentication, and registration to take place. The ISIM contains:

- Private Identity in the format of a URI
- Home domain name in the format of a URI
- Public identities, each in the format of a URI
- Mode of operation (normal, type approval operation, specific facilities, etc.)
- Access rules reference
- Service table that includes indicators for service availability
- P-CSCF addresses, in priority order
- GBA with:
  - Associated Bootstrapping Transaction Identifier
  - Lifetime of the GBA bootstrapped keys
  - List of Network Application Functions that may request keys on behalf of the user
  - AKA Random challenge (RAND)
- HTTP Digest details, although HTTP Digest security requests do not apply to a terminal using a 3GPP access network or a 3GPP I-WLAN

5.5.1.4 Soft Clients

Laptops and PCs have soft clients that can be downloaded to any compatible device. The soft client traditionally identifies the user via password and username. The soft client can generate a user ID, but this is application specific and is not subject to standardization.

For IMS, the ISIM is used within the soft client to access IMS services. This means that such soft clients utilize the same authentication mechanisms as described for IMS.

Advanced biometric technologies enable user identification through unique biological features, such as fingerprints or iris retina. These methods assist in associating a user with an ID and a subscription.

Mechanisms that provide firmware identifications are considered safer. Therefore, the idea of providing UICC with SIM identification for laptops, PDAs, and Fixed IP phones is becoming more popular. Other ideas include a smart-card, a memory USB stick or a dongle that can uniquely identify the user.

5.5.2 Mobile Station Identifiers

The Mobile identifiers described here are already in use in 2G Mobile networks. They still form a basis for interworking with CSNs and for generating identifiers for communications with the PS network and IMS, particularly in the absence of the ISIM.

5.5.2.1 Mobile Station Integrated Services Data Network (MSISDN)

MSISDN is an ISDN Number allocated to an MS. Blocks of unique international ISDN numbers are allocated according to the ITU-T Recommendation E.164 numbering plan. The composition of the MS international ISDN number should be such that it can be used as a global title address for routing messages to the Home location register of the MS.

The format of MSISDN is shown as follows:

MSISDN		
	National Number	
Country Code	National Destination Code	Subscriber Number

The Country Code identifies the country for international routing. For GSM/UMTS applications, the National Destination Code represents an operator's PLMN, where there can be several in one country. If further regional routing information is required, it is contained in the most significant digits of the Subscriber Number (SN).

One or more numbers of the ISDN numbering plan must be assigned to an MS for use in incoming call routing. The assignment of at least one MSISDN to an MS is mandatory in GSM/GPRS. A sub-address, with a maximum length of 20 characters, can be appended to an ISDN number for use in call setup and in supplementary service operations where an ISDN number is required.

5.5.2.2 International Mobile Subscriber Identity (IMSI)

IMSI is used for cross-operator routing. This must be a globally unique identifier that is allocated to each MS in the GSM/UMTS system. It is used in IMS for generating identifiers when there is no URI or ISIM.

IMSI consists of decimal digits only (0 through 9), not exceeding 15 digits in total. The IMSI components can be represented in the following format:

IMSI		
	NMSI	
MCC (Country Code)	MNC (Network Code)	MSIN (Subscriber ID)
3 digits	2 or 3 digits	Up to 9 or 10 digits
15 digits		

The IMSI parts comprise:

- The MCC (Mobile Country Code) consists of three digits. The MCC uniquely identifies the country of the Home Network of the Mobile subscriber. The ITU-T administers the allocation of MCCs. The current country code allocation is given in the ITU-T Recommendation complement E.212.
- The MNC (Mobile Network Code), consisting of two or three digits, identifies the Home PLMN of the Mobile subscriber. Within a single Country Code, the MNC must uniquely identify the PLMN. This code is normally allocated regionally by the regulatory body.
- The MSIN (Mobile Subscriber Identification Number) uniquely identifies the Mobile subscriber within the Home PLMN. This number is allocated by the Home Network operator.

The NMSI (National Mobile Subscriber Identity), as shown in the IMSI table, consists of the MNC and the MSIN.

IMSI is used in user verification in several ways. Authentication in the IMS Domain is usually based on a corresponding Private User Identity; but when the authentication bootstrapping is based on information from the CS Domain, the IMSI is used to derive the user's identity.

5.5.2.3 International Mobile station Equipment Identity (IMEI)

IMEI consists of digits only. It is composed of the following elements:

- TAC (Type Allocation Code) with length of eight digits. This is allocated by a central authority.
- SNR (Serial Number), which is an individual serial number uniquely identifying each device. Its length is six digits. This number is allocated by the equipment manufacturer.
- SVN (Software Version Number), to identify the on-board software.

The SVN may not be present, but instead there may be one spare digit.

IMEI-SV		
TAC (Type Allocation Code)	*SNR (Serial Number)*	*SVN (Software Version Number)*
8 digits	6 digits	2 digits

The IMEI is complemented by a check digit. The check digit is not part of the digits transmitted when the IMEI is checked. The check digit is intended to avoid manual transmission errors, for example, when customers register stolen ME at the operator's customer care desk. The check digit is calculated on the first 14 characters; that is, it not applied to the SVN.

5.5.2.4 Temporary Mobile Subscriber Identity (TMSI)

To support the subscriber identity confidentiality service, the VLRs or SGSNs can allocate TMSIs to visiting Mobile subscribers. The VLR and SGSN must be capable of correlating an allocated TMSI with the MS IMSI. To ensure that allocated numbers do not overlap, the allocation time of the temporary number can be included in the TMSI. Because the TMSI is confidential, it must be allocated in ciphered form only.

The MS is instructed to use the allocated TMSI. The UE does not check or validate this temporary number (which is in ciphered form) but uses it, whatever its value. The TMSI can be used for both CSN and PSN handsets; therefore, it includes a flag to denote to which network it is referring.

The P-TMSI is temporary subscriber ID and is conditionally stored in the SGSN. The P-TMSI is accompanied by the P-TMSI Signature. The P-TMSI Signature consists of three octets and is allocated by the SGSN and stored in the SIM.

5.5.2.5 Routing Numbers: MSRN and IMRN

Temporary numbers are also allocated when roaming. MSRN (Mobile Station Roaming Number) is used to route calls directed to an MS. The Gateway MSC can allocate an MSRN via the HLR when requested by the VLR that handles the roaming MS. The MSRN may be identical to the MSISDN on the MS. The Visited MSC and the assigning VLR will then use the MSRN to reach the MS in the same way as for MSRN. More than one MSRN can be assigned simultaneously to an MS.

An IMRN (IP Multimedia Routing Number) is a routable number that points to IMS. In a roaming scenario, the IMRN has the same structure as an international ISDN number. The Tel URI format of the IMRN is treated as a PSI. The IMRN can be used in services such as VCC (see Chapter 3).

5.5.3 PSTN Numbering

PSTN numbering is needed for routing from the IMS Domain toward the PSTN and for ENUM translations. PSTN numbering is based on ITU-T Recommendation E.164. The E.164 numbering ensures that each number is unique and can be connected from any point worldwide. This recommendation defined the structure of the international PSTN numbering as well as setting rules for numbering reservation, allocation, and reclaiming numbers when unused.

E.164 PSTN (International)		
Country Code (CC)	*National Destination Code (Optional)*	*Subscriber Number (SN)*
1–3 digits	Remaining digits up to 15	
Maximum 15 digits		

The number structure is restricted to a maximum of 15 digits (without prefixes) written with a "+" prefix. Prior to 1997, the number length was 12 digits (per E.163) but this was extended to 15 digits to allow unique global numbering. Each country can set further rules, for example for allocation of block of numbers to carriers.

5.5.4 SIP Identifiers in IMS

IMS is using SIP as its major signaling protocol. This means that SIP identifiers, SIP URI or Tel URI, must be used in the signaling messages. As a rule, IMS requires the SIP URI to be used in routing within the IP networks and resorts to the use of MSISDN or IMSI in Tel URI only when routing to external CS networks.

For generic descriptions of SIP URI and Tel URI, see Appendix A, Protocols.

5.5.4.1 Domain Name

A URI contains a domain name portion and a username portion. A URI of "http://www.Homewebsite_example.com/index.html" contains a domain name of "www.homewebsite_example.com," where "homewebsite_example.com" is a registered domain name. The domain name includes a gTLD (Generic Top-Level Domain) extension appended, such as ".com" or ".org." The gTLD is allocated by the IANA (Internet Assigned Numbers Authority).

Domain names are logical names representing host servers registered by their owners. Such hostnames can be meaningful, using both alphabetic and numeric characters. Therefore, they are more memorable than the full IP address, which is a string of digits delimited by dots. Maintaining logical names also enables changing the IP address temporarily, so as to use alternative servers (e.g., for maintenance work). The Home Network domain name is used for setting up SIP addresses to route sessions to the Home Network or from it.

When the ISIM application is present in the IMS handset UE, it contains the default name for the Home Network, securely stored. It should not be possible for users to modify this name. When the UE initiates a session, the ISIM inserts the stored domain name in the "FROM" header address. If there is no ISIM application to insert the domain name automatically, the UE client needs to derive the Home Network domain name from the IMSI as follows:

- Parse the IMSI into the MCC and the MNC according to the digit assignation in the particular network, that is, considering whether two or three digits are used for the MNC. If the MNC is two digits, then a zero is added at the beginning.
- The MCC digits are prefixed with ".mcc" and the MNC with ".mnc."
- The label "ims" is inserted in front.
- Add the label "3gppnetwork.org" at the end.

The derived Home domain is therefore "@ims.mnc[MNC value].mcc[MCC value].3gppnetwork. org."

As an example, a derived Home Network domain name from IMSI = 792 42 1234567899. The IMSI is parsed as follows: MCC = 792, MNC = 042, MSIN = 1234567899. With the inserted labels, the derived Home domain name will be:

"@ims.mnc042.mcc792.3gppnetwork.org"

5.5.4.2   IMS SIP URI

Routing of SIP signaling within the IMS is performed using SIP URI. URI is the identifier format used in the Internet and IP networks in general. SIP defines a SIP-specific address by inserting "sip" in the protocol element of the URI, expressed as "sip:username@domain." Alternatively, non-SIP addresses in the "AbsoluteURIs" format can be used. Routing of SIP signaling using AbsoluteURI (non-SIP) is intended only for routing messages from IMS users to external IP networks or the Internet.

The URI syntax is typically

http://username@domain.com:999/pathname?query=anchor#fragment

This syntax is parsed as shown below:

Every URI begins with the scheme that defines its namespace, purpose, and the syntax of the remaining part of the URI. For SIP routing in IMS, the protocol part is "sip:"

http:	*//username*	*@domain.com*	*:999*	*/pathname*	*?query*	*=anchor#fragment*
Protocol	Login	Host	Port	Path	Query	Anchor and fragment

5.5.4.3   The "Tel" URI

IMS does not route by telephone numbers. However, telephone numbers can be used in IMS to identify the users taking the form of "tel" URI, as described in RFC 3999.

The telephone numbers must be in the form of E.164 format, prefixed with a "+" sign, that is, " + (country code) (operator code) (user number)". The inserted "tel:" denotes that the username is not merely a string of digits, but it is a valid telephone number that is internationally recognizable and globally unique, and therefore there is no need for a domain name.

This format allows for shorter identifiers and less processing than the "sip:+number@domain. com".

5.5.4.4   URI for Different Protocols

IMS can make use of URI formats including "sip:" (SIP Secure), "sips:", and "tel:". In addition, URI to address Instant Messaging and Presence can be associated with IMS sessions. This enables IMS to deal with non-session-based calls as well as SIP session messages. Although there are other methods for these services, it is easier to correlate them to sessions under a single IMS umbrella.

The following is a table of URI formats with their specification documents:

Format	URI Designation	Specification
sip:	SIP signaling URI address	RFC 3261
sips:	Secure SIP signaling URI address	RFC 3261
tel:	Telephone number and dial strings	RFC 3999
pres:	Presence resource	RFC 3861
im:	Instant messaging	RFC 3861

5.5.5   IMS User Identity

5.5.5.1   IMS Subscription

As a rule, the reference to an IMS Subscription defines a user overall account, that is, "The User" as viewed by the operator's network. This IMS Subscription can be used by multiple devices referred to as UE, in both Mobile and Fixed communications.

5.5.5.2   IMS Private User Identity (IMPI)

Within a network, an IMS user can have more than one instance that can be activated or deactivated, and billed. This user identity is said to be "Private" because it is used only within internal network processes and is never divulged to the user or anyone else. Multiple such Private IDs are useful to segment Enterprise business, for example, or support a user with different billing schemes.

The Home Network operator is responsible for the assignment of Private User Identities. These identities are permanent subscriber data, stored in the HSS. The IMPI format follows the NAI form of username@realm as specified in IETF RFC 4282.

The IMSI can be included in the username if desired to help relating it to existing subscriptions. If there is no ISIM application, and the IMPI is not known, then the IMSI is used to derive the Private ID. This is performed utilizing the entire string of digits calculating the realm from the Country Code and Network Code into the format of:

ims.mnc[MNC].mcc[MCC].3gppnetwork.org

For example, a derived Private ID from an IMSI of +792421234567899 will be parsed as follows: MCC = 792, MNC = 042, MSIN = 1234567899. With the inserted labels, the derived Private ID will be:

1234567899@ims.mnc042.mcc792.3gppnetwork.org

5.5.5.3   IMS Public User Identity (IMPU)

The Public User ID (IMPU) is a published contact name or number by which the UE is addressed. Several IMPUs can exist per IMS subscription — for example, both a telephone number and a URI address. The Home Network operator is responsible for assigning IMPUs. These IDs are permanent subscriber data items, stored in the HSS and access network admission nodes such as the BSF.

IMPUs can be shared across multiple IMPIs within the same IMS subscription. Hence, a particular IMPU can be registered simultaneously from multiple UE that use different IMPIs and different contact addresses. All IMPUs of an IMS subscription must be registered at the same S-CSCF to effect proper coordination.

If an IMPU is shared among the IMPIs of a subscription, then it is assumed that *all* IMPIs in the IMS subscription share that IMPU. Subscription data should indicate which IMPUs within a subscription are shared and which are not shared.

The IMPU sharing mechanism is not intended to support the sharing of identities across large numbers of IMPIs (e.g., for large IP-PBX extensions) because this would result in all these users being forced to associate with the same IMS subscription and hence be served from a single S-CSCF.

The IMPU takes the form of either a sip URI or a Tel URI. The format is sip:user@domain. If there is no ISIM application to provide prestored IMPU, a temporary IMPU can be derived from the known IMSI. The temporary IMPU is built to the format of: "sip:user@domain," with the prefix label "sip:" plus the IMPI. The IMPI is derived as described above; for example, the IMSI of +792421234567899 makes a private identity: "sip:1234567899@ims.mnc042.mcc792.3gppnetwork.org."

In the example shown in Figure 5.16, the IMS Subscription consists of two IMPIs and several IMPUs. One IMS Subscription has two IMS IMPUs. IMPU-3 and IMPU-4 belong to both IMPI-1 and IMPI-2.

5.5.5.4 Implicit Registration Sets

An IMS user can have several IMPUs associated with several IMPIs is an implicit set. Upon registration of one IMPU (as described in Chapter 4), several other IMPUs can be registered — "implicitly" — at the same time. This implicit registration is performed for those identities that have been predetermined as the "implicit set," as illustrated in Figure 5.16.

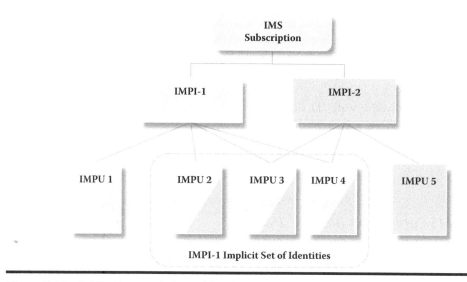

Figure 5.16   Public ID association with Private IDs and IMS subscriptions.

An Implicitly Registered Public User Identity Set contains several instances of IMPUs for the same IMS Subscriber. Several Implicitly Registered Sets can be configured for a given IMS user, with the following rules:

- An IMPU can be included in no more than one Implicitly Registered Set.
- One Implicit Set can contain IMPUs that belong to different IMPIs.
- A single IMS subscription can include more than one Implicit Set.
- Only one Implicit Set can be stored per IMPI.
- There must be a Default IMPU indicator that marks the IMPU to be used as default in each Implicitly Registered Public User Identity Set.
- There is only one Default IMPU per Implicitly Registered.
- It is not necessary for an IMPU to be associated with any Implicit Set.

5.5.5.5 Temporary Public User ID

As shown above, it is possible to access the IMS service, even when the terminal does not contain the ISIM function, by generating a temporary ID from an IMSI that is stored on the client. The Home domain name is derived from the MCC and MNC, and the username is set to the SN (Subscriber Number). This temporary IMPU takes the form of a SIP URI and is used in SIP registration procedures to the IMS domain. This temporary IMPU is made available to the CSCF and HSS nodes.

The temporary IMPU should be set to a "barred" status for any *non*-registration SIP procedures. Being "barred," the temporary IMPU is not displayed to the user, and should be secured from user tampering. Because it is "barred" for anything else, this temporary identity can be used only for registration, re-registration, and deregistration procedures.

On registration, all associated IMS Public IDs tagged for "implicit registration" are deemed registered. Once registered, these "implicitly" registered identities can be used for SIP session initiation and subsequent messages. In effect, the temporary IMPU is used to register the other IMPUs associated with the same user subscription.

5.5.5.6 Anonymous User ID

When originating users elect to remain anonymous, their IDs should not be revealed to the called party. The Anonymous User Identity takes the form of a SIP URI where the user part consists of the string "anonymous" and the domain part is set to the string "anonymous.invalid." The full SIP URI for Anonymous User Identity is therefore "sip:anonymous@anonymous.invalid."

The recipients can activate a service to reject all such anonymous calls when encountering this format.

5.5.5.7 Emergency Public User Identity

Emergency calls are treated differently from ordinary calls. Users do not need to be registered to originate an emergency call, and therefore there may not be a valid Caller ID. However, such calls must be recognized and given priority over other traffic. Therefore, the UE needs to set up a special user ID as originator in the "FROM" header. The Emergency Public User Identity is derived from an existing Public User Identity as follows:

■ The UE retrieves a Public User Identity from ISIM.
■ If there is no ISIM, the UE generates a temporary Public User Identity from IMSI as described above.
■ The UE prefixes the string "sos." to the beginning of the domain part of the selected Public User Identity.

For example, if the SIP URI for the selected Public User Identity is "sip:user@domain," the corresponding Emergency Public User Identity is "sip:user@sos.domain."

5.5.5.8 Alias Identities

The HSS can provide, on request, all the aliases available for a particular user. When the ALIAS_ IDENTITIES are requested, the HSS sends a list of all non-barred IMPUs that share the same service profile and are included in the same Implicit Registration Set. The MSISDN user identity is not considered an alias.

5.5.5.9 Globally Routable User-Agent URI (GRUU)

The concept of the GRUU is to relate a specific UE with a specific Public User Identity for a duration, although the user may have multiple public IDs and multiple terminals.

SIP and packet networks support handset-independent users that can be associated with several terminals and several Public IDs. SIP enables an efficient way of alerting such a user on multiple registered terminals all at once via SIP "forking." However, this is not always desirable. In fact, there is a requirement to provide a one-to-one relationship between a Public User ID and the UE, when a predictable result is required, e.g., for session continuity (VCC).

To operate with GRUU, the UE client application must be able to support a unique combination of specific Public User Identity with that UE. At registration time, the UE indicates that it supports GRUU for a given Public User Identity. A single ID is then assigned to this UE, which can be used in any IP-CAN to access IMS. If the Core IMS servers do not support GRUU, the request for a single ID will be rejected. A UE supporting GRUU must be able to distinguish a GRUU from ordinary IMPUs. It should be able to communicate with other UE supporting GRUU and other IMS networks with GRUU capabilities.

When a GRUU is registered, and the IMPU is in an Implicit Registration Set, the IMS registration process will generate GRUUs for all implicitly registered IMPUs belonging to it. This means that all the implicit set members will be associated with the same UE. This also necessitates that all GRUU messages associated with a single IMPU will be directed to the *same* S-CSCF.

Figure 5.17 shows how GRUU associations can be created within the structure of the Identity tree. Note that one IMPU can be linked with two separate GRUUs as long as they are associated with one IMPI. For example, GRUU 3 and GRUU 4 are associated with IMPU 4 and IMPI 2. Also evident is the ability of one UE to associate with two GRUUs (e.g., UE2), and to an implicit set.

When a GRUU is cited in a SIP INVITE message, IMS nodes supporting GRUU will identify the "gr" parameter with the "instance" header, and use it to identify the address contact for

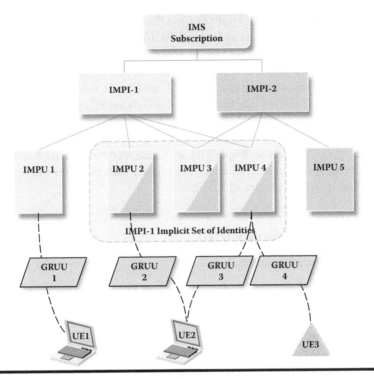

Figure 5.17   GRUU ID associating IMPUs with different UE.

the associated UE. This IMPU must correspond to the Sender's URI inserted in the SIP Contact header. The message will be rejected if the URI is different from the derived IMPU, or if the UE has no GRUU registration. IMS nodes supporting GRUU should be able to route the session according to the GRUU parameters and rules and only route to the selected UE address. Even when there are other UE instances registered for the user, only the UE named in the GRUU will be alerted.

The GRUU Public Identity can be associated with Filter Criteria for triggering services, just as any other IMPU. The triggering of a service for a GRUU Identity may need to take note of the GRUU status and trigger only applications that are applicable to the nominated UE.

GRUU users may be able to receive terminating services that can be triggered for unregistered subscribers, according to network policy. Because the association between the UE and the IMPU is only made during registration time, this may entail keeping historical data for such users to predict their wishes, or allowing users to elect a GRUU combination to use when not registered.

Two special GRUUs are defined:

1. *T-GRUU (Temporary Globally Routable User-agent URI)*. This type of GRUU is assigned temporarily and the actual URI remains unknown to the user. The T-GRUU remains valid until the contact is explicitly deregistered or the current registration lifetime expires.
2. *P-GRUU (Public Globally Routable User-agent URI)*. P-GRUUs are GRUUs similar to IMPUs in that the assigned Public User Identity is revealed to the user. The P-GRUU is a long-term assignment.

The IMS UE is able to obtain both T-GRUUs and P-GRUUs when performing IMS registration, exchanging GRUUs using SIP messages and using GRUUs to address SIP requests to specific UE.

Any Public User Identity can have one or more associated GRUUs. A pair of generated T-GRUU and P-GRUU is assigned at registration and a lifetime is associated with a single Public Identity. In subsequent re-registration, the P-GRUU remains the same but the T-GRUU is generated afresh. After a re-registration, the previous T-GRUUs are all still valid. The UE can retain some or all of them, or replace them with new T-GRUUs. A GRUU set is made of the current P-GRUU and all T-GRUUs that are valid during the registration period.

If the UE registers (explicitly or implicitly) with multiple Public User Identities, a separate GRUU set is associated with each. If different UE register with the same Public User Identity, a different GRUU set is associated with each UE.

5.5.6 Public Service Identity (PSI)

5.5.6.1 Public PSI and Private PSI

The PSI identifies a service or a specific resource created for a service on an AS. The PSI can be used to access service components and service enablers, such as Presence or Messaging components. PSIs are published addresses that associate the user with a specific service on a hosting AS. Such applications can be used within the IMS network or from outside it. PSIs can be used to support non-IMS sessions.

The PSI is similar to the IMPU concept that is applied to resources and enablers — the AS name can be associated at different times with different Private Service Identifier (internal server addresses) to suit network requirements, without having to change the globally routable published identifier. Like the IMPU, this makes it easy for anyone to request connection to this service AS. The Private Service ID is similar to the User Private ID. It is allocated by the operator and takes the form of an NAI.

The PSI consists of the domain it belongs to and an allocated unique address within this domain. It can be identified as a unique identifier or as a wildcard PSI, but there is always a single Private PSI for all the wildcard Public PSI values. Both Private Service Identity and Public Service Identity are regarded as permanent data that is stored in the HSS and downloaded to the assigned S-CSCF.

Further information about routing to PSI and its session control is found in Chapter 3.

5.5.6.2 PSI Provisioning

PSIs can be created, modified, and deleted in the HSS by the operator via O&M mechanisms. PSIs can also be created and deleted by users using the Ut interface between the AS and the UE, which also may transfer an XML document with the service configuration.

Subdomains can be created to isolate the PSI servers from the trusted zone of IMS servers, because access to these PSI servers is often open to external users and is handled without the constraints and protection of the IMS session control.

5.5.6.3 PSI Identifier Formats

The PSI format can take the form of the SIP URI or Tel URI within the form of a NAI. The domain part is predefined by the IMS operators and the IMS system provides the flexibility to create the user part of the PSIs dynamically.

The PSI can be expressed as a distinct PSI or as a group of PSIs with a wildcard identifier. A distinct PSI contains a globally routable name of a specific server platform, while the wildcard format allows grouping of a collection of PSI names according to wildcard rules. The grouping of PSIs enables better network management of these servers, for operations and maintenance purposes. A group PSI is expressed with a wildcard that contains a delimited Regular Expression. This wildcard is inserted in the "userinfo" portion of the SIP URI or in the telephone-subscriber portion of the Tel URI. Where the destination is expressed as a wildcard, the actual address must be resolved by a local DNS lookup or similar network management facility.

5.5.6.4 Originating PSI Services

The PSI information for originating services is permanent data and is stored in the HSS and downloaded to the assigned S-CSCF. Users who need to use a service hosted by a PSI application server will have their subscription profiles changed to include the PSI in their list of service Filter Criteria, in the same way as for other applications. Therefore, the PSI can also be used as a service to originating callers, prioritized in a predetermined sequence.

5.5.6.5 Terminating PSI Services

For terminating PSI services, there are three ways to access the PSI hosted services:

1. *Via S-CSCF*: The PSI is created on the HSS as a special "PSI-User" with an assigned S-CSCF address and ISC Filter Criteria information. In this scenario, when the I-CSCF receives a SIP request destined to the PSI name, it queries the HSS owning the "PSI-User" and directs the request to the assigned S-CSCF. The S-CSCF invokes the service by connecting the calling party to the PSI server.
2. *Via HSS*: The HSS maintains the address and routing information for the PSI AS itself, not the S-CSCF. In this scenario, when the I-CSCF queries the HSS, it retrieves the actual address of the AS hosting the PSI service in the location query response. Then the I-CSCF can forward the request directly to the AS hosting the PSI, without the help of an S-CSCF. In this way, the PSI can use HSS service data for non-IMS communications.
3. *Via DNS:* For users outside the Home Network of the PSI service, the I-CSCF needs to resolve the domain name given as part of the PSI. The DNS inquiry, in this scenario, returns the full address of the PSI server, so the session can be connected directly to the PSI AS without involving the S-CSCF or the Home HSS. This scenario permits connection to the PSI AS when the user is not an IMS subscriber, and avoids opening the HSS to external parties.

5.5.6.6 PSI Data

The PSI data associates "PSI User" details with a PSI service hosted on a PSI AS. This is a particular instance of a service, such as Conferencing or Instant Messaging that has been assigned to a particular hosting server.

The table below shows the data required for this facility:

Function	Parameter	HSS	S-CSCF	AS
User Identity	Private Service Identity	Y	Y	—
	Public Service Identity	Y	Y	Y
	Display Name	Y	Y	—
Service Profile	Initial Filter Criteria	Y	Y	—
	Shared iFC Set Identifier	Y	Y	
	Services related to Unregistered State	Y	—	—
	Subscribed Media Profile Identifier	Y	Y	—
Routing Data	S-CSCF Name and AS Name	Y	—	—
	Diameter Client Address of S-CSCF	Y	—	—
	Diameter Server Address of HSS	—	Y	Y
Service Data	Registration Status	Y	—	—
	Server Capabilities	Y	—	—
	PSI Activation State	Y		Y
	Application Server Information	Y	Y	—
	Service Indication	Y	—	Y
	Transparent Data	Y		Y
Charging	Primary Event Charging Function Name	Y	Y	—
	Secondary Event Charging Function Name	Y	Y	—
	Primary Charging Collection Function Name	Y	Y	—
	Secondary Charging Collection Function Name	Y	Y	—

Source: Based on 3GPP TS 23008.

5.5.7 WLAN Identifiers

5.5.7.1 Root NAI for 3GPP Interworking WLAN

Root NAI is in the standard format for NAI that is defined as username@realm. The Home Network realm is in the format of Internet domain name (e.g., "mynetwork.com"). The WLAN UE can derive the Home Network domain name, when it is not preconfigured as such, from the IMSI, using the Country Code and the Operator Network Code.

The username also can be derived from the IMSI and used for either authentication method, EAP SIM or EAP AKA, where the Subscriber Number is preceded by "0" for EAP AKA authentication and by "1" for EAP SIM authentication. The format (in the case of AKA) is as follows:

"1<IMSI>@wlan.mnc<MNC>.mcc<MCC>.3gppnetwork.org"

5.5.7.2 Decorated NAI

The Decorated NAI format allows transferring information necessary for roaming. The format allows for the Home realm to be sent with the username.

A Decorated NAI is constructed in the form of an NAI but with the format of "homerealm!username@otherrealm." The "homerealm" is the Home Network for the user, while "otherrealm" is the Visited Network identifier.

The Decorated NAI format for EAP AKA authentication is built in the following format (for EAP AKA):

> "wlan.mnc<HomeMNC>.mcc<HomeMCC>.3gppnetwork.org !1<IMSI>@wlan.
> mnc<visitedMNC>.mcc<visitedMCC>.3gppnetwork.org"

5.5.8 Summary of Data Retrieval by Identifier

The following table provides a summary of the identifiers used for extracting data stored on various elements in 3GPP/WLAN networks:

Data Item	From:	IMSI	MSISDN	TMSI	P-TMSI	IMPU	IMPI	PSI	GRUU	IMRN
IMS Subscription	HSS					Y	Y			
PSI Service Data	HSS							Y		
IMS Roaming number	HSS									Y
GRUU UE/IMPU	HSS					Y			Y	
MS Subscriber	HSS	Y	Y							
	VLR	Y		Y						
	SGSN	Y			Y					
	GGSN	Y								
	AAA Server	Y	Y							
	AAA Proxy		Y							
	WAG		Y							
	PDG	Y	Y							
	WLAN UE	Y								

Source: Based on 3GPP TS 23008.

Note that some servers accept more than one identifier type; for example, IMSI and MSISDN are both used to retrieve data from the HSS, the AAA Server, and the PDG.

5.6 Server Profiles and Triggers

5.6.1 IMS Service Profiles and Triggering Data

5.6.1.1 Service Profiles Associations

The IMS Service Profile is a collection of data for services for a particular user. Multiple Service Profiles can be defined in the HSS for a subscription. The relationship of the Service Profile to the user ID is independent of the Implicit Registration Set that can exist. For example, Public User Identities with different Service Profiles can belong to the same Implicit Registration Set. Each Public User Identity is associated with one and only one Service Profile, but each Service Profile can be associated with one or more Public User Identities.

Figure 5.18 maps out associations between service packages (Service Profiles) and the IMPU. A single Service Profile can be shared across several Public Identities (IMPUs) if desired. This means that this service set can operate in the same manner on several types of equipment or for multiple different users. In this case, the IMPUs are said to be "alias."

A set of filter criteria can form a service profile, which is a package of services with a pre-determined order of priorities. Such service profiles are associated with each user. Multiple users can then share the same service profile. The service profile identifier is stored in each subscriber record that shares the same service profile, thus reducing the volume of data per user record.

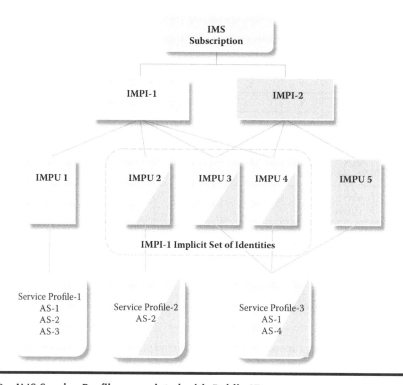

Figure 5.18 IMS Service Profiles associated with Public IDs.

5.6.1.2 Service Data Contents

Data contained in the filter criteria includes the following information:

- AS addresses or logical names to resolve into addresses
- AS priority (i.e., order of priority in the sequence in which the criteria is applied)
- Trigger Points and Optional Service Information
- Subscribed Media
- One or more SPT Filter Criteria linked by means of logical expressions (AND, OR, NOT)
- Default Handling and Error Handling
- Possibly additional Service Information to be sent to the AS (e.g., the IMSI for the IM-SSF)

5.6.1.3 Types of Filter Criteria

The HSS contains data per service per user that enables activation of services at special trigger points and in a predetermined sequence, as is necessary for their appropriate execution. Filter criteria is the information that an S-CSCF receives from the HSS or an AS, which defines conditions for service invocation.

Each set of Filter Criteria includes the AS address, application precedence, default handling, subscribed media, trigger points, and optional service information.

There are two types of Filter Criteria:

1. *iFC* are the conditions and parameters for invoking applications on the S-CSCF. These filters are stored in the HSS as part of the user profile and are downloaded to the S-CSCF upon user registration. The iFC are stored for each user, for each application that the user request might invoke.
2. *sFC* are downloaded from the AS (SIP AS, OSA SCS, or IM-SSF) directly to the S-CSCF when initiating the application. They affect this application logic flow and allow for dynamic definition of the relevant trigger points during the execution of that application.

For a PSI that is routed according to basic IMS routing principles, iFC are mandatory in order to route toward the AS hosting the PSI.

For greater efficiency, the iFC can be shared across groups or classes of IMS subscribers, as a shared service profile. The set of shared iFC is marked by an identifier stored in the user profile. The S-CSCF obtains it from the HSS (or caches it) and interprets the shared iFC into a set of application triggers, according to the operator's configuration.

For a description of processing AS triggering, see Chapter 3.

5.7 Service and Service Identities

5.7.1 Access Point Name (APN)

In the process of generalizing access networks, data for different access networks must be retained. The concept of the Access Point can cover a node in a particular access network, or the access server within a multi-access node. The name can refer to alternative instances of the function on different servers. For example, in GPRS, an APN can be a particular GGSN; or in WLAN, the W-APN corresponds to a PDG.

The APN identifies a service by associating it to an FQDN (Fully Qualified Domain Name) and a server specific address or by using labels and reserved labels that identify that server. The operator may allow use of wildcard values for the APN if the UE has a choice of servers to which it can attach. The APN contains:

- Network Identifier
- Operator Identifier within that network

An APN can be constructed of one or several "labels," using a dot as a separator (e.g., APN = "label1.label2.label3"). This enables the use of more meaningful names.

5.7.2 IMS Support of Local Service Numbers

Local Service Numbers are traditionally telephone numbers not in the international format, and are normally in the form of a short code. Such numbers can allow access to local services in the Visited Network. These local services can be, for example, directory inquiry, obtaining last incoming call number, reporting service faults, and network-based voicemail.

The requirement to support Local Service Numbers arises from their current widespread use in CSNs and the need for seamless interworking across traditional networks and the new packet-based networks.

In IMS, short numbers in a session request must be accompanied by a flag to indicate that a Local Service Number is used. This is similar to the flag denoting numbers of a private plan (e.g., in an Enterprise numbering plan). This indication is included in the Request URI of the SIP request.

The Local Service Number must be translated to a globally routable number format or a PSI (Public Service ID) URI. The S-CSCF routes the SIP request toward the Home Network Application Server, based on filter criteria that are triggered by the information in the "local indication" received from the UE.

The AS in the Home domain expands the short code to a routable address and passes the session request back to the S-CSCF with the Request URI that contains either a globally routable SIP URI or a Tel URI with a number in the international format. The message is then routed to the network hosting the required service.

The SIP request must contain information to enable the hosting network to identify the intended service (usually by the dialed short service code). Roaming users can be served by the Home Network special services even for services that are relevant to their location.

5.8 Routing Data Support Elements

5.8.1 Routing Database

5.8.1.1 Internal Routing Tables

In traditional Fixed Networks, the routing tables determine routes to each line using number analysis and algorithms to improve performance and avoid having to keep large tables with a route per line. This is no longer necessary in IMS because subscribers are not permanently residing on preconfigured servers, but are assigned a serving session controller only at registration time. This still means that the address of the subscriber's current S-CSCF must be fetched from the central database, the HSS. Assigning processing servers dynamically, when needed rather than permanently, is advantageous in growing networks because it is flexible and scalable. In contrast, static routing tables require a high maintenance level.

In a PSTN Emulation system, routing tables are still used for Call Servers that control access of legacy terminals, usually using MGCP or H.248 protocols. However, IMS-based PSTN Emulation, via an Access Gateway Controller Function that performs the SIP to H.248 conversion, allows users to be registered on IMS as normal. Session routing information is then based on the address of the access gateway. This provides the mobility aspects of a dynamic registration to Fixed Networks' nomadic users.

5.8.2   Electronic Numbering Mapping (ENUM)

The S-CSCF supports translation of E.164 format addresses contained in a Request-URI (in the "Tel:" format) to a routable SIP URI using ENUM and DNS facilities.

Although the ENUM conversion algorithm is well known, the standards currently do not specify the interfaces to the ENUM or DNS database. ENUM and DNS can be a public service open to any operator, but many operators choose to maintain an internal facility. When an internal ENUM or DNS servers are maintained, there is no need to adhere to the Internet top-level domain name of "e164.arpa" and an alternative proprietary name can be configured.

ENUM is instrumental in routing data, Number Portability, and Carrier Pre-Selection. Addresses in the packet network can be generated from a telephone number using the ENUM algorithm. This was conceived as a way of achieving unique addressing when migrating from telephone numbers to SIP URIs and IP addresses. It can be applied to dialed numbers in the CNS or when Tel-URI is entered as a destination address.

ENUM is a set of procedures that allows interworking and interpreting the Telephony E.164 numbering format with Internet IP addressing schemes. This involves a conversion formula to generate an IP address from the E.164 number and lookup tables to obtain NAPTR (Naming Authority Pointer Resource) Resource Records from a DNS server.

The ENUM algorithm reverses the number and inserts dot delimiters to get an IP address, with the Country Code in the least significant position. This is appended with the string of "e164 arpa," which is a domain name specially created for public service ENUM but could be any other convention for internal network use.

The ENUM conversion algorithm is illustrated here:

Algorithm Step	Result
The given telephone number:	+449876123456
The number is reversed:	654321678944
Dot delimiters are inserted:	6.5.4.3.2.1.6.7.8.9.4.4
Top level domain name added:	6.5.4.3.2.1.6.7.8.9.4.4.e164.arpa

The assumption is that having created an IP address via ENUM, the session can be routed directly there, as in peer-to-peer communication. However, algorithms that are more complex are required to cope with circumstances where PSTN does not behave in the same manner in interpreting the numbers or in choosing how to route the call.

Networks with wholesale and transit services choose the next hop, depending on the subscriber ID and the service provider or the enterprise from which the call has originated. It also depends on where and when the session is initiated, that is, where "Least-Cost-Routing" is deployed. A further consideration is the way to optimize network routing to minimize the transitions between IP and CS networks. For these reasons, straight ENUM conversion is not necessarily useful but needs to be supplemented by additional rules that are operator specific.

5.8.2.1 DNS Database

The global DNS system consists of a hierarchical set of DNS servers. Each domain or subdomain has one or more authoritative DNS servers that publish information about their own domain and the identities of DNS servers for any subdomains "beneath" them.

At the top of the hierarchy there are root servers that can be queried for top-level domain names. Within the root domains, there are groups or "zones" of domains — for example, country-level zones. The search of DNS servers follows this structure to find the relevant user server address.

The first port of call for user inquiries is the nominated DNS server provided by the service provider. This DNS server is usually configured either manually or via a DHCP server. This DNS "Resolver" usually has a cache containing recent lookups to speed up popular domain name resolution so that it needs only to search the global DNS structure if there is no entry for the particular domain name.

There are several types of DNS inquiries:

- *NAPTR.* This is the NAPTR DNS Resource Record type of query. To provide scalability, load balancing, and offline maintenance facilities, a group of servers can match an identifier on the DNS database. To manage multiple results for a server search, the use of regular expression rules has been added. The search can be made using Regular Expressions, which define the matching criteria, expressed by standard delimiters and special search logic.
- *SRV Query.* An SRV record (Service record) is a category of data returning information on available services, as defined in RFC 2782. The results of the DNS SRV inquiry can provide a list of alternative servers with associated priority levels. Clients, receiving multiple answers, use the SRV record with the lowest priority value first and only fall back on other records if the connection with this host fails.
- *E2U.* The result of an ENUM algorithm is an address containing the "e164.arpa" domain name. This is sent to the DNS to be resolved into a server address. This query is flagged to the DNS as type "E2U," that is, requesting E.164 number resolution into a URL.
- *NPDB.* A DNS query can be set to initiate a Number Portability lookup (see below) to determine the final destination at this point, rather than first attempt to deliver the session.

5.8.3 Number Portability Database (NPDB)

NP service is defined by regulatory bodies in many countries to make the competitive second operator's offerings more attractive to users. NP is mandated for Fixed Line networks as well as for Mobile networks, and now also for broadband-based IP networks.

NP enables users to move to another service provider without changing their number. This is considered a measure that supports competition because most users prefer to keep their familiar numbers when opting to use another service provider.

Because the telephone number includes the network identification, the "donor" network still receives the call request but it must forward it to the recipient network. In effect, NP makes subscribers numbers independent of the Home Network or the Serving Network, at the expense of the donor.

Two different solutions are defined to support NP:

1. *Intelligent Network (IN)-based solution.* The Number Portability Database (NPDB) is maintained by the network operator for the ported-out and ported-in subscriptions. The NPDB keeps the association of the user subscription number with the new network, within a portability zone or cluster (e.g., within a country) that permits portability. The IN server responds to NP inquiries based on the user subscription number and returns the new network identifier. In some territories, the IN NP service is countrywide, shared and maintained by all operators.
2. *Signaling Relay-based solution.* The Signaling Relay Function is based on internal routing data rather than an independent external database. This is performed by routing to the would-be MSC or PSTN local exchange, obtaining the new route information and then relaying the information to the destination network. In GSM, the location is stored in the donor network's HLR but also conveyed to the recipient network.

NP in IMS can follow the same two options. Ported numbers must be recognized for routing to Mobile 2G networks and PSTN networks, as well as within the new IMS domains. In many scenarios, a session initiation process requires DNS resolution or ENUM conversion. Therefore, it may be expedient to associate any address query with an NPDB lookup so that the ENUM or DNS query will automatically spawn an NPDB query, in case the number is ported.

Chapter 6

Policy and Quality-of-Service Control

6.1 This Chapter

This chapter describes policy decision management, resource allocation control, and quality-of-service management. First, this chapter explores general concepts, measurement and control mechanisms of QoS, as deployed in CSNs as well as in IT and on the Internet. It also explains how QoS, resource management and QoS network policies can be enforced in the IP bearer, at service level and at bearer level.

Based on these concepts, the IMS QoS principles are introduced. These concepts are largely shared by Mobile and Fixed Networks, including considerations for roaming and end-to-end QoS across multiple networks. Finally, this chapter provides a detailed description of IMS QoS management according to the 3GPP and then according to TISPAN, describing their different terminologies and separation of functions.

Note that the standards for Fixed and Mobile QoS management have not merged entirely and functions are not wholly integrated. This may be due to disparity in bearer handling, and using different parameters and values. Hence, the different function names and reference points are explained here separately.

6.2 Principles for QoS

6.2.1 Definitions of QoS

6.2.1.1 Types of QoS

QoS (Quality of Service) can mean different things. QoS can be defined as "the measure of the degree of user satisfaction of the user of the system." This suggests a rather subjective approach that depends on the user's perception. However, in telecom traffic engineering, this term refers to measurable and qualitative techniques that enable to select, control, predict and measure the level of QoS and to guarantee a predictable behavior.

QoS must not be confused with performance — it is not about high throughput. It is also not the same as resilience or even service availability. QoS in telecom covers a multitude of measurable attributes of media flows.

Dynamic QoS control, in QoS-aware service delivery, can be served in two ways:

1. *Guaranteed QoS:* service delivery that defines the bounds of QoS parameters, such as throughput, latency, jitter, and packet loss. This is achieved by strict admission control in the access network and enforcement via throughput control and traffic policing. This requires management of resources, such as available bandwidth. The enforcement of throughput control and traffic policing is performed in the IP edge, and the CPE or Access Node. It can entail rejection of new call attempts when the resources are exhausted, so that existing sessions will not deteriorate.
2. *Relative QoS:* this model implies traffic class differentiation by applying QoS mechanisms such as prioritizing packets (using RSVP and DiffServ), so that higher QoS is achieved for higher priority services. QoS differentiation is performed at the IP edge, using the DiffServ Edge functionality as defined in IETF specifications for Differentiated Services or a similar technique. The CPE also may be able to provide QoS differentiation (e.g., by applying DiffServ marking).

6.2.1.2 Timing of Establishing QoS Levels

In a network that distinguishes between QoS levels per service, the desired level of QoS must be requested, and if resources are available, this QoS request will be granted. This process occurs at the following points in time:

- *In advance.* Where there is a prior agreement for a class of users (e.g., users belonging to a particular service provider), the level of service is agreed according several criteria. The network must reserve enough resources to allow for the appropriate execution of the SLA.
- *Dynamically at call setup.* Negotiation of resources for a session takes place dynamically at session establishment time. Each session is assessed individually for requirements of QoS and resource availability.
- *At mid-call, responding to events.* In multimedia services, mid-call events may require a change of media type (e.g., video instead of voice), addition of a media flow (e.g., a third party to the session), or the release of a media flow. The session controller is notified of such changes by user-initiated events. These events cause a reevaluation of the QoS requirements.

6.2.1.3 Setting QoS in Service Level Agreement (SLA)

SLA is a generic term for agreement to perform a service under a set of performance criteria. It entails taking steps to ensure that the network can accommodate the SLA, and measuring the agreed upon performance parameters to report adherence to the SLA.

In computing, an SLA is often signed between a managed network service provider and a corporate customer. In a telecom, Fixed Network operators have contracted with their corporate clients to assure them of a level of network connection availability and reliability, service availability and ubiquity, throughput of calls, connectivity extents, etc. Such agreements are also made between service providers and the hosting network carrier.

An SLA in packet networks is likely to be an important mechanism of specifying the level of service contracted between two parties. QoS classification plays a major role in such agreements where commercial guarantees are given. While previously an SLA was managed retroactively by reviewing statistics, there is now a demand for proactive management of the resources according to the SLAs in force. This necessitates installing a measurable QoS management system that is enforced by the "Traffic Plane" according to the service QoS profile.

6.2.1.4   Resource Allocation

Resource allocation is often considered part of QoS. A resource allocation must be performed for every connection, while QoS management means allocating adequate resources to ensure that the agreed level of QoS is achieved.

The term "network resources" means anything on the network that is required to deliver a network service and that has a finite supply. This includes not only the bandwidth, the bit rate on a particular connection, but also hardware and virtual ports, including specialized ports that can be assigned to certain protocols. The term "resources" can also refer to the pool of IP addresses that are globally routable and can be temporarily assigned to a far greater number of inactive users behind a NAT gateway.

6.2.2   QoS Networks and Non-QoS Networks

6.2.2.1   Best-Effort Networks

The quality of transmission of packets is impaired when the packets do not arrive in sequence, arrive late, or do not arrive at all. This is caused by network congestion. Where there is large enough capacity with spare headroom, or where excess of session admission is prevented, congestion problems will not occur.

Best-Effort network service, by definition, means that there are no guarantees of delivery except that the best effort will be made to deliver. There is no filtering of service request because there is no notion of exhausting finite resources, and there is no resource reservation or allocation.

Unlike call sessions in CSN, Best-Effort Networks cannot protect existing sessions when the congestion level is raised during the call. Such networks are "non-blocking" — they allow sessions in even when the network is overloaded. The QoS level of ongoing sessions in this case will deteriorate accordingly.

In a blocking network, the protocol used (X25 or ATM) rejects session requests when a threshold of used bandwidth has been reached. The circuit that was reserved for the session remains intact, and the call is not interrupted or degraded with a surge of new session requests.

The Internet thrives on the simplicity of routers that implement the Best-Effort principles, while X25 and ATM are deemed too expensive and too complicated. Equally, UDP which has no error correction is used in preference to TCP, because it is more permissive and flexible. By excluding reliability features such as acknowledging receipt or error correction, the Best-Effort Network is faster, more efficient, and less expensive. However, the "Best Effort" Internet service lacks predictability and QoS differentiation.

For real-time interactive sessions such as conversational voice and video sessions, good quality cannot be achieved without the ability to prioritize VoIP packets. The "Best-Effort" model can satisfy low expectations for free, or nearly free service; but for mainstream SLA-driven contracts, better service assurance is required.

6.2.2.2 Over-Provisioning for Non-QoS Networks

Best-Effort Networks can maintain a good quality of service if they are over-provisioned so that the bandwidth capacity within the network is greater than any demands on it from the access side or the network-to-network links. This method avoids elaborate QoS control and is relatively simple, but at a considerable cost for largely unused capacity. The Internet can provide good-quality VoIP while ISPs (Internet Service Providers) provide large capacity for any Internet traffic, not distinguishing between QoS-sensitive services (for example, user-to-user voice or video) and QoS-tolerant services (e.g., FTP file download or near-real-time messaging).

Ensuring that there is much higher capacity than actually used can get expensive. It is also inefficient because the required headroom is applied also to services that do not need it. Creating two tier networks — data network and voice/video network — may provide a solution, but detracts from the goal of a unified bearer and requires extensive network engineering and dedicated equipment. It has been estimated that a network would need about four times the capacity as provisioned headroom over a finely tuned QoS network to ensure adequate quality for the same traffic volumes. The exact amount of over-provisioning required to replace QoS depends on the number of users and their traffic demands, that is, service profile and volumes.

Over-provisioning works for packet networks that carry mostly non-time-critical application data, which tolerates irregular transmission rates, high latency and jitter. For example, file transfer (music download, Internet browsing) or applications with large buffering capabilities (streaming media) are not time-critical.

The over-provisioning method can be used as an interim strategy when voice and video services generate only light traffic and share the network with high volumes of non-real-time data. So far, this strategy has worked for VoIP when it is delivered to users who have low expectations of quality (e.g., when the VoIP service is "free").

6.2.2.3 QoS Considerations for the Internet

Not everyone believes in the need for QoS control in general, and especially for the Internet. There is a large lobby for the Internet style of traffic management, which is said to be more elastic and permissive.

The main reasons that QoS has not been introduced to the Internet include the following:

- QoS mechanisms are needed only when there is not enough capacity. All that is required is to increase bandwidth capacity.
- Increasing capacity may prove cheaper than building and maintaining QoS control.
- QoS only matters when the network operates at nearly full capacity, and thus the benefits are related only to the excessive traffic at near-congestia state.
- QoS management appears to be over-engineered and entails high management costs.
- Where QoS is achieved by prioritization of services, telco staff may not know the real customer's preference for prioritization, and thus may miss the ultimate goal.

■ Negotiation of end-to-end QoS with other network operators is fraught with difficulties. It is easier to provide a guaranteed pipe with a certain capacity and let the customer determine the service priorities.

6.2.3   QoS Attributes

6.2.3.1   Throughput and Bandwidth

The term "bandwidth" is widely associated with QoS to mean the capacity that needs managing to ensure a given QoS level. In general, bandwidth is the capacity of a particular network link or channel; that is, the physical channel capacity available to transmit bytes of data per unit of time. However, ensuring capacity for end-to-end transmission involves the available bandwidth at every hop, between routers and between network nodes, such as media gateways and border gateways.

Throughput in packet networks defines the ability to transmit bits per second (bps), where the transmission unit is a "bit" (1/8 of a byte). Throughput is measured by two main attributes:

1. Peak rate of bps, at the maximum level possible, according to the capability of the physical ports and speed of processing packets
2. Mean rate when packets are flowing at a steady, even rate, i.e., the mean value of gaps and bursts of packets

6.2.3.2   Lag Time and Latency

Communication lag time is the time taken for a packet, containing any type of media, to make the journey from the sending application to the receiving application. This time lag includes encoding the packet for transmission (i.e., packetizing), transiting across the network between the nodes, and the processing time at both the source and the destination. Latency is directly related to the distance between the network nodes.

Latency in a Packet-Switched Network is measured in two ways:

1. *One-way:* the time from the source sending a packet to the destination receiving it. To be more precise, this will measure start of transmission to start of receiving, excluding processing time at the destination.
2. *Round trip:* the time from the source to destination plus the return trip back to the source, excluding the processing time at the destination. Round-trip latency is easier to measure — it is done at the source point only, and therefore frequently quoted. Most networked nodes offer a "ping" service that allows the source point to measure round-trip latency.

In most networks, a packet will traverse several nodes, making several "hops." The minimum latency is then the sum of one-way latency plus the processing time (transmission delay) in each node.

6.2.3.3   QoS Issues Characteristics

The following are the characteristics of the bearer that affect the QoS:

■ *Dropped packets.* When a router in the network is handling high volumes of packets, the router buffers fill up and new packets cannot be accepted; that is, they are "dropped." The missed packets can belong to any session, resulting in quality deterioration but not necessarily in lost connection altogether.

◼ *Packet loss.* In Best-Effort Networks (using UDP), the missed packets are not recovered; that is, they are "lost." Other protocols (such as TCP) or the recipient application can apply error detection and correction methods, for example, check sum and retransmission to recover lost data packets. If dropped and packets are not recovered, excessive packet loss results in poor service. However, the process of retransmission can slow the stream of media received at the other end, causing distortion of the voice or video.

◼ *Latency.* The time it takes for packets to traverse the network before reaching the destination is the latency, or delay. This is a major factor in the perceived quality of the connection. The one-way latency between two network elements is primarily a function of the distance between them, but the transmission delay on the particular network element depends on the queue on this node and the speed it can process the packets.

◼ *Jitter.* In IP networks, packets can travel on different paths, even when they belong to the same media stream. This enhances resilience and performance but may cause variable delays for packets in the same stream. Jitter results from the variation in time delay between packets in the same stream, as they arrive in irregular time intervals.

◼ *Packets out-of-order.* Due to taking different routes traveling to the target node, packets that belong to a single media stream can incur variable delays and congestion. Therefore, packets arrive at the destination in a different sequence than the original sent media. Rearranging packets into the original sequence requires special techniques that restore the isochronous state. This is particularly important for video and audio media streams.

◼ *Transmission errors.* Errors can occur in processing packets, where packet content is corrupted, or packets are wrongly combined or misdirected. Such errors are detected by means of check sums, for example. Errors are corrected by retransmissions, if the protocol supports a self-correcting mechanism.

6.2.3.4 Congestion

Non-QoS networks are considered more elastic in their treatment of large volumes of traffic, while high-QoS networks are considered inelastic. Therefore, arguably, high-QoS networks lack flexibility and can reach a state of congestion sooner. Where there is very high congestion, a "congestion collapse" can occur; that is, the network functions well below its capability. Congestion collapse can be caused by the very mechanisms that enhance reliability. Congestion causes a retransmission of lost data, but a large number of retransmissions can overload the network even further to a point of a "meltdown," where the network starts to under-perform and the throughput is well below its normal capacity.

To avoid congestion collapse in QoS networks, session requests that exceed network bandwidth capacity must be rejected. This is preferable to dropping packets in existing sessions, which causes QoS degradation and breaches of the SLA.

Traffic shaping, using algorithms such as Token Bucket or Leaky Bucket, helps reduce the dangers of congestion in a non-QoS network by allowing in only what the network can cope with, at the expense of dropping excessive packets. However, this can decrease their ability to carry media for VoIP or IPTV applications, which need a constant bit rate, and have low tolerance to packet loss, latency, and jitter.

Congestion collapse can be avoided by disciplined behavior of the endpoint software client and the network routers. The endpoint can still request repetition of the transmission of missed packets but it does so more slowly, giving the network time to recover gracefully. When routers encounter an imminent congestion collapse, they begin to drop packets at random (using RED, the "Random Early Detection" procedure) to reduce the pressure on the network. Other methods include various types of queuing, such as "fair queuing." The fair queuing procedure buffers packets for certain destinations in separate queues, and sends out the packets from queues nearest to an estimated finishing time of delivery, thus allowing for urgent delivery of packets to occur first and reducing the user's perception of a delay.

6.2.3.5 Reliability and Data Integrity

Quality means, among other factors, that the data is reliably transmitted, with the correct contents in the original sequence and without any duplication, omission, or corruption. Some protocols used for the transmission are congestion-aware while others need additional mechanisms.

TCP was enhanced (in 1988) to support methods of avoiding congestion but UDP is not capable of doing so. Real-time transmission of voice and streaming of media has low tolerance for packet loss, and therefore necessitates a different type of QoS control, at the service level.

Data can be transmitted with full error detection and correction, causing retransmission when errors have been detected. This high-integrity method has some drawbacks — for example, retransmissions slow the response time, even when the meaning of the data can still be understood despite the errors. Retransmissions can also swamp the network that is reaching its top capacity, and therefore drive the network into congestion collapse. This means that maximum data integrity is not necessarily what should be chosen for every type of service.

The table below describes classes of reliability (Maximum/Medium/Minimum) for typical services with the shown reliability characteristics:

Required Data Integrity Profiles	Example Services	Error Sensitivity	Error Tolerance	Lost Packets Tolerance	Duplicate Packets Tolerance	Out-of-Sequence Packets Tolerance	Corrupt Packets Tolerance
Maximum	Voice, conversational video	High	Limited	Low	Low	Low	Low
Medium	Interactive data services	High	Good	Medium	Medium	Medium	Medium
Minimum	Non-interactive data services	Low	Very good	High	Medium	Medium	High

Source: Compiled from multiple sources.

Note that deploying error correction may be inherent for the type of service or by choice.

6.2.3.6 Media Transmission Handling

The following are some methods of handling media:

■ *Buffering.* Applications that use large buffers can alleviate QoS issues and deliver a smooth media flow to the recipient. The buffer fills at irregular rates of input but provides a steady stream of output media. The application, not the network, is then providing a level of QoS.

However, buffering creates a time lag between originating and receiving the packet. If this time lag is greater than a few seconds, it will not be acceptable for voice conversation or videoconferencing.

■ *Streaming media.* Streaming of media means that media is sent as a continuous stream from the originating source and the recipient end user continues to receive it. This is in contrast to the delivery of media that is based on short bursts, such as speech communication. Streaming media is typically used for sending digitized video or music over IP networks.

■ *Cascading with BitTorrent.* The BitTorrent procedure is a cost-efficient way of distributing the same media to a large number of end users. It allows cascading of transmitted data, where recipient nodes perform further transmission to other parties. Thus, the originator does not service each and every user but relies on other network nodes to complete the distribution of the media. In addition, packets do not overload the same path, because they take many more routes to reach all the destinations. In essence, BitTorrent is a facility to broadcast data to a large audience, similar to broadcasting over the air. This can be used for Internet events, where mass downloading of data occurs at a specific time.

6.2.4   User Perception of QoS

6.2.4.1   Perceived Quality Issues

User experience is the ultimate yardstick for successful real-time interactive communication. The human perception of QoS is affected by a diverse range of defects, faults, and annoying features.

■ For audio:
  – Response time from dial to connect
  – Echo
  – Interrupts
  – Crossed line, cross-talk
  – Noise on the line ("white" noise, buzzing, etc.)
  – Too loud or too quiet speech
  – Stutter effect, pauses, breaks in sound
■ For video:
  – Broken movement
  – Frozen picture
  – Untrue colors

6.2.4.2   Measuring User Perception: The Mean Opinion Score (MOS)

In the effort to quantify QoS, a way of capturing user experience is sought. This is particularly needed to assess the differences between different compression rates and codecs. Because human perception is highly subjective, the method must attempt to clarify what the level of quality means.

To arrive at a consensus, the perception of levels of quality is recorded by a number of testers and averaged out to give a single rate. As with all such statistics, the "mean" rate should be averaged out over a sufficiently large sample.

The ITU-T has published a recommendation (P.800) for a structure of the measurement tool. This is the Mean Opinion Score (MOS), based on the concepts and algorithms of the previously defined "Perceptual Evaluation of Speech Quality" and the "Perceptual Analysis Measurement."

MOS is expressed as a single number in the range of 1 to 5, where 1 is lowest perceived quality and 5 is the highest perceived quality, but 4 is regarded as acceptable.

MOS defines the level of QoS, as follows:

MOS	Quality	Level of Impairment
5	Excellent	Imperceptible
4	Good	Perceptible but not annoying
3	Fair	Slightly annoying
2	Poor	Annoying
1	Bad	Very annoying

Source: Based on ITU-T recommendation P.800.

Using the MOS tool, responses from numerous users are collected and the mean value is calculated for each type of transmission tested. MOS is used to assess the effect of codecs (compressors/decompressors) and DSP (Digital Signal Processing) used for media transmission, to find how best they can be configured to conserve bandwidth with least degradation of QoS.

The following table is an example of measurement results using MOS (1 = worst, 5 = best) for a number of codecs:

Codec	Data Rate k (bps)	MOS
G.711	64	4.1
G.729	8	3.92
G.729a	8	3.7
G.726	8	3.85
G.723.1	6.3	3.9
G.723.1	5.3	3.65

Sources: Compiled from multiple sources.

The table summarizes a user's experience for G.711 codec as best, but the compression rate is low, and therefore throughput is not as effective. Looking at G.729, even at the highest throughput of 8 K bps, it is still achieving second-best MOS, while G723.1 is the lowest throughput and only medium MOS. This example demonstrates how codecs can be evaluated from a QoS perception point of view.

Although MOS measures only subjective assessment of QoS, proponents of MOS claim that other methods, attempting to produce user-independent automated measurement system, cannot provide adequate evaluation of QoS because it is, in fact, a matter of human perception.

6.2.5 Classifying Services

6.2.5.1 Class of Service

Classifying services according to their requirements for QoS simplifies resource negotiation at the time of session establishment. Several classification methods exist in different networks. This means that mapping the Class of Service must occur at border points between domains and different carriers' networks.

In general, to identify the Class of Service for a specific service, the network's switches and routers examine the call based on several factors, including the technology used; type of media (text, voice, media); level of precedence; and requirements for real-time response versus *near*-real-time.

The following are factors in defining the Class of Service:

- The type of service
- Priority due to precedence (e.g., emergency service)
- The identity of the initiating party
- The identity of the recipient party
- The path and involved IP-CANs

6.2.5.2 Grade of Service

In CSNs, calls are "gapped" (i.e., delayed) or blocked (i.e., rejected) to ensure free-flowing traffic. In broadband networks, similar techniques are deployed to protect the network and avoid congestion. IP data services deploy QoS at the packet level of granularity, but managing call attempts is applied at the service level (e.g., IMS session and application control).

"Grade of service" is a method of measuring the health of the network by assessing the probability ratio of calls to encounter a certain threshold level of delay or outright rejection during the "busy hour" when the traffic is at its highest. This ratio can give different results for the incoming leg or outgoing leg, and between different combinations of originations and destinations.

When a user attempts to make a telephone call, the routing equipment handling the call has to determine whether to accept the call, reroute the call to alternative equipment, or reject the call entirely. By rejecting calls when the level of congestion dictates it, the network is protected from a complete "meltdown." By delaying, more time is given for calls to proceed.

The "Grade of Service" can be measured according to lost calls and delayed calls:

- Measuring by lost calls calculates the proportion of calls lost due to congestion in the busy hour (i.e., during the highest load condition). This is calculated as:

Grade of service = Number of lost calls/Number of calls offered

- Calculating Grade of Service by delay provides:
  - Average waiting time for a connection for delayed calls
  - Average waiting time including calls that are not delayed
  - Ratio of delays over a configurable time delay variable

The Grade of Service is a useful tool for network management, applying it to different sections of a network. It can be applied between the access point and the serving core node, or end-to-end, crossing several nodes. This tool helps fine-tune the configurable thresholds and time delays at various points of the network.

6.2.5.3 Prioritizing by Type of Service

In Packet Networks, QoS is an important issue, as packets can take variable paths and get reconstituted back into speech, video, image, or text at the other end, in a coherent manner. Real-time interactive voice services demand a higher level of QoS to reassemble the correct packets fast enough for the human hearing experience.

Video-based services require large volumes of packets to traverse the packet network and still arrive promptly to reflect smooth movement and changing colors. While streaming video can tolerate small delays due to buffering, user-to-user interactive videoconferencing is highly demanding. Requirements for QoS levels therefore depend on the type of service.

6.2.6 Enforcing QoS at the Bearer

6.2.6.1 Resource Management

Call Admission Control (CAC) has been used for voice call admission in connection-based links such as ATM. CAC is associated with connection-oriented resource requirements with standard media attributes and assured connection between the two ends. With connectionless communication, using protocols such as IP and UDP, the media requirements are varied and the actual route is not predefined; therefore, a mechanism such as RSVP (Resource reSerVation Protocol) is needed to reserve resources along the path of the packets.

Resource management is a function of session control that must be performed as part of network admission and service authorization, whether there is QoS management in place or not. Resource management for Best-Effort connection involves only allocating a resource to start the session, with no guarantees to maintain it and ensure adequate quality. However, "resource management" is often used as a synonym for QoS management because it is instrumental in realizing the required session QoS Profile.

Resource management functions include:

- Reservation of resources
- Commit resources
- Reservation and Commit in one step
- Refresh reservation of resources
- Modification of resources
- Release unused reserved resources
- Event notifications of resources changes

Assured QoS is the process that confirms the ability to support the level of QoS end-to-end, and the process of reservation of suitable resources. Reserving resources and allocating them to a particular session ensures that this particular session will not suffer from QoS problems that are often caused by insufficient network resources, such as bandwidth and media processing.

Resource reservation for "assured QoS" may not be required in the following scenarios:

- The originating UE needs to confirm the fulfillment of pre-conditions relating to the resource requirement end-to-end, up to the destination party. However, if reservation is not required all the way (e.g., with a PSTN connection), only local IP-CAN Assured QoS is required.
- Although the process of Assured QoS is utilized, the particular session does not require end-to-end confirmation; therefore, both endpoints report that pre-conditions are fulfilled with no further action.
- The session does not require reservation and the endpoints do not use the mechanism of Assured QoS.

The process of reserving resources entails communicating to media bearing servers requests to set aside resources. This can be accomplished using RSVP and DiffServ Edge Protocols. To appreciate how this is performed, see the descriptions of these protocols in Appendix A.

A set of restrictions is defined as a pre-condition for the session resourcing during the session admission control process. These pre-conditions must be met across the path up to the target recipient. A minimum QoS requirement per direction (uplink or downlink) is defined to enable initial local access. The Reservation is performed within network segments. Endpoints initiate resource allocation in their local IP-CAN. The endpoint suggests the desired QoS level but this must be negotiated with every network segment, through a multi-step dialog, until agreement is reached.

Assured QoS, by its nature, means that when there are no sufficient resources, the session will be rejected. When the session is rejected, depending on the reason, the UE logic can lower the QoS requirement, and try again after a short time interval or notify the user to do so.

6.2.6.2   Policy Enforcement Functions (PEF)

The PEF is contained within the IP-CAN nodes, that is, in the GSN nodes or BRAS for DSL or other access nodes. PEF also performs QoS control in CN bearer nodes and in the border control across to other networks.

As policies for charging and policies for resourcing tend to be related and are processed in the same way, later standard definitions have combined charging rules enforcement with resource management. The combined charging and resourcing enforcement function is the PCEF. The role of the PCEF is to police packet flow according to Class of Service or QoS markers. This entails "opening" or "closing" the gate to packets of certain flows. Only session media flows that are authorized and allocated for this session will be allowed through in the bearer network nodes. In addition, packets will be prioritized according to the given rules and Class of Service.

6.2.6.3   Media Flow Management

6.2.6.3.1   Shaping Traffic

Traffic shaping, or traffic engineering, is the means taken to control the media flows and engineer a network that optimizes the use of its capacity (bandwidth). Such means include:

- Service classification
- Packet prioritization

- Routing optimization
- Queue discipline
- Enforcing network-wide policies
- Congestion management
- Buffering and gapping bursts of traffic
- Rate-limiting channel thresholds

Traffic shaping deploys many mechanisms on different network elements to achieve its aims. Such techniques include smoothing out bursty traffic using "Token Bucket" and "Leaky Bucket," as described in the next section. Other techniques are based on separating the flows and prioritizing them according to network policies.

6.2.6.3.2 Achieving Steady Rate

Regulating traffic peaks and troughs into an even flow is one of the aims of traffic shaping. This reduces the occurrence of unwarranted delays and jitter. There are two popular methods of shaping lumpy traffic: (1) Token Bucket and (2) Leaky Bucket. Despite the similar names, their effects are different: while the Leaky Bucket imposes a hard limit on the data transmission rate, the Token Bucket allows certain amounts of uneven transmission within the limits of an averaged transmission rate.

- *Token Bucket.* The Token Bucket method works on the principle of buffering incoming traffic flow and releasing it at a steadier rate, according to configuration of parameters. This conceptual bucket slowly fills with tokens at a constant rate; that is, token replenishment is time based.

  The rate at which the bucket fills with tokens is configurable per unit of time, for example, 1000 tokens per second. The release of packets into the network is governed by the existence of tokens in the bucket. When there is data in the buffer, the packets are sent out one by one, reducing the contents of the bucket by the equivalent number of tokens. A token can be configured to be worth a particular SDU (Service Data Unit), one or more packets or frames.

  When the bucket is empty, the packets will not be transmitted until new tokens arrive, that is, until a period of time has passed. When the bucket is full, tokens are not added to it, thereby defining the maximum "depth" of it. If the bucket is full, all packets can go at once, up to the capacity of the bucket. Thus, the rate at which tokens arrive dictates the average rate of traffic flow, while the "depth" of the bucket, or the bucket capacity, determines how "bursty" the shaped traffic can be.

  In this way, the rate of filling the bucket and the rate of emptying the bucket are independent of each other, and packets can be released at a configured maximum peak rate, equal to a full bucket. The traffic is still lumpy, within the limits of the total capacity (depth) of the bucket.

- *Leaky Bucket.* The Leaky Bucket method is used to control tightly the rate at which data is injected into the network, producing a smooth flow of packet traffic. The algorithm works like a bucket that is continuously but erratically filled, and is emptying through a hole at the bottom that lets out a steady flow.

  The algorithm that controls the release of SDUs allows arriving packets to fill the bucket and queue them for "first-in-first-out" (FIFO). The bucket has a predefined size, which is a function

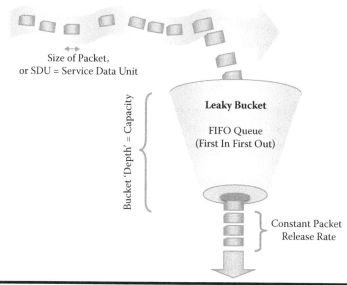

Bursty Flow of Incoming Packets Over the Time Line

Size of Packet,
or SDU = Service Data Unit

Bucket 'Depth' = Capacity

Leaky Bucket

FIFO Queue
(First In First Out)

Constant Packet
Release Rate

Figure 6.1   Leaky Bucket algorithm for traffic management.

of available memory for buffering. When it is full, packets are discarded and lost. The bucket drains at a constant rate of packets per second, which represents the given bandwidth.

While achieving a steady out-flow of packets from a specific port, regardless of the fluctuating data in-flow, this is not necessarily the most efficient usage of resources for the network as a whole, where higher rates of flow could still be accommodated through the capacity of other "buckets." Additional methods can be applied to remedy this, such as "Fair Queues," to provide further means of shaping the traffic.

In Figure 6.1, the incoming stream of SDUs, or packets, is uneven. The packets go into the bucket with a configurable depth. When the bucket is full, packets will be rejected. There is a steady outflow of packets, at a regular rate through the hole at the bottom of the bucket, which represents packets released into the network.

6.3   Generic IMS QoS Management

6.3.1   Subscriber QoS Profiles

In IMS, the level of QoS is generally associated for each session according to the user subscription and the service type. Each subscriber can have a set of parameters describing the QoS for several services, according to terminals and access networks. The QoS Profile can contain values for parameters such as service precedence (priority), reliability, delay and throughput, etc. The total of these parameters constitutes a QoS Profile.

The QoS Profile is held for each subscriber in the user database, that is, within the User Profile in HSS (3GPP) or in UPSF (TISPAN). The QoS profile can be assigned for an agreed contractual period.

Service requests are validated against the subscriber's QoS profile, and network ability to deliver it. The request is assessed against the QoS profile for the particular service for this user.

6.3.2 QoS Profile Parameters

In IMS, subscribers are assigned a QoS Profile with a set of parameters including negotiated and default values. The following can be elements in the QoS Profile:

- *Service Precedence.* This parameter defines the service priority level; that is, when the network is congested, packets for the highest priority service will be allowed through while lower-priority packets may be discarded. Priority levels are defined as High, Normal, and Low.
- *Reliability.* This parameter defines the transmission attributes required by a certain service or an application. The reliability parameter sets up the probability of packet loss, duplication, out-of-order packets, data corruption (undetected), and transmission errors. This parameter is selected according to the type of service and transmission protocols, for example:
  - Sensitivity to errors
  - Incorporated error detection and error correction
  - Tolerance of network faults
  - A maximum holding time for packets, after which they are discarded.
- *Delay.* Session delay is defined as the sum of the one-way latency in each hop between routers plus the transmission delay on each of these network elements that must process the packet on its way to the destination. The voice/video transmission is expected to occur at real-time with no delay, although some buffering can be employed to smooth out peaky traffic. The aim is to eliminate or minimize delay, but delays still occur due to queuing at nodes and congestion in the path. The delay parameter sets a maximum value for permitted mean delay, defined for end-to-end transmission.
- *Throughput.* The throughput parameter defines volumes per period of time required for the service. This parameter consists of two variables: (1) maximum bit rate and (2) mean bit rate. The maximum bit rate defines top capacity while the mean bit rate defines the average rate between the peaks and troughs (i.e., the ideal steady rate). The maximum and mean bit rates can be negotiated to a value up to the maximum transfer rate.

6.3.2 Session QoS Management

6.3.2.1 The Basic Concepts Evolution

QoS management means that resources are allocated in a controlled fashion, under instructions from an operator-defined network policy. The QoS Profile and the scope of the resourcing is set for the session at signaling level. In the process of negotiating the session particulars, the bearer nodes that will carry the media must commit to deliver the required level of QoS for the duration of the session. When the media flow is established, the QoS attributes are enforced at the Traffic Plane by the IP-CAN and Core Network media bearing elements.

Figure 6.2 shows the QoS management functions in the Session Control layer interacting with the QoS enforcement function in the transport nodes in the IP bearer network.

The QoS procedure involves an agent that requests the session with a certain QoS Profile, a decision maker that considers the session particulars against a set of rules provided by the network operator, and a media bearing node that applies packet gating to achieve the agreed QoS. The AF

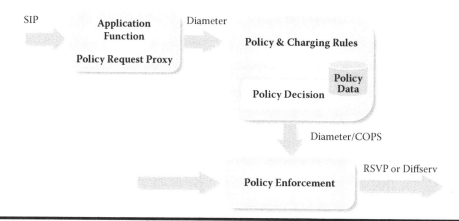

Figure 6.2   QoS control basic elements.

is a proxy user agent that initiates QoS requests. It receives session requests in SIP and creates QoS requests using Diameter.

The Policy Decision Point, as defined by the IETF, provides a decision of what resources will be allocated to the requested session, based on stored rules and network availability. The decision is received in the PEP (Policy Enforcement Point), which is located in the IP-CAN bearer nodes, where it is enforced. The 3GPP has adopted these terms although they are referred to as *functions* (soft modules), not "points" (nodes), to avoid association with physical implementation. Hence, in the 3GPP, a PDF and a PEF are defined rather than PDP and PEP.

Due to the need to distinguish QoS levels between services, the concept of SBLP (Service Based Local Policy) was developed. This is similar to the TISPAN approach that identified an SPDF (Service Policy Decision Function) that interacts with the RACS.

The concept of defining rules for the bearer to execute was extended again to include Charging rules, which are closely related to decisions regarding QoS levels and priority resourcing. In Release 7 of the 3GPP, the PCC (Policy and Charging Control) is defined, with the PCRF as the decision maker, and the PCEF as the bearer enforcer. The PCRF stores PCC rules in the SPR (Subscription Profile Repository). For descriptions of PCC components and functions, see later in this chapter.

6.3.2.2   Single-Phase and Two-Phase QoS Control

QoS establishment can be performed in a single phase or via two phases:

1. *Single-phase approach.* In this mode, the network allows for a single transaction that identifies the QoS settings, requests resources, reserves resources, and activates them.
2. *Two-phase approach.* In this mode, QoS resources are requested and reserved prior to session requests. If, and only if, adequate resources are authorized, then resources for each session initiation request can be committed. In this way, resources are reserved by the IP-CAN and the Core Network before the session is offered to the destination.

6.3.2.3   Setting and Negotiating QoS Sessions

In determining which resources are used and what QoS filters are applied, QoS signaling must allow for session control procedures that enable the correct setting of the session, considering the connecting parties, connecting networks, and the involved applications.

Prior to the connection of the parties, the bearer attributes are negotiated. The originating call leg requests resources to match a QoS profile according to the desired settings for the service type. The originating network servers apply local policy and negotiate the bearer characteristics with the user agent. Similarly, before the destination user is alerted, the capability of the destination endpoints and those of the target network, along with all interim transit networks, must be established to obtain end-to-end commitment of network resources. It is imperative that no network resources are made available until the QoS negotiation is complete.

Where applications are involved, the relevant applications in both the originating network and the terminating network must be able to set QoS requirements for the session. These requirements, including restrictions and filters, are then taken into account when the QoS parameters are set.

If there is already an established connection for a media flow, it should be possible to use the allocated resources for the new session, if it needs the same QoS profile as the existing one. This is only possible within the IP-CAN rules and in agreement with network policies.

Setting QoS for IMS users is best performed through a dedicated signaling IP-CAN bearer. This enables the operator to define a standard QoS Profile that delivers consistent and predictable user experience. However, a multi-service network sharing a single IP bearer is more cost effective, provided services can be identified and differentiated in terms of QoS and charging.

Resource reservation, which is initiated by the UE, takes place only after successful authorization of QoS resources. Resource reservation requests from the UE contain the binding information that enables the IP-CAN to match correctly the reservation request to the corresponding authorization.

When a UE combines multiple media flows onto a single IP-CAN bearer, all the binding information related to those media flows must be provided in the resource reservation request. The authorization is "pulled" from the Policy Decision function by the Policy Enforcement function in the IP-CAN.

6.3.3   Resourced-Based or Service-Based QoS

6.3.3.1   Bearer-Level QoS

QoS can be defined at the bearer level only, or at the service layer as well as at the bearer level. Where the service requires no higher-level control, the QoS management is said to be resource based rather than session based. Bearer control is closely associated with the type of bearer protocols used, e.g., IP, MPLS (Multi-Protocol Label Switching), or PBT (Provider Backbone Transport).

The bearer resources are allocated to a session by the bearer nodes. This can be performed based on preconfigured, provisioned information on the nodes that provide the anticipated level of resources available for allocation. More informed information of current resource availability requires an interface to a real-time monitoring of the network. This is yet to be set into the standards and is not easy to manage.

At the bearer level, media flows can be set up to be shared by several sessions, as long as the QoS requirements are the same. Such services may be carrying large volumes of data packets, for multiple sessions. To vary charges according to usage at the bearer level, Flow-Based Charging (FBC) controls have been introduced.

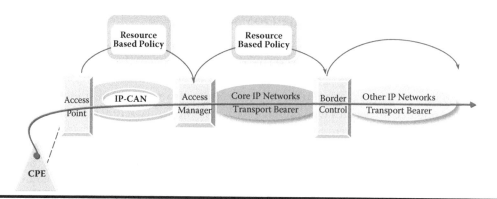

Figure 6.3   Resource-based QoS.

FBC relies on the bearer accepting policies from a policy decision server that determines the level of QoS as well as charging for the particular flow.

Figure 6.3 shows session connection where resource allocation is performed within the access or bearer layer. The access network nodes GGSN (Gateway GPRS Support Node) or BRAS (Broadband Remote Access Server) determine the requirements and forward them to the Core Network (CN) routers, which in turn negotiate QoS along the path and across the border.

6.3.3.2   Service-Level QoS

For QoS purposes, IMS is perceived as an application. IMS needs the differentiation between Data services with low QoS requirements and QoS-sensitive services, such as telecommunication sessions that have particular high QoS requirements. Therefore QoS control must be based on the service policy. The level of QoS is defined by information that is stored or generated in the service layer, i.e., at the IMS nodes or on the application servers.

The user's QoS Profile is found in the CN data repository (HSS or UPS). The network policies are kept on a CN element capable of making policy decisions (PDF) or service-based policy decisions (SPDF). These decisions and QoS specifications are mapped by the SPDF into resource management instructions for the bearer nodes to perform.

Figure 6.4 illustrates generically how resources are allocated via the service-based policy function, according to the TISPAN vision for NGN resource management. The A-RACF manages the initial resource allocation to enter the access network. The C-RACF (Core – Resource and Admission Control Function) negotiates the passage through the CN and if the session is bound for external domains, the C-RACF interacts with the border nodes to establish QoS levels end-to-end.

6.3.3.3   The Principles of Service-Based Local Policy (SBLP)

In data networks, the quality of the transmission is based on the ability of the IP bearer to reliably transmit all packets to their destination, using appropriate error detection and correction techniques to ensure data integrity. The Internet has introduced a requirement for greater flexibility and speed at the expense of absolute data fidelity, which is achieved using the more permissive UDP and relaxing rules of error handling.

With the advent of demanding services such as Voice and Video over packet network, the QoS requirements dictate different solutions again. It necessitates segregating media flows that

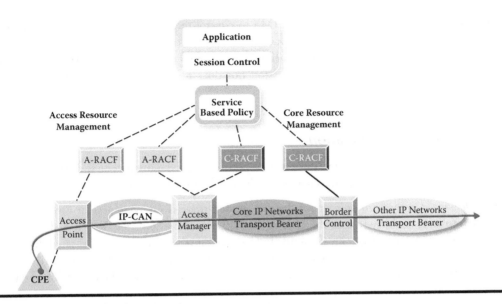

Figure 6.4   Service-level QoS.

require different levels of QoS, ensuring priority for services with low tolerance for network delay or packet loss.

The standards define the SBLP that supports resource authorization in the IP connectivity network, and correlation of charging for resources with the charging for IMS session, to prevent duplicate charging.

Resource is allocated by two types of mechanisms:

1. *Reservation based*, in which the granularity of authorization and charging correlation is a single resource reservation (i.e., a session)
2. *IP flow based*, in which the granularity of authorization and charging correlation is an IP flow that can consist of several sessions

The underlying network must have the following capabilities to support SBLP:

■ Relay requests of resource reservations to IMS.
■ Act on an authorization decision received from the IMS to reserve and allocate the appropriate level of resources within the currently available resources.
■ Provide IMS with correlated information on resource usage and the appropriate charging policy and charging data.
■ Act on revocations of the authorizations sent by the IMS and withdraw the resource.
■ Inform IMS when a resource reservation for allocation is revoked by the network.
■ Provide individual IP flows configured according to policy information received from IMS.
■ Provide charging information and correlation information for individual IP flows.

SBLP, if supported, can operate in several modes:

■ *Authorization and charging correlation.* SBLP performs resource authorization and correlation of charging information.

- *Authorization only.* SBLP allows the IMS to control resource authorization in the IP connectivity network but there is no support for charging correlation.
- *Charging correlation only.* SBLP allows correlation of charging for IP connectivity network resources with the IMS session but does not support any control of resource authorization.

6.3.4 Policy Dissemination

Policy can be generated centrally but there may be policies also preconfigured on the UE or in the IP-CAN bearer. The following scenarios have been identified for the way QoS is provisioned:

1. *QoS via a proxy with policy-push.* In this model, there is no awareness of QoS classes or parameters at the user end, the UE, or the CPE; therefore, policies are "pushed" from the network. QoS is not specifically requested by the user when the session is initiated. It is up to the QoS proxy to evaluate the type of requested service and the IP-CAN that carries it, and formulate the appropriate QoS request. The session's QoS needs are then determined within the session control layer.

2. *User-requested QoS with policy push/pull.* In this mode, the UE is able to determine the QoS requirements and sends a specific QoS request. The session controller relays the request to a resource manager, which obtains a network authorization token. The token is sent back to the UE. Now the UE, using the token, can request a particular session with the specific QoS attributes.

3. *User-requested QoS with policy-pull only.* In this mode, no prior authorization is required. The UE is capable of defining the QoS requirements. It initiates a service request toward the IP-CAN, without the involvement of the session controller. QoS allocation is handled directly in the IP-CAN nodes, but they activate sessions according to policy downloaded ("pulled") from the resource controller. In this way there is no need for the network resource controller to retain knowledge of the UE relationship with its IP-CAN.

6.3.5 Resource Allocation

6.3.5.1 Reserve-Commit

Resourcing is done either by Reserve or by Reserve-Commit operations. The resources can be reserved for a session before the actual session initiation. This means that service authorization and QoS Profile determination are executed first, and resources can be committed later, as and when the INVITEs arrive. This is necessary, for example, for multi-party conferencing, where the parties call in to join the bridged media.

By definition, reserving resources results in underused network capacity. It may be possible for advanced algorithms in the Resource Manager, based on variable bit-rate flow, burstiness, and other characteristics of the media flows, to calculate what overbooking levels will be permitted.

6.3.5.2 Monitoring of Resource Availability

The transport nodes must recognize when they are unable to admit more sessions to avoid congestion. This can be performed by configuring the total capacity these resources have, so that the a

Resource Manager can calculate their availability, reduce it when they are reserved and increase it when they are released. When changes occur in the network, the Network Manager should provision changes in these volumes though a management interface.

This is not fully satisfactory because it does not reflect real network status and cannot react to failure fast enough. To obtain dynamic availability, all resources must be reported to the Network Manager in Real Time. The QoS reporting information is extracted from all participating elements, such as IP Edge devices, access nodes, and border gateways. Network probes that sample the network performance can be used to add credibility to the QoS reports and can also be used independently.

6.3.6   Multi-Domain QoS Control

6.3.6.1   Inter-Domain QoS Negotiation

As carriers are gradually upgrading their IP networks to carry voice and video, concerns about QoS increase. In particular, they are concerned about end-to-end QoS; that is, the ability to guarantee a level of service even when the destination is outside the operator's own network. While great improvements in quality can be made in controlling the access domain (IP-CAN), the access bearer aggregation networks, the internal backbone, and the Core Networks, the quality is not assured across the border.

E2E (end-to-end) QoS relies on the ability to:

■ Control the call admission, obtaining authorization.
■ Negotiate with external networks to reserve resources along the path.
■ Control the flow of packets according to priorities and block unauthorized ones.

QoS enforcement works at both bearer level and service level. SBLP need not be implemented at the terminating network for QoS to be controlled, as long as the router network is able to exercise reservation of resources via the widely implemented protocols such as RSVP and DiffServ. For descriptions of these protocols, see Appendix A.

Figure 6.5 shows a series of service control domains (IMS) and their transport domains through which the media flow travels. A service-based QoS request for a session is forwarded to the local IMS domain. From there it is carried through intermediate IMS domains to the final destination network.

At the request of the originating IMS server, the destination IMS server finds out the capability of the destination UE to establish its ability to support the requested type of service. If this is acceptable, then each network must confirm the QoS Profile for the media flow and commit to enforce it. This process includes some negotiation of the agreed QoS level between these networks. This can happen only if all participating networks support the same QoS signaling standards.

Transporting high-priority session media remains a matter for operator implementation. It takes some discipline to avoid marking one's own traffic as a top QoS class, even when it does not need it. Operators must be in charge of the bearer network to manage the traffic; therefore; they must be able to block unwanted traffic, and downgrade or upgrade desired traffic. Tools to do this should be installed at the borders of the network, in the BGF (Bordering Gateway Function).

The media transport servers can support different mechanisms to achieve the agreed QoS within their domain. Media generated by a server rather than a user (i.e., a Media Server), is also

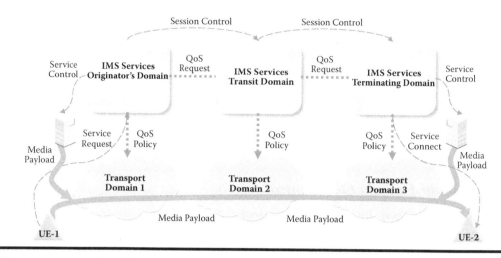

Figure 6.5   Inter-domain QoS management principles.

subject to the QoS definition. Note that such media can be generated in the destination domain (e.g., local announcements or other content).

E2E QoS solutions are still undergoing research, where various options are discussed for QoS signaling to be carried in Layer 1, 2, or 3. Also investigated is the mechanism of defining some level of QoS even traversing networks that have only Transport-level media flow management.

6.3.6.2   *Inter-IMS QoS Signaling*

A session destined for an external domain must be managed at the signaling or control plane and at the traffic plane. QoS must be established and agreed to by all participating domains and the participating terminals. Signaling occurs between the originating and terminating domains, and it can transit through other networks along the way.

QoS signaling can be transferred along the signaling path for session-based services as parameters inserted into the SIP messages. Because the session is "offered" to the destination, QoS parameters are suggested by the originating domain. The destination terminal and its access network can reject or counter-offer with different QoS settings, according to the capabilities and resource availability of the receiving network.

At both the originating and terminating domains, the session controllers (CSCFs) seek policy decision from the Policy Decision functions (PDF, PCRF, or SPDF), which examine what particular conditions are fulfilled by this session and match them against a predetermined set of rules.

At the traffic plane, both access nodes and border gateways can have locally preconfigured rules. These rules contribute to the central policy so that the final decision is made considering the traffic plane capabilities and the rules. The exact way in which QoS parameters are established is not yet identified or standardized.

Enforcing the final decision occurs within each network. At the border, each operator applies its own traffic management. However, where the resource reservation process has preceded the connection of the parties, the bearer along the path is already committed to the delivery of the session.

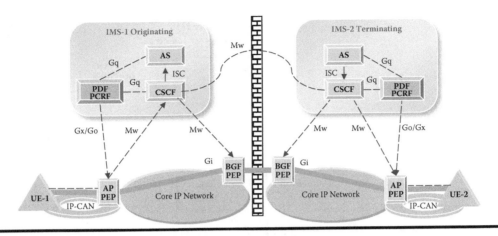

Figure 6.6   Negotiating QoS between operator domains.

Figure 6.6 depicts a scenario of a session between the originating IMS-1 and the terminating IMS-2. The originating UE-1 requests a session to UE-2. The request is forwarded by the AP to the CSCF, which consults the PDF/PCRF, over the Gq interface, when setting up the session QoS parameters. Also shown is the media flow (solid line) that passes directly through the media bearing nodes.

6.3.6.3   Multi-Access QoS

Although seemingly far apart, the TISPAN and 3GPP concepts of resource management and QoS already have converged considerably. However, major differences remain in implementations, particularly in the IP-Access bearer procedures.

The AF (Application Function) is accepted for both Fixed and Mobile networks. This function can be incorporated into the P-CSCF. It can also be incorporated into the AGCF, allowing for fixed and traditional terminals to be treated in the same way as SIP terminals via IMS.

Converged Charging and Policy Rules function can serve all types of domains, even when separate rules might be drafted for the different scenarios. The combined functional entity will be responsible for all rules, QoS and charging (and possibly security) to be managed together, thereby simplifying the IMS functional structure. Thus, the same PCRF (Policy and Charging Rules Function) can be utilized for the charging rules of FBC, that is, non-session-based services, as well as for Service-based Policy Decision Function (SPDF), which is used with IMS sessions.

TISPAN specifies an additional resource management function in the A-RACF (Access-Resource and Admission Control Function) in conjunction with the network admission subsystem, the NASS. Similar functions are already performing within the GPRS nodes, GGSN, and SGSN. Aspects of resource allocation over the air interface in Mobile Networks, by definition, remain separate functions.

Figure 6.7 provides an example of a session between a wireline terminal and a wireless handset in the same IMS network. The AF that serves a SIP-based soft client on a PC can also handle the Mobile UE. The SPDF and the PCRF may well be governed by logic in a single product. However, the A-RACF is responsible for the interaction with the xDSL bearer only.

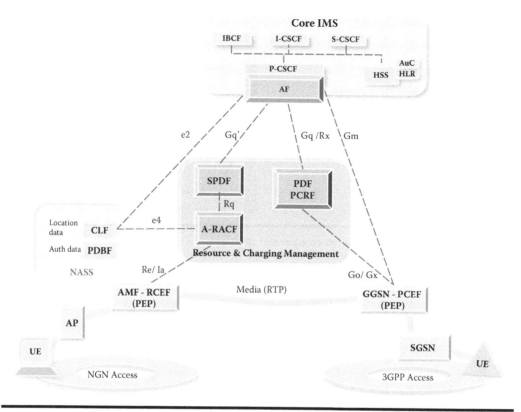

Figure 6.7   Enforcing (PEP) service-based QoS.

6.4   The 3GPP QoS Architecture

6.4.1   General 3GPP Concepts

6.4.1.1   UMTS Bearer and IP Bearer

Setting up an IMS session involves two incremental levels of QoS negotiation and resource allocation:

1. *The UMTS Bearer level.* This is the process that conveys the QoS requirements from the UE to the RAN, the CN, and the IP Bearer Service Manager. When UMTS QoS negotiation mechanisms are used to negotiate end-to-end QoS, the translation function in the GGSN coordinates resource allocation between the UMTS Bearer Service (BS) Manager and the IP Bearer Service Manager. UMTS-specific signaling, such as RAB QoS negotiation and PDP Context setup, are performed at the UMTS bearer level.
2. *The IP Bearer level.* This is the process that conveys QoS requirements at the IMS application level to the UMTS and performs any required end-to-end QoS signaling by interworking with the external network. This process is performed between the IP BS (Bearer Service) Manager at the UE and the GGSN. QoS signaling mechanisms such as RSVP are only used at the IP BS level.

When UMTS QoS negotiation mechanisms are used to negotiate end-to-end QoS, the translation function in the GGSN needs to map and coordinate resource allocation between the UMTS

BS Manager and the IP BS Manager. The resource allocated at the UMTS Bearer can be utilized for multiple sessions at the IP BS level, but the same GGSN that is used for QoS control negotiation between the IP BS level and the UMTS BS level must be used for QoS class selection, mapping, translation, and reporting of resource allocation.

6.4.1.2 GPRS/UMTS Resource-Based and Service-Based QoS

QoS in GPRS/UMTS refers to both delivery of IP IMS traffic (service based) and for traffic delivered entirely by the GPRS (resource based). IP-based data services can include broadcasting and streaming services, which are controlled by application servers in conjunction with 2G service controllers.

When creating PDP Context, resources for the end-to-end session should be allocated to enable session connection. The session may be using either a bearer-only resourcing procedure, (such as MBMS (Multimedia Broadcast/Multicast Services) or service-based resourcing (i.e., IMS-based sessions).

The UE can request access to GPRS/UMTS without service-based local policy or with service-based local policy. The GGSN determines the need for service-based local policy, based on preconfigured data or based on the AP of the PDP Context.

- *Resource-based policy.* This may or may not be required for the basic GPRS IP service. The QoS function in the GPRS node examines the user's subscription as well as any preconfigured local operator's rules. Certain QoS specifications depend on the admission control process and access network details, and on roaming agreements. The resource-based QoS policy is applied to the bearer and enforced by the GGSN and SGSN nodes. The authorized resources provide an upper bound on the resources that can be reserved or allocated for the defined set of IP flows. These parameters are expressed as a maximum authorized bandwidth and QoS class. The QoS class identifies a Bearer Service, which has a set of associated bearer service characteristics.
- *Service-based policy.* This is always required for a UMTS session driven by applications through CN session controllers (e.g., IMS). The decision of the level of QoS requirement per session is defined at CN nodes, taking into consideration the type of service, its priority level, etc. to make a decision. The policies are stored centrally. These network-wide service-based policies must be mapped to UMTS bearer instructions for creating media flows of the appropriate attributes for the requested service. Service-based local policy decisions are either "pushed" (downloaded) to or "pulled" (requested) by the GGSN.

6.4.1.3 IMS Service-Based Local Policy (SBLP) QoS Procedures

To support IMS with SBLP, the GPRS procedures are enhanced with QoS related procedures:

- Authorize QoS resources
- Resource reservation with service-based local policy
- Enable media and disable media
- Revoke authorization for GPRS and IP resources
- Authorize PDP Context
- Modify PDP Context

■ Indicate PDP Context release
■ Indicate PDP Context modification

6.4.1.4 Policy Decision in GPRS/UMTS

The IMS media components carried by the PDP contexts in GPRS are controlled by the PDF, which now is converged with Charging Rules in the PCRF. The PDF/PCRF becomes involved in QoS negotiation between the IP BS level and the UMTS BS level. The resource allocated at the UMTS Bearer can be utilized for multiple sessions at the IP BS level.

In GPRS, the PDF originally was integrated into the GSN node, for mapping QoS parameters and reporting on the resource allocation. In this case, the same GGSN used for QoS class selection must also be used for QoS negotiation. For this reason, the PDF is better implemented on a node external to the GSN nodes.

The early 3G PDF was assumed to co-reside with P-CSCF at the entry point to IMS. However, in multi-access IMS, the policy decision functions can be utilized by other proxies (e.g., the AGCF). Therefore, the PDF/PCRF should be implemented as a separate function.

The P-CSCF makes its decisions according to instructions from the PDF within the same network, that is, in the Visited Network for roaming UE and in the Home Network for non-roaming UE. The PDF declares a maximum authorized QoS class for the set of IP flows. This information is mapped by the Translation/Mapping function in the GGSN, to generate the authorized resources for UMTS Bearer admission control.

6.4.1.5 IMS Signaling in the PDP Context

The network must be notified of the intention to use the PDP Context for IMS signaling. A Signaling Flag for IMS is used to signify that the packets belong to a dedicated signaling PDP Context that carries IMS signaling. Otherwise, the packets are deemed general-purpose PDP Context. This IMS flag also indicates to the network elements (radio and core) that service-based signaling will be deployed. The IM CN Subsystem (IMS) indicator is enforced by the GGSN, regardless of the contents of the UE REQUEST message, because the user device cannot be trusted.

There is also a QoS Signaling Indication that identifies QoS requirements for IMS sessions. It provides prioritized handling over the radio interface for IMS signaling. The QoS Flag can be used independently of the IMS Signaling Flag. Both the IM CN Subsystem Signaling Flag and the QoS Signaling Indication can be used in the PDP Context activation procedure at the same time.

6.4.1.6 Media Flows on PDP Context

The IMS media components carried by the PDP Contexts are controlled by the IMS network Policy Decision Function (PDF/PCRF). The AF acts as the QoS agent on behalf of the user for service-based sessions. The AF, which can be incorporated into the P-CSCF in 3G implementations, indicates to the UE whether a separate PDP Context is required for each requested IMS media component. This media component level indication is transferred in SIP/SDP signaling of the session initiation messages. All associated IP flows (RTP/RTCP) used by the UE to support a single media component must be carried within the same PDP Context.

It is possible for some media components to share a single PDP Context, but subsequent media flows cannot then request a dedicated PDP Context (e.g., in the case of subsequent answers received due to forking). If the UE receives no instruction to open a separate PDP Context from the P-CSCF, it can open a separate PDP Context anyway or use an existing PDP Context, modifying it if necessary.

If an existing PDP Context is used, the UE can also decide to carry media components from different IMS sessions in the same PDP Context, as long as none of the bundled media components are marked for a dedicated PDP Context.

If an existing PDP Context is used without modification, it cannot be subject to SBLP. Therefore, where SBLP will be applied, the P-CSCF must include the indication to the UE to request the establishment of a new PDP Context.

6.4.2 The AF

The function that requests resources from the network is the AF. The AF can reside within the:

- P-CSCF, for resources toward the IP-CAN requested by the UE
- AGCF, for connecting CPE over H.248 to multimedia services in IMS
- IBCF, for resources toward external IP networks or incoming from external networks

The AF within any of these functions must be recognized by the PDF (or the PDF element within the PCRF) and get authorized to request resources. The AF initiates a request to be authorized, furnished with suitable server details and authentication keys. When the authentication is successful, the PCRF sends an authorization token that is subsequently used by the AF when requesting QoS settings for individual sessions.

When the UE initiates a session request toward the AF, the AF responds by setting up a request for resource reservation at a defined QoS level, and forwards it to the appropriate bearer node. For any session that implements SBLP, the AF consults the PCRF.

The AF can provide the PCRF with the following session-related information:

- Subscriber identifier
- IP Address of the UE
- Media type
- Media format and associated parameters
- Bandwidth
- Flow description (e.g., source/destination, IP Address, port numbers, and protocol)
- AF application identifier
- AF communication service identifier (e.g., for IMS), provided by the UE
- AF application event identifier
- AF record information
- Flow status (for gating decision)
- Priority indicator, used by the PCRF to differentiate service QoS parameters
- Emergency indicator

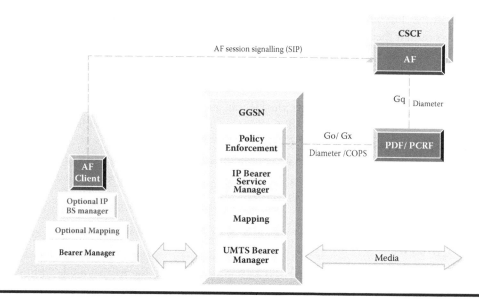

Figure 6.8 The AF client and server.

As Figure 6.8 illustrates, there is an AF client in the UE to initiate and participate in the QoS setting dialog. The AF in the proxy server contacts the PDF/PCRF to retrieve network-based prestored policies, which can override the UE instructions. The dialog between the PDF and the P-CSCF is governed by the Gq interface, which is the Diameter Protocol.

6.4.3 The Policy and Charging Control (PCC)

6.4.3.1 The PDF Evolution to PCRF

The PDF has been replaced by the PCRF as the function that provides policy decision for both Charging and Resource management. The PDF is still referenced in numerous documents and diagrams, and therefore merits mentioning here. Where the functionality of the PDF is described, it must be assumed that the same functionality is incorporated into the PCRF in the PCC system.

6.4.3.2 The Scope for PCC

PCC is a combined system that controls both QoS and Charging for services that need dynamic determination of level of service. Resource allocation is closely associated with charging for these resources. For this reason, later standardization sought to combine these functions and simplify the architecture.

Policy control means performing the following:

- *QoS control.* Authorize and allocate resources to match a QoS profile for a service data flow.
- *Event triggering.* Send and receive event notifications that change the behavior of the network or the handling of a particular media flow.

- *Establishing media flow in IP-CAN bearer.* Initiate procedures that establish and manage the media flow in the Traffic Plane, according to given parameters.
- *Gating control.* Block packets that do not fit the QoS Profile, or have lower priority when the network is congested, and allow through packets that belong to authorized, high-priority media flows.

PCC is a generic function intended to work with any IP-CAN. It is currently defined for UMTS/GPRS and I-WLAN, and is also examined for Cable DOCSIS (Data Over Cable Service Interface Specification).

PCC should be enabled for each Packet Network, defined by the AP and the published IP Addresses of its servers. PCC also can operate for a Visited IP-CAN, where the final QoS settings take into account the QoS capabilities of all participating networks.

The main role of PCC is to control which packets get through and which ones are discarded. Packets that do not match any of the service data flow that has been authorized are rejected. Packets belonging to a media stream with assigned high priority are sent forward before lower-rated packets. For example, PCC should be able to handle Emergency services as top priority.

PCC rules can be preconfigured and stored locally, or provisioned dynamically, at session initiation time. PCC rules can be applied to a group of IP flows, selected by a wildcard descriptor. Thus, one QoS Profile will be used for a number of sessions of the same type, although the PCC architecture allows control charges to be applied on a per-service data flow basis, independent of the policy control. This means that PCC makes it possible to relate the Charging particulars and the QoS settings for a single session.

6.4.3.3 PCC Components Evolution

In PCC, the concept of PDF evolves to include Charging as a type of policy control, and the function is now called PCRF. The PCRF maintains the policy data in the SPR (Subscription Profile Respository). The enforcing function in the bearer is also renamed, from PEF to PCEF (Policy Control Enforcement Function).

The AF is the PCC user agent that requests resources on behalf of the user or the driving application. The AF requests can be rejected, upgraded, or downgraded during negotiation with the PCRF. For some service types, the AF will delegate full responsibility to the PCRF to act on its own. The AF also relays events that can trigger a change in the requirements during the session for the PCRF to reevaluate the QoS settings.

Note that the PCC, PCRF, and PCEF charging functions are described in Chapter 8.

6.4.3.4 Policy and Charging Rules Function (PCRF)

The PCRF makes admission decisions according to the resourcing rules it contains. The PCRF information can be provisioned dynamically in response to a PCEF request. Alternatively, the PCRF can download the PCC rules to the PCEF in "push" mode, when these rules have been changed, responding to operations needs. The PCEF combines the predetermined rules and any dynamic rules in setting up the QoS setting for the session.

The PCRF resolves conflicts between different rules and the QoS level requested. The PCEF can, as a result of traffic conditions, suggest a different QoS Profile, and the PCRF will adjudicate in consideration of that. The PCRF has the final word in case of conflict. As a rule, dynamically provisioned parameters take precedence over pre-installed rules. The PCEF must accept the final decision and enforce it.

The PCRF can use an assigned service priority level where the session falls within a cumulative QoS band authorized for a subscriber's multiple-session service. This means that when the last arriving request will exceed the guaranteed bandwidth, for example, the PCRF can downgrade another ongoing session with lower service priority to allow in the new session. If it is not possible to downgrade some of the PCC rules for lower priority services, the new session will be rejected.

The PCRF, at any time, can modify an active, dynamic PCC rule, although not all rules can be modified. The PCRF conveys PCC rules either by their ID or downloads the entire set of QoS attributes. The Gx reference point defines this interface.

6.4.3.5 PCRF Resourcing Functions

The interfaces of the PCRF to IMS are made via the AF. The following are the resourcing- and QoS-related activities that are managed via the AF–PCRF interface:

- *Authorization of resources.* The PCRF receives details of the session as expressed in the SIP message headers and SDP. The PCRF also can receive specific requests of a level of QoS or service classification. All these details are input to the resource control logic to match the relevant prestored conditions and rules, and arrive at a decision for the level of resourcing to offer. The authorized resourcing may not be the same as requested, and a dialog may take place to arrive at an agreed level of QoS. The authorized resourcing is expressed as media flow parameters, such as the upper limits for the media flow (peaks) and rate of flow, or restrictions on IP destinations and port usage.
- *Resource reservation.* Having authorized the maximum level of resources, the request for resource reservation is made for a particular session. The resource reservation request is linked to the binding information that associates the session with the IP address and the IP-CAN path, which has been generated during the call admission procedures. The resource reservation request from the UE must identify not only the individual media flow, but also all other related media flows if they are combined under a single QoS policy. The PCEF, within the IP-CAN bearer, "pulls" the policy authorization details from the PCRF. It can then determine whether or not the new reservation request falls within the given parameters of the authorized QoS.
- *Committing to QoS Resourcing.* Responses for reservation requests are returned to the PCRF. If there is more than one response for the same session, indicating that the session has been forked in the network, the PCRF will permit either one to be committed. This is established when the PCRF receives the final response from the PCRF and commits to resources according to the session-established indication. Until the final commitment of resources is given (QoS-Commit), the IP-CAN continues to restrict any use of the IP resources for the session, by keeping the gate closed.
- *Removing a QoS Commitment.* The PCRF can remove a previously approved QoS-Commit. The PCRF communicates this decision to the PCEF for execution. Once received by the PCEF, it will restrict any flow of packets relating to this media flow (i.e., close the gate).
- *Revoking Authorization for IP-CAN and IP Resources.* The UE, via the AF, initiates session release (termination) when the parties terminate the sessions. When the UE is unable to do so, or there are administrative reasons, a network server can initiate a session release on

behalf of the UE. The AF notifies the PCRF, which then provides an indication to the PCEF in the bearer to release previously authorized, allocated, or activated resources.

- *Indication of IP-CAN bearer release.* Under certain circumstances, a session can be released by the IP-CAN bearer itself. For sessions managed by the PCRF, this event is communicated by the PCEF to the PCRF. This indication is forwarded to the P-CSCF via the AF and can be used by the CSCF to initiate a session termination (BYE) toward the remote endpoint.
- *Authorizing of IP-CAN bearer modification.* Existing media flow for a session may need modification. This can involve requirements for additional resources or a connection to a previously unauthorized path for which there is no binding information. The PCEF may attempt to cater to the modification request, but if this exceeds the previously authorized QoS levels, a new authorization must be requested from the PCRF.
- *Indication of IP-CAN bearer modification.* The PCRF decision to authorize or reject the modification is communicated back to the AF. This can be used by the P-CSCF to initiate a session release toward the remote endpoint or passed on to an application for further processing.

6.4.3.6 Subscription Profile Repository (SPR)

The PCRF uses the SPR for stored subscription QoS data. The PCRF uses subscriber ID and PDN identifiers to retrieve information from the SPR. Once that information is retrieved, the PCRF will retain the relevant PCC data for the subscriber until the IP-CAN media flow for that subscriber terminates.

The SPR can be updated with SLA information for the subscriber or with new or modified PCC rules. The SPR can be set to notify the PCRF of any PCC rule changes or any modification to subscriptions. The PCRF will then apply the decision process to the changes and "push" it to the PCEF as unsolicited charging information.

6.4.3.7 Policy and Charging Enforcement Function (PCEF)

The PCEF manages the media flows according to the agreed PCC rules and QoS settings for the media flows. The PCEF activates, deactivates, and modifies PCC rules. The rules are provided to the PCEF by the PCRF, either by the "push" method from the PCRF when a new rule or modified rule is installed, or dynamically by the "pull" method when a session request for a media flow is initiated. The PCC information is transferred from the PCRF to the PCEF via the Gx reference point.

The PCEF initiating media flow establishment uses either preloaded PCC rules or solicited QoS Profile decisions from the PCRF, for IP-based or service-based sessions, respectively. The PCEF, if configured to do so, can activate local PCC rules that are not stored in the SPR for the PCRF. Such local rules are activated if there is no UE-provided traffic mapping information related to the IP-CAN bearer.

The PCEF can terminate a session request if the QoS attributes required for it cannot be satisfied. It can also terminate ongoing media flow, as a result of PCC changed rules that can no longer be satisfied.

The PCEF has two methods to execute its tasks:

1. *Gate control.* Based on authorization of media flow, the gate will let through packets belonging to it, regardless of their associated service. If packets do not have an identifier of a media flow that is recognized by the gate, they will be rejected.

2. *QoS enforcement* by the following means:
   - *QoS Class identifier.* The PCEF interprets the QoS Class ID into a set of QoS attributes, as appropriate for the IP-CAN.
   - *PCC rules for QoS.* The PCEF enforces the active PCC rules for the required service data flow (e.g., enforcing uplink DiffServ Code Point marking of packets).
   - *IP-CAN Bearer QoS.* The PCEF manages QoS for a media flow that contains more than one service data flow. The PCEF ensures the sum of the service data flows does not exceed the maximum settings that have been authorized for that bearer.

6.4.3.8 PCC Rules

The data transferred between the PCC elements consists of rules that are updated by the operator to reflect external events and are downloaded to the bearer functions. The tables below list the PCC rules. Only QoS affecting resourcing rules are shown here, while charging rules are described in Chapter 8.

The following table relates to rules governing the detection of service data flows:

Service Data Flow Detection	
Rule Name	*Description*
Precedence	Determines the order in which the service data flow templates are applied at service data flow detection.
Service data flow template	A list of service data flow filters for the detection of the service data flow.

Source: Based on 3GPP TS 23203.

The table below lists PCC rules for the control of service data flows:

QoS Policy Control for Service Data Flow	
Rule Name	*Description*
Gate status	The gate status indicates whether the service data flow, as detected by the service data flow template, can pass (gate is open) or will be discarded (gate is closed) at the PCEF.
QoS class identifier	This is an identifier for the authorized QoS parameters for the service data flow. It allows for differentiating QoS in all types of 3GPP IP-CAN. The value range is expandable to accommodate additional types of IP-CANs.
Up Link maximum bit rate	The uplink maximum bit rate authorized for the service data flow.
Down Link maximum bit rate	The downlink maximum bit rate authorized for the service data flow.
Up Link guaranteed bit rate	The uplink guaranteed bit rate authorized for the service data flow.
Down Link guaranteed bit rate	The downlink guaranteed bit rate authorized for the service data flow.

Source: Based on 3GPP TS 23203.

The rule identifier is a mandatory field that defines the rule to use within the IP-CAN session. The rule identifier is known to both the PCRF and the PCEF. Not all rules can be modified by the PCRF, but all those shown in the tables above can be changed.

6.4.3.9 Binding PCC Rules

There are three steps for binding proceedings:

1. Session binding
2. PCC rules binding
3. Bearer binding

Session binding is performed by the UE when requesting entry into the network. PCC binding associates an appropriate PCC rule to the requested session. Bearer binding confirms that bearer acquiescence has been obtained or binding the session with bearer policies has taken place.

Bearer binding is specific for each type of IP-CAN. The bearer binding process associates a PCC rule to an existing PDP Context that has matching QoS attributes, or a rule that can be modified to match these attributes. If there is no suitable PDP Context, the PCC rule will be attached to a new PDP Context with the appropriate QoS class. The PCC rule will be set to pending status until the PCEF reports the successful establishment of a PDP Context that fulfills the QoS level.

UMTS bearer binding is performed according to the following scenarios:

- *UE-Only mode.* Due to lack of capabilities in the UE or the network to request and negotiate resources, the PCRF alone is consulted.
- *Network-Only mode.* The PCEF alone makes the resourcing decision (e.g., for session-less services).
- *Network-User mode.* Both the PCRF and PCEF are involved in resourcing decisions, where the PCRF binds rules for user-controlled services and the PCEF binds rules for network-controlled services.

6.4.4 UMTS QoS Mechanism

6.4.4.1 UMTS Bearer QoS Management Functions

At the UMTS Bearer Control, the following functions are performed:

- *Service Manager.* This function coordinates establishing, modifying, and tearing down services. It manages the peer service interfaces between multiple instances of the bearer nodes. It can interpret the required service attributes for other functions and seek permission for service provision from other control functions in the network.
- *Translation.* This function translates generic protocols to logic understood internally by UMTS Bearer servers. This includes conversion of QoS parameters between external networks and UMTS-specific attributes.
- *Admission control and capability.* This function maintains a view of available resources for the network entity and the UMTS Bearer service, enabling resource allocation for a new service request. This function also checks that there is no administrative bar to admitting this service request.
- *Subscription authentication.* This function checks on the rights of the requesting user to use the UMTS Bearer service.

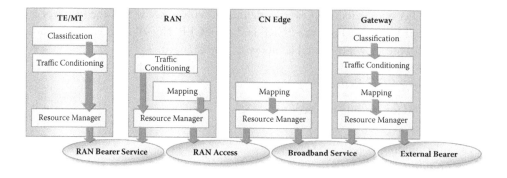

Figure 6.9 QoS functions within UMTS network elements.

At the User Plane Control, these functions ensure that the negotiated QoS attributes are being maintained, for both signaling and user data traffic:

- *Classification.* This function assigns data units to an established service, when an MS has multiple service data flows, each with a different set of QoS attributes, that is, a different class of service. The classification of data units is based on either header information or traffic characteristics.
- *Traffic Conditioner.* This function involves traffic "shaping" to police the traffic flow according to the required QoS attributes. This function lets through or drops packets according to their QoS marking and the available resources.
- *Mapping.* This function ensures that data units are marked according to the intended QoS level assigned for the service.
- *Resource Manager.* This function distributes and schedules the load of the established services across the shared UMTS Bearer resources. Such resource management can be scheduling, managing bandwidth, or controlling power for the radio bearer.

Figure 6.9 shows the QoS functions as they are deployed within the various network elements of the UMTS network. The Resource Manager, which resides in all these elements, interfaces to the physical network segment.

6.4.4.2 UMTS Bearer QoS Classification

The UMTS Bearer distinguishes four classes of service: (1) Conversational (real-time), (2) Streaming (near-real-time), (3) Interactive Best Effort, and (4) Background. These classes segment services according to their sensitivity to network delay and jitter and tolerance to errors. The Conversational class is for service traffic that is most delay sensitive while the Background class is the least sensitive.

The table below describes the attributes and usage of the UMTS four classes:

	Class	*Key Attributes*	*Utilization*
1	Conversational	Preserve time relation (variation) between information entities of the stream (i.e., minimum delay). Conversational pattern (stringent and low delay).	Voice, video telephony, video conferencing

	Class	Key Attributes	Utilization
2	Streaming	Preserve time relation (variation) between information entities of the stream but allow lag for starting point. One-way stream, relying on buffering and time alignment performed at the receiving end.	Streaming video (e.g., VoD, IPTV)
3	Interactive Best Effort	Request response pattern Preserve payload content	Web browsing, polling for data collection, telematic data retrieval
4	Background Best Effort	Destination is not expecting the data within a certain time Preserve payload content	Background download of e-mails, file transfer, SMS

Source: Based on 3GPP TS 23107.

6.4.4.3   GRX Interworking QoS Classification

Despite commercial rivalry, the telecom industry has reconciled itself to the necessity of cooperation between competing business entities. This extends not only to roaming and service interworking, but also to the delivery of predictable QoS levels. Agreements between carriers are achieved through industry groups concerned with interoperating, such as the GSMA (GSM Association).

The GSMA IREG 34 is a specification for the GRX (GPRS Roaming eXchange). This is an agreement between the GSMA members on how DiffServ bits are interpreted into the four QoS classes and implemented over the inter-PLMN backbone (GRX).

The following table shows the mapping of UMTS QoS classes to DiffServ PHB (Per-Hop Behavior). For further details about the DiffServ protocol, see Appendix A.

GSMA Specification for QoS Mapping in GRX							
3GPP QoS Services		*DiffServ*		*QoS Requirement on GRX*			
Traffic Class	**Example Services**	**DiffServ PHB**	**DSCP**	**Max. Delay (ms)**	**Max. Jitter (ms)**	**Packet Loss (%)**	**SDU Error Ratio**
Conversational	VoIP, video conferencing	EF	101110	20	5	0.5	10^{-6}
Streaming	Audio and video streaming	AF4	100010	40	5	0.5	10^{-6}
Interactive (but "near"-real-time)	Transactional services	AF3	011010	250	N/A	0.1	10^{-8}
	Web browsing	AF2	010010	300	N/A	0.1	10^{-8}
	Telnet	AF1	001010	350	N/A	0.1	10^{-8}
Background	E-mail download	BE	000000	400	N/A	0.1	10^{-8}

Source: Based on GSMA IREG 34 and 3GPP TR 23836.

The table above defines ratios for each type of service:

1. *Conversational:* voice and video in real-time interaction, requiring immediate response, no jitter.
2. *Streaming:* voice and video in a continuous stream that allows slight delays in reaching the target but delivers a contiguous, reliable stream.
3. *Interactive:* data online or in near-real-time for texting, text chats, banking transactions, etc., including Web surfing, faxing, etc.
4. *Background:* file transfer that is tolerant of packet loss, delay, and jitter, but may require good data integrity controls.

The table above also identifies the DiffServ PHB per service type, such as EF (Expedited Forwarding), AF (Assured Forward), and BE (Background). The DSCPs (DiffServ Code Points) identify the class of service in DiffServ flags. The values for maximum delay, jitter, packet loss, and transmission errors are set by agreement between the members of the GSMA and implemented on the GRX.

6.4.4.4 UMTS QoS Attribute Indicators

The QoS parameters for the media flow are communicated via a set of indicators described below. These indicators are used in the UMTS Bearer as well as the RAN, although in a different mix and different valued.

- *Maximum bit rate (Kbps):* the upper limit of SDUs (Service Data Units, i.e., packets or data frames) to and from the UMTS within a period of time, per time unit. This is based on a Token Bucket algorithm where the token rate equals the maximum bit rate and the bucket size equals the maximum volume of SDU.
- *Guaranteed bit rate (Kbps):* the assured rate of data flow (bits) that can be delivered by the UMTS per defined period of time. This is based on a Token Bucket algorithm where the token rate equals the guaranteed bit rate and the bucket size equals the maximum volume of SDU. The UMTS BS attributes (e.g., delay) are guaranteed up to the guaranteed level of traffic, thus facilitating admission control.
- *Delivery order (Y/N):* an indication of whether or not the SDUs are guaranteed to be delivered in the correct sequence. This information cannot be extrapolated from the Traffic class. This indicator determines whether out-of-sequence SDUs are dropped or reordered, in line with the specified level of reliability.
- *Maximum SDU size (octets):* the upper limit on the SDU size allowed for the given QoS guarantees. The size of the SDU can affect the optimization of the transport, especially at the RAN. However, this must not be confused with the Maximum Transfer Unit of the IP layer.
- *SDU format information (bits):* a list of possible exact SDU sizes. This is important for enabling the RAN to operate in transparent Radio Link Control Protocol mode, which is beneficial to spectral efficiency. Thus, if the application can specify SDU sizes, the bearer is less expensive.
- *SDU error ratio:* an indication of transmission errors of SDUs that are lost or corrupted. This is independent of network congestion where resources are reserved, but is used as the target value for Best Effort.
- *Residual bit error ratio:* an indication of the ratio of undetected bit error that gets through into the delivery of SDUs, where error detection is not used.

- *Delivery of erroneous SDUs (Y/N):* an indication of whether erroneous SDUs, even when detected, are to be delivered or discarded. This is used together with error detection where the delivery of the corrupted SDUs may still be useful, or is considered superfluous.
- *Transfer delay (ms):* the upper limit allowable for the 95th percent of the distribution of delay in all delivered SDUs during the lifetime of a bearer service. Delay is defined as the time from a request to transfer an SDU at one node to its delivery at the other node. Delay calculation that takes into account bursts of traffic since the arrival of the *first* SDU corresponds better to the perception of delay than calculation on the arrival of the *last* SDU. This attribute allows RAN to set transport formats and parameters.
- *Traffic handling priority:* an indicator that defines the relative precedence for handling of SDUs compared to the SDUs of other services. As a relative measure, it provides an alternative mechanism to segment high-quality traffic from Best Effort.
- *Allocation/retention priority:* an indicator that helps in admission control because it defines precedence for allocation and retention of the UMTS Bearer for a subscriber. This is not negotiated from the Mobile terminal, but the value can be changed either by the SGSN or the GGSN Network Element.
- *Source statistics descriptor (speech/unknown):* an indicator that specifies characteristics of the source of submitted SDUs. The information that the SDUs are generated by a speech source, for example, can be used by a GSN node to calculate a statistical multiplexing gain for use in admission control on the relevant interfaces.
- *Signaling (Y/N):* an indicator sent by the UE in the QoS request for an interactive service, with a high Class of Service, asking for higher priority and lower delay. An example use of signaling indication is the flag for IMS signaling traffic.

6.4.5   3GPP QoS Reference Points

6.4.5.1   Gq/Rx Reference Point (between AF and PCRF)

The Rx reference point runs between the AF and the PCRF. This interface incorporates the functionality of both Gq and Rx interfaces as defined in 3GPP Release 6. This enables the PCRF to interwork with previous releases of the AF and interact with existing PDFs.

The Rx is used for both charging and resource management purposes. The Rx transfers information from the AF to the PCRF. This information includes filters that enable identification of packets belonging to the particular service data flow. It also transfers requirements for QoS settings, such as bandwidth and level of service. The Rx also enables the AF to subscribe to notifications from the IP-CAN over the signaling path. Such events can prompt an AF action to vary the session or terminate it.

6.4.5.2   Go/Gx Reference Point (between PCEF and PCRF)

The Gx reference point is located between the PCEF and the PCRF. The Gx incorporates the previous functionality of both Go and Gx, but unlike the Go (using COPS) it is based on Diameter. Like the Rx, the Gx is used for both charging and resource management purposes. The Gx reference point enables a PCRF to have dynamic control over the PCC behavior at a PCEF. The Gx supports both SBLP and FBC decisions for resourcing and charging. The Gx supports the following functions:

- Initiation of a media flow for a connection
- Termination of a media flow under certain conditions
- Initiation of requests for PCC decisions from the PCEF to the PCRF
- Response, sending PCC decisions from the PCRF to PCEF
- Negotiations of bearer media involving:
  - UE only
  - UE with Network
  - Network only

6.4.5.3 Sp Reference Point (between SPR and PCRF)

The Sp reference point lies between the SPR and the PCRF. Through this interface, the PCRF obtains details of users' subscriptions, including their network policies for resourcing and charging. Data is retrieved according to given subscriber ID and possibly further IP-CAN session attributes. Optionally, through the Sp, the PCRF can subscribe to changes in the subscribers' data, so that it will be notified by the SPR of such changes and request downloading the subscription details again.

6.5 The NGN QoS

6.5.1 NGN QoS Architecture

6.5.1.1 NGN QoS Architecture

The NGN architecture aims to fit multiple access technologies, multiple administrations of CNs and ANs, and multiple types of user terminals. The NGN architecture supports:

- Resource request, direct or via a proxy
- Resource reservation, allocation, and activation
- SBLP
- Network policy control
- "Push and pull" policy distribution
- Gate control for policy enforcement

This architecture provides an abstract method of interaction between the Service layer and the Transport layer, allowing for any kind of service to interface to any kind of transport. The resource management is independent of the session control. It also accommodates both setting up QoS sessions via the session controller or directly with the transport access node.

Figure 6.10 shows the NGN architecture elements, where the AF acts as a proxy to the UE and can access the SPDF. It shows multiple instances of the RACF, which are positioned at the edges of the access network and the core network. Note that the AN resources are managed by the A-RACF and the CN by the C-RACF.

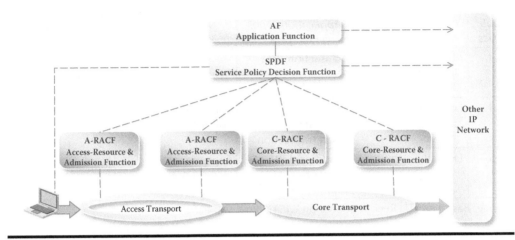

Figure 6.10　NGN QoS architecture.

6.5.1.2　Types of QoS Resourcing

In Mobile Networks, the rate of media flow is always constrained by the air link — the radio network, which is limited by the available bandwidth and the permitted spectrum. In NGN, there are no such constraints, other than the physical local loop. Support of QoS-unaware ("Best-Effort") networks as well as support of networks that have statically provisioned QoS differentiation does not require any NGN resource management via the RACS functionality.

To share the IP bearer, rather than segregate voice/video traffic, NGN needs a dynamic resource management. QoS can be served as Guaranteed QoS or Relative QoS. When it is guaranteed, strict admission control is applied with precise enforcement via a deterministic traffic policing. For Relative QoS, prioritization of traffic types is made based on service type and other factors. The IMS QoS architecture supports both models, Guaranteed and Relative, allowing access providers to select the most suitable QoS architecture for their needs.

6.5.1.3　NGN QoS Classification

TISPAN standards follow classification recommended by ITU. The following table provides guidance for IP QoS classes according to the ITU recommendation:

QoS Class	Attributes	Application Types Examples	Node Processing Mechanism	Network Routing Technique
0	Real time Highly interactive Jitter sensitive	VoIP Video conferencing	Separate queue with preferential services, traffic "grooming"	Constraint routing and distances
1	Real-time Medium interactive Jitter sensitive	VoIP Video conferencing		Less constraint routing and distances
2	Highly interactive	Transaction data Signaling	Separate queue, drop priority	Constraint routing and distances

QoS Class	Attributes	Application Types Examples	Node Processing Mechanism	Network Routing Technique
3	Medium/low interactive	Transaction data		Less constraint routing and distances
4	Low loss only	Video streaming Bulk data Short transaction	Long queue, drop priority	Any route path
5	Default	Traditional IP network application	Separate queue, lowest priority	

Source: Based on ITU Recommendation Y.1541.

6.5.2 NGN Application Function (NGN AF)

6.5.2.1 Application Functions

The AF is the function that holds a dialog with the RACS on receiving a network admission request, with or without special resources. The AF operates as the resource requesting agent on behalf of UE, CPE, AGCF, or other applications initiating sessions. The AF can be located in a different administrative entity, and therefore enables nomadic users to reserve resources and be admitted to the Visited Network.

AF behavior is influenced by the applications it supports. The RACS may not be aware of each application that the AF represents but it manages the media flow that can contain streams of media for multiple sessions. It is up to the AF to manage these sessions. The AF capabilities allow this function to:

- Identify the media flow for the SPDF, with the desired QoS levels and a service classification, and what transfer media type is required.
- Indicate whether BGF or A-RACF or both are to be in the path.
- Indicate whether gates should be opened only when resources are committed, or earlier, when they are merely allocated as in the case of "early media". Early media is used for playing tones and announcements prior to the end-to-end media connection.
- Modify existing reservation, release existing reservation, and "refresh" the time limits on ongoing reservation.
- Request address mapping and support NAT functions, including hosted NAT.
- Support overload management and reduce QoS requirements (if possible) when congestion is detected by the RACF.
- Support mode of operations with a single or multiple media flows per application per request, and multiple reservation requests per application session.

6.5.2.2 AF Service Request Process

The AF receives SIP messages requesting a session initiation for a particular user or an application on behalf of a user. Next, the AF needs to discover the location of the RACS entry point, that is, the active SPDF server. The address for the active SPDF, if not already configured, is obtained from the NASS (CLF) as an FQDN or as an IP Address. The AF identifies the media flow to the

RACF by the Subscriber ID or the IP Address. The additional information provided to the RACF is conditioned on the type of the application requesting the resources.

The AF formats the resource request, using a session's specific information, such as user ID and perhaps the identity of the Access Point node, the access network ID, etc. The type of requesting application is identified (e.g., Voice, Video, real-time gaming, Video-on-Demand). This is used to request a level of QoS required for this Class of Service and to request an appropriate service priority. The AF also generates a unique identifier for the resource reservation request. This request is then forwarded to the RACS (SPDF and RACF) over the Gq' interface.

The AF can request modifications of QoS parameters of existing media flows. The RACS will reevaluate the resource requirements, which can result in reestablishing a new media flow. The RACS also can initiate a change to an existing media flow, as a result of administrative operational instruction. If the existing session is affected, the RACS will notify the AF.

6.5.3   The RACS Subsystem

6.5.3.1   The RACS Principles

The RACS is a set of functions of the Resource and Admission Control Subsystem. The subsystem comprises multiple nodes implementing the RACF at the ingress and egress of networks, IP-CAN, and Core Bearer Network. The RACF allocates resources according to the policy decisions from the SPDF. The terms RACS and RACF are sometimes used interchangeably, although strictly speaking, the RACF refers to the functionality while the RACS refers to the combined elements of SPDF and RACF in their individual roles. The RACS elements include:

- Traffic Plane bearer functions, with A-RACF, at the access end of the network and C-RACF at the edges of the CN
- Service layer control function, with SPDF

6.5.3.2   The RACS Functions

The RACS provides a facility for applications (including IMS) to request resource reservation at the level of QoS that is suitable for the type of service. The RACS can receive and process requests for QoS control from IMS via the AF. The RACS can also process QoS control requests that originate from a particular application on behalf of the UE, for example, using SIP 3PCC.

The RACS supports session-by-session resource reservation, but also can provide resource reservation functions to non-session-based services performed transparently to the Service Control layer. The RACS, therefore, can offer the following functions:

- Admission control
- Resource reservation
- Policy control

Figure 6.11 shows the functional elements of the RACS and their interfaces by their reference points to other functions. In the RACS, the two functions are shown: SPDF and A-RACF.

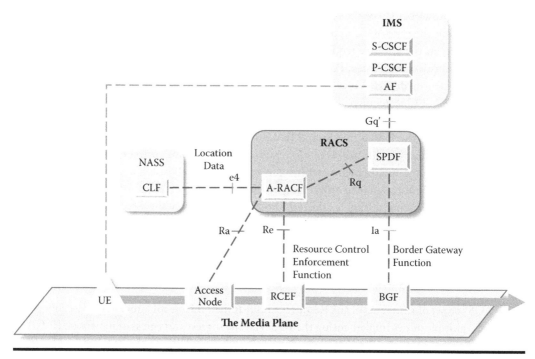

Figure 6.11   The RACS components.

The RACS is intended to provide support for all subsystems (e.g., a video broadcasting subsystem), and therefore should support non-sessional services such IPTV, interactive games, and video streaming.

The RACS supports several scenarios of resource reservation, including:

■ Where admission control is required for the "last-mile" AN segment but not for the Aggregation Network segment behind it
■ Where admission control is required for the Aggregation Network segment but not for the "last mile"
■ Where admission control is required for both the "last mile" and the Aggregation Network segment

The RACS should be able to support multiple instances of the AF. It also supports different kinds of AFs, as long as they use the standard interface defined by the Gq'. For security purposes, because the AF can reside in a device outside the control of the network operator, the RACS should be able to recognize the AF and authenticate it.

6.5.4   *The Service Policy Decision Function (SPDF)*

6.5.4.1   *SPDF Roles*

The SPDF is an element within the RACF that is responsible for *service-based* network policy decisions for sessions.

The SPDF makes policy decisions using policy rules defined by the network operator. The SPDF contains network-wide policies but local decisions can also be made according to local policies, thus affecting the A-RACF. The SPDF selects the appropriate policies to apply to the requested session, based on session details and the predefined set of rules.

The SPDF is also responsible for interpreting the policies into instructions for enforcement by the media bearing elements. It maps the policy decision into the parameters to send to the A-RACF or BGF.

In summary, the SPDF performs the following functions:

- It checks if the request information received from the AF is consistent with the policy rules defined in the SPDF.
- It authorizes the requested resources by calculating what is needed for the media components, at the requested QoS level.
- It finds the address of the local A-RACF or BGF that is able to support the requested session.
- It requests resources from the A-RACF or more services from the BGF.
- It responds with the details of the RACS to the AF.
- It hides the details of the transport layer from the AF.
- It performs resource mediation by mapping requests from the AF toward an A-RACF or BGF.
- It initiates media flow termination when necessary.

6.5.4.2   SPDF Decision Making

In analyzing policies for a specific session request, the SPDF takes the following into consideration:

- Requestor name and ID
- Service class, type of media (e.g., voice, video, text)
- Service priority (e.g., Emergency, near-real-time text)
- Reservation class

The decision-making process considers the local circumstances, local node locations and capabilities, and available bandwidth along the path.

At the RACS, the received resource request is examined against predefined rules by the SPDF. This enables the SPDF to derive the following:

- The network elements that must be involved in the resource reservation signaling path, that is, the need for the involvement of an I-BGF (Inter connection Border Control Function)
- The network elements that must be involved in the resource enforcement (e.g., which A-RACF to connect), whether or not a BGF is involved
- QoS parameters that define traffic attributes and shape, and are used by the SPDF to generate IP filters (packet marking rules) to be sent to the A-CRF or the BGF.

These policy-based parameters are inserted by the SPDF into the resource requests that are forwarded to the RACS and the BGF. The request parameters are also conveyed back to the requesting agent, the AF.

6.5.4.3 SPDF Interfaces to Other Functions

The SPDF has an additional role in NGN. It coordinates message exchange between the AF, the BGF, and the A-RACF when they are all involved in the connection. For example, the SPDF reports on abnormal state of the service data flow. If a problem arises in the PCEF, the SPDF can decide to initiate a release of resources as an administrative action. The SPDF, in this case, will instruct the A-RACF and BGF to action it, and then notify the AF.

By designing the SPDF as a module separate from the AF, the ability to apply SBLP in another network is facilitated, and a common view of the network, hiding its topology, is presented to the AF requesting the session.

The SPDF interfaces to other functions are defined in the following reference points:

- *To the AF:* the Gq′ reference point (similar to Gq in 3GPP).
- *To the A-RACF:* the Rq reference point.
- *To the BGF:* the Ia reference point (H.248).

6.5.5 Access RACF (A-RACF)

6.5.5.1 A-RACF Functions

The role of the A-RACF is twofold:

1. It performs Admission Control by checking whether resource requests can be met against QoS specifications received from the SPDF.
2. It also assembles Network Policy by obtaining the QoS decision from one or multiple instances of the SPDF. This can involve authenticating the SPDF, which can be in an external network. It also involves matching local policies with the received network policies from a SPDF.

The A-RACF is the element that looks after AN resources, to distinguish from the CN. The A-RACF provides resourcing information as follows:

- *Explicit traffic descriptors selected for the session.* This applies to policies such as DiffServ settings.
- *Attached predefined traffic profile ID for the media flow, based on bearer node information.* This ID is translated into the precise set of traffic policies at the RCEF, which applies it to that media flow. This is applicable to both Data Link Layer and Network Layer policies.

The A-RACF activates the following control functions in the RCEF:

- *Gating control.* This function enables or disables the flow of IP packets, applying instructions from the A-RACF. The A-RACF makes Gate Control decisions according to the details of the session request and a set of preconfigured conditions, and policies received from the SPDF.

■ *Packet marking.* This function is used to apply QoS differentiation mechanisms involving the DiffServ Edge Function. Where the associated parameters for the DiffServ Edge Function (classifiers, meters, packet handling actions) can be statically or dynamically configured on the RCEF.

■ *Traffic policing.* This function involves inspection of each packet to enforce the decision of the A-RACF. This inspection can lead to a reassessment of the packet classification, which will result in packets being forwarded or discarded.

6.5.5.2   A-RACF Procedures

The A-RACF receives information of the user's access into the network from the CLF (Connectivity session Location and repository Function) in the NASS. This includes the user's attachment information, with physical and repository, logical access node ID, type of access network, and the globally routable IP Address.

Optionally, the A-RACF also receives a QoS Profile that defines a set of QoS parameters, including the subscribed bandwidth, rate of priority, media types, etc. Also delivered to the A-RACF are initial gate settings, such as a list of allowable destinations and uplink/downlink bandwidth default rates, when QoS cannot be negotiated.

The A-RACF receives from the SPDF for the particular subscriber ID, the service class, media description, and service priority. With this information, the A-RACF is able to bind the user ID and IP Address to a particular path, and to Access and Aggregation Network nodes.

The initial QoS settings and QoS Profile are used to determine initial policies for the enforcing function, the RCEF. These settings can be activated for the subscriber when a new service request is received. When a new request arrives via the SPDF, the A-RACF can reject it if it does not fall within the limits of the QoS Profile parameters, in line with local policies.

6.5.5.3   C-RACF Functions

The A-RACF (Access side) and C-RACF (CN side) are specialized versions of the RACF. The A-RACF needs to establish the subscriber's QoS Profile before making a policy decision, and it obtains this information through the NASS. The C-RACF deals with CN resources and has no requirements to consult the user's profile. The C-RACF can handle resource allocations for aggregated media, that is, not on a per-subscriber basis. While the A-RACF performs session admission, the C-RACF merely maintains QoS control over CN segments.

6.5.6   Resource Control Enforcement (RCEF)

6.5.6.1   RCEF Functions

The RCEF is the function in the Transport layer that controls the media flow and enforces the traffic policies received from the Control layer (i.e., the A-RACF). This function can be embedded in the transport nodes. The RCEF is similar to the PCEF for GPRS/UMTS.

The RCEF main functions are to:

- Obtain policies to enforce for the session.
- Control the gating functions to allow through only authorized traffic.
- Mark packets of authorized service data flow according to filters received from the A-RACF.
- Ensure that service data flows conform to the policy limits and QoS parameters.

The RCEF can perform in either "push" or "pull" mode. The "push" mode implies that the request comes from the AF via the SPDF, pushing it to the RACS that resolves policies and resource availability and instructs the RCEF to allocate the negotiated resources. However, the RCEF also can "pull" when a transport-based service requires resources. The RCEF subsequently will request the RACS to provide network admission control and policy decision.

6.5.6.2 RCEF Data

The RCEF receives information of unidirectional service data flows from the A-RACF. This information includes data that defines the flow, such as:

- Source IP address
- Destination IP address
- Source port
- Destination port
- Protocol

These values can contain a "wildcard" (i.e., an open range of values) to allow the RCEF some flexibility in managing them. A media session identifier can be addressed independently, and that address can be changed during the session. This allows for sessions that get connected first to one destination or an application server (e.g., playing announcements), followed by reconnection to the destination.

The media flow is described by the media type, media ID, and media priority:

- The *media type* is defined by the requested service, such as voice, video, or data.
- The *media ID* is a unique identifier for the specific session media.
- The *media priority* is used by the A-RACF to determine call admission parameters. This is a separate parameter from the service priority that determines the precedence of the service (including signaling).

Within media flows there are traffic flows with common or unique characteristics. The following table itemizes these characteristics:

Traffic Flow Parameters	Traffic Flow Description of Media
Direction	Direction of the unidirectional flow, from or toward origination or destination
Flow ID	Identifier for the specific media flow, which could be sharing a bearer
IP addresses	Source and destination IP addresses and the address realm. The addresses may be compatible with IPv4 or IPv6 (using the prefix of IPv6 for IPv4 implementations)
Ports	Source and destination port numbers. The values for port numbers can be expressed as wildcard, range of numbers, or within limits (maximum and minimum)
Protocols	The media handling protocol ID, such as UDP or TCP
Bandwidth	The maximum requested bit rate
Reservation class	Identifier for a set of traffic characteristics, e.g., burstiness and packet size
Transport service class	Parameters for the forwarding behavior assigned to a particular media flow. This is part of the QoS profile used in NASS for call admission

Source: Based on TISPAN ES 282003.

These characteristic descriptors are used to request resources over the Rq interface between the SPDF and the A-RACF.

6.5.7 The NGN Border Gateway

The NGN Border Gateway is discussed here in relation to resourcing and policy management only. See Chapter 7 for border security description.

6.5.7.1 Border Network Elements

The BGF is located anywhere in the transport network. It can be installed between an AN and a CN (C-BGF) or between two CNs (I-BGF). An A-BGF can be installed between the Access Network and the customer's equipment. The qualified identification of the BGF is intended to highlight the exact scenario rather than to indicate different operations.

Figure 6.12 shows the interaction of bearer border functions with the NASS and the RACS. In this example, the NGN RCEF is accompanied by an L2TF (Level 2 Termination Function) server at the IP Network Edge node. The C-BGF and the I-BGF are shown together in the Border Control node, although they serve different parts of the network: the C-BGF provides border control before entering the CN, and the I-BGF protects the inter-operator border with other IP networks.

6.5.7.2 The NGN Level 2 Termination Function (L2TF)

The RCEF and the L2TF are two separate functions that frequently share the same platform, that is, in the IP Edge nodes, where the session resourcing is determined by the RCEF and a tunnel is set by the L2TF. The L2TF is the function that communicates with CPE, that is, terminates the signals from the UE at this point and sends generated messages in a tunnel.

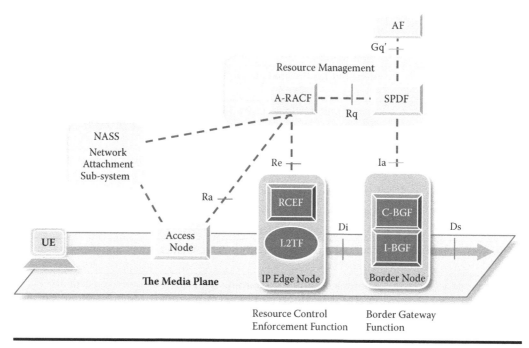

Figure 6.12   NGN border functions.

6.5.7.3   NGN Resourcing in the Border Gateway

The BGF enables media flows between two terminations expressed as IP addresses and ports, where the IP addresses can be translated addresses. The BGF allows requesting NAT binding and address latching for UE behind NAT gateways. These requests can be for unidirectional or bi-directional connections.

The BGF resides in media bearing nodes in the transport network. It can be positioned between an AN and a CN (C-BGF) or on the border with another IP Network (I-BGF). It marks packets according to received instructions from the SPDF. It monitors packet flows and shapes the traffic using the gating function. The BGF is a packet-to-packet gateway that performs policy enforcement as well as address translation and interworking conversion functions, as and when required.

The BGF functions are summarized as follows:

- Performing NAT to convert to a globally routable IP address or converting between IPv4 and IPv6 addresses.
- Marking packets for QoS classes according to received QoS parameters and limiting the bandwidth to fall within the rates agreed to for the session media flow.
- Performing address latching, to enable media flowing to and from a translated address or port, for UE behind a NAT gateway.
- Providing mid-session modifications of media flow where NAT control and QoS parameters have changed.
- Measuring usage.

The BGF receives information of the service data flows from the SPDF over the Ia reference point. This interface is realized using the H.248 Protocol or COPS. This interface provides

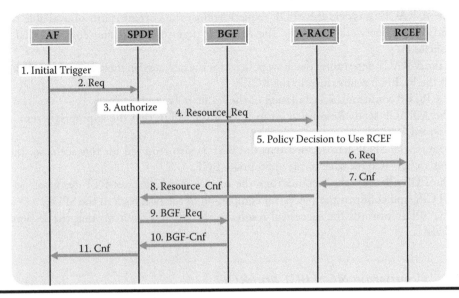

Figure 6.13   Call flow to select RCEF and BGF.

information that includes the source IP Address, destination IP Address, source and destination ports, and the type of protocol. Where the value for these items is not fully known, a wildcard range can be sent to the BGF. The SPDF also sends various filters to "shape" the traffic flow according to the QoS parameters and limits. The interface carries requests of bandwidth for specific media flows, including modifications to existing ones. Also transferred on this link is the status of the BGF, to enable recovery from different failure scenarios or avoiding overloading.

To enable protocol-specific processing on the BGF, the interface enables defining the media transport protocol per media flow, such as RTP, T.38, and MSRP (Message Session Relay Protocol). The protocol-specific actions include, for example, reserving dual ports for RTP or RTCP, or collecting statistics per protocol.

There are several scenarios for procedures to involve border elements in a session. The session may or may not need the involvement of certain border elements (for on-net destinations), or a number of permutations of them. This depends on the requested service, the type of media transport, and the network topology.

Figure 6.13 details a scenario in which the AF requests resources, and the SPDF establishes the need for RCEF to be involved as well as BGF. The sequence of messages is as follows:

1. The AF receives a request to set up a session. The AF ascertains what kind of service is required and what transport layer resources are needed.
2. The AF sends service request information to the SPDF.
3. The SPDF checks and authenticates the particular AF and also checks that the network policies allow for resources for the dialog between the AF and the SPDF, as requested by the AF.
4. Next, the SPDF determines what is needed to serve the request (the level of resources required to satisfy a class of service) and the need for RCEF, BGF, or both. In this scenario, it sends a Resources-Req to the A-RACF to allocate resources, or a BGF-Req to the BGF. The SPDF uses the local policies and the parameters in the request to make the decision.

5. The A-RACF considers the SPDF request against the current status of available resources and internal network topology. The A-RACF performs admission control based on AN policies.
6. If the A-RACF determines that new policies for the new session media flows must be installed on the RCEF, it sends them to the RCEF.
7. The RCEF confirms the installation of the traffic policies.
8. The A-RACF sends Resource-Cnf to inform the SPDF that the appropriate resources are reserved.
9. Because the SPDF has decided that the BGF is also required for this scenario, the SPDF sends a bgf_Req message to the appropriate BGF.
10. The BGF allocates the resources for the service (e.g., allocates RTP resources as well as RTCP), and confirms the successful completion of the task back to the SPDF.
11. The SPDF forwards the successful results to the AF and confirms that the session can go ahead.

6.5.7.4   Comparing NGN BGF and RCEF

The functions of the BGF, both the Inter-network and Core BGF, overlap the RCEF, as described above. This is necessary to provide these functions in different scenarios and different parts of the network.

The following table compares these functions:

Functionality	RCEF	C-BGF	I-BGF
Open and close the gates	Y	Y	Y
Packet marking	Y	Y	Y
Resource allocation for a flow		Y	Y
Address translation		Y	Y
Hosted NAT traversal		Y	
Police uplink and downlink traffic flow	Y	Y	Y
Usage metering		Y	Y

Source: Based on ETSI ES 282 003.

Because these functions are similar, they can be contained in a single platform, however, they may be needed at different points in the network topology.

The RCEF and the BGF are notionally different functions but they are assumed to co-reside in one node. The interfaces between them are not specified, as they may well be internal connections.

6.5.8   End-to-End Resource Negotiation

The negotiation methods of QoS parameters across the border are yet to receive wide consensus. Each network operator must have some autonomy on how it admits sessions and ensures packet delivery. Each network operator wants to retain control of resource management and reserve the right to override a QoS Profile assigned by another network.

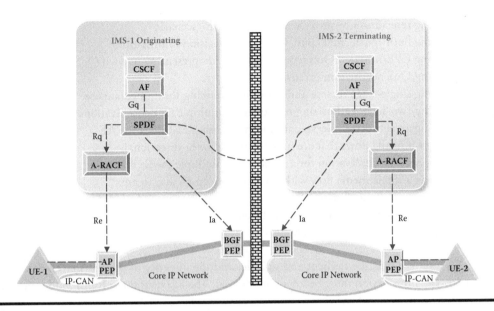

Figure 6.14   End-to-end QoS negotiation via SPDF-to-SPDF.

There are a number of potential solutions, each with its own merits and risks:

- SPDF-1 to SPDF-2
- RACF-1 to RACF-2
- SPDF-1 to RACF-2

The interfaces between the RACS components across the border are under investigation.

The Transport layer products can provide internal resourcing and anti-congestion facilities; therefore, the concept of the final resourcing decision being dictated from the Service Control layer is not acceptable to network management people.

Figure 6.14 presents an option for cross-border QoS negotiation, one in which the SPDF of the originating network is communicating requirements to the SPDF of the terminating (or intermediate) network. The terminating network consults its Transport layer via the RACS, and returns its available level of resourcing. The originating SPDF can accept it, or it can reject the offer and refuse the call admission request.

6.5.9   NGN QoS Reference Points

6.5.9.1   Rq Reference Point (between SPDF and A-RACF)

The Rq interface is between the SPDF and the RACF. The Rq enables management of an individual media stream that is associated with a specific application. The Rq can create, modify, and release each media flow or handle the collective media stream.

The Rq interface is capable of:

- Requesting resources from Access and Aggregation Networks
- Allowing for SPDF and A-RACF in different administrative domains

- Supporting reservation by proxy with policy "push" scenario
- Supporting both "push" and "pull" transfer of decision information
- Supporting CPE-initiated QoS requests via the AF, presenting its own QoS requirements
- Supporting QoS negotiation extensions
- Supporting single-stage reserve and commit resourcing for immediate use
- Supporting two-stage authorizing resources for multiple service activations
- Supporting an authorize-reserve-commit model, with SBLP via the AF
- Allowing for media flows that are uni-directional or bi-directional
- Allowing resource reservation for a session with multiple media flows
- Supporting a single media flow with multiple session media streams
- Modifying an individual media flow or service data flow in mid-session
- Releasing an individual media flow or the entire service data stream
- Extending given time limits parameters

The Rq interface must be resilient and reliable, and the integrity of the transferred data must be ensured. The information must be kept confidential. The SPDF and the RCEF nodes must be able to authenticate each other.

6.5.9.2   Gq' Reference Point (between AF and SPDF)

The Gq' reference point lies between the AF and the SPDF. It handles resource requests from the AF to the SPDF, and notifications from the SPDF to the AF. The resourcing requests are forwarded, together with policy decisions, by the SPDF to the RCEF on the Rq, and to the BGF over the Ia, if the destination is in an external network.

In fact, the Gq' combines both signaling functions of the Rq' and Ia. Although all these interfaces operate in the same way, the actual parameters values needed at service execution may be different across each interface, as determined by the SPDF, according to requests coming from AFs and according to the operator's local policies.

6.5.9.3   Ia Reference Point (between BGF and SPDF)

The Ia reference point carries messages for dialogs concerned with establishing sessions across network egress to other networks. The Ia interface is based on H:248.

The Ia supports the following:

- Requests of binding addresses through NAT, i.e., keeping an internal IP address and port as a matching pair with the external IP address and port
- Request allocation of resources (i.e., bandwidth)
- Relay information about the destination media parameters, the media direction (bi-directional or uni-directional and in which direction), and the IP address and port latching for specific terminations
- Set up the media transport protocol (RTP, T.38, MSRP, etc.) for the media flow in each NAT binding request to enable specific resourcing (e.g., dual port for RTP and RTCP)
- Allow modifications of media parameters in mid-session, possibly requests for new IP address and port latching in mid-call, including changes to bandwidth in mid-call

- Convey the selected policy for the session
- Transfer QoS marking values (e.g., DiffServ DSCP) to be applied at the transport gateways
- Reports from the BGF to SPDF on BGF status (booting, offline) and resource usage at session termination or mid-term

6.5.9.4 Re Reference Point (between A-RACF and RCEF)

The Re reference point is used by the A-RACF for transferring policy-decision information that controls the Layer 2/Layer 3 transport plane. The RCEF will then perform the necessary functions of gating, packet marking, traffic shaping, etc. to enforce the policies. The Re interface is also used to send mid-session updates from the Transport layer toward the A-RACF.

The Re reference point is used for both non-functional configuration purposes and functional QoS management purposes. This means that the Re can transfer traffic policies to be installed on the RCEF. Installing a new policy for a particular flow or a group of flows can result in the replacement of a policy previously installed.

The Re interface controls the following functions:

- *Gating* is performed by the RCEF. The open/close gating decision essentially is based on instructions from the A-RACF to enable or disable IP flows.
- *Packet marking* is performed to apply QoS differentiation involving the DiffServ Edge. The DiffServ Edge parameters (i.e., classifiers, meters, packet handling actions) can be statically or dynamically configured on the RCEF.
- *Traffic policing* is performed by the RCEF to enforce the decision of the A-RACF. It involves inspection of each packet, classifying it for forwarding or discarding it.
- The *removal of policies* is initiated by the A-RACF. This message causes the RCEF to release all the resources associated with an existing traffic policy.
- The *revoking of policies* is initiated by the RCEF. This is required when an external Transport layer event signifies that the access information is no longer valid. The RCEF notifies the A-RACF and releases all the resources associated with this reservation.

Chapter 7

The IP Border and Security

7.1 This Chapter

This chapter deals with aspects of the ingress or egress of the operator's IP network. This encompasses management of the IP border with another IP network, as well as toward the access network — that is, with access gateways, CPE devices, and ultimately the handsets. In all these cases, the network must take certain measures to protect itself, to ensure protocol interworking and to enable routing by global addresses.

This chapter first identifies the required functions that should be performed at the border, such as security filtering and encryption or media conversion. It describes security measures (such as IPSec tunneling) to protect the IP network signaling and provide privacy and data integrity, and to enable address translation between private network addresses and public addresses and between different versions of IP. It also describes general firewall techniques used at the network edge, including Deep Network Inspection.

Next, the converged network border functions are described for both NGN and 3G, with their network positioning and functionality. As most functions are now common to both Mobile and Fixed NGN, there is a short description of some aspects that are particularly typical of 3G Mobile and of NGN borders, followed by a detailed description of the border control elements, in particular the IBCF and the BGF.

Finally, this chapter describes the commonplace Session Border Controller (SBC), comparing it with the defined standards. It also gives a brief description of State Security Lawful Intercept mechanism required by regulation.

7.2 Border Management

7.2.1 Network's Border Points

Unlike the case in most CSNs, in IP, different layers of the OSI model can belong to different commercial entities, thus creating borders that must be monitored at various levels. In IP communications, the media flow travels in a different path from the signaling route, and therefore crosses borders at different checkpoints.

A network border point is where signaling and media packets cross over from a network administrated by one entity to another network managed by another entity, that is, the "Business Border." This demarcation point is handled by specially designated servers that "publish" or "advertise" their addresses, therefore they are "well known," and can be routed to by external servers.

Whenever a communication is crossing a border between networks, there are several border functions to be performed. In fact, borders exist even between segments of the same network that span several regions, cross country borders, and may have a level of autonomy. Each network can provide a border server that performs various security checks. The border is, in fact, a logical De-Militarized Zone (DMZ) stretching between two border servers on either side.

Figure 7.1 demonstrates the multiple types of borders for the operator's IMS network, with UNI (User to Network Interface) and NNI (Network to Network Interface). Each border is managed by a set of specific functions. These functions include a wide range of signaling functions, different types of media bearing gateways, security filtering gateways, and interworking facilities.

A different mix of functions is needed for each type of border:

- NNI gateways are placed between the operator's network and other networks, including the CSN, PDN, and external IMS
- UNI gateways toward access networks that can be Fixed or Mobile

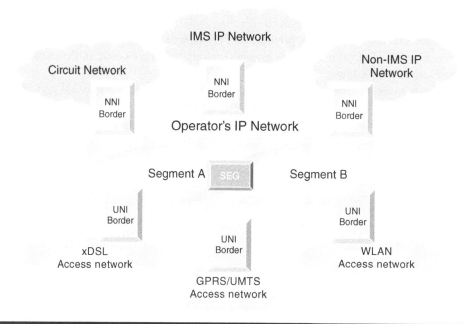

Figure 7.1 Various network borders.

■ Internal gateways between segments of the same network

The flow of packets is intercepted at both the signaling level and the media payload level when crossing the border. Even where there is no need to convert the transport mechanism (i.e., to packet), the border must be monitored for the following purposes:

■ Unlike CSNs, IP connection is deemed to bring more risks and less control. Packet networks can be flooded by unwanted traffic that deteriorates the quality of service to *bona fide* paying customers.
■ Unlike IP data connection (e.g., Web browsing or e-mail), IMS and VoIP present many more interworking issues that must be resolved in real-time to enable global communications. Such issues include not only protocol conversion, but also signaling adjustments when different SIP profiles or different versions are used.
■ Like Web services, IP communications need to resolve differences in IP addressing that occur between private and public addressing, and between IPv4 and IPv6 networks.

7.2.2 Border Management Requirements

7.2.2.1 Border Control Requirements

Border management, operates at both the signaling level and the media flow level. It affects Layer 2, the Data-Link layer, where packets are forwarded according to packet headers through "gates" that open and close, and controlled by the signaling elements. Border management is also concerned with Layer 3 protocols (the Network-Link layer) where network addresses are managed and converted (e.g., between IPv4 and IPv6), and where certain types of security attacks can be detected.

Border management is applied for verifying, even asserting, the identity of the senders and the destination parties that wish to communicate across operators' boundaries. At the border, security checks are made, as well as network admission. Security functions ensure that media flow is not subverted or used maliciously, and that the carrier or the user is not defrauded or harrassed.

The IP border functions also include interworking between domains and technologies. IP protocols must be translated (H.248, H.323, and SIP) or merely adjusted for different implementations (e.g., variants of SIP profiles or different release levels).

The network borders exist not only between two administrative entities, but also between the network and the user. This access border exists whether the equipment used by the user is a handset or a CPE. User devices can be tampered with, and therefore deemed outside the perimeter of the trusted network. The border is defined as the perimeter around a "trust" zone where communication within the zone can afford lower security, and communication toward the untrusted zone requires additional measures.

Finally, border functions play a major role in mobility. As users can be mobile or nomadic, they often need to cross operator boundaries to receive any service from the network. This entails access edge security and network-to-network border control even at registration time, let alone every session request.

7.2.2.2 Border Functions

The functions required for the IP border include the following:

- Controlling transport plane functions, including QoS (addressed in Chapter 6)
- IP Address and ID management (addressed in Chapter 5)
- Mutual authentication (addressed in Chapter 4)
- Address translation between private and public address realms (see below)
- Converting between IPv4 and IPv6 for devices and servers
- Software and protocol version adjustments
- Hiding network configuration to prevent disclosure of internal information, such as identity, capability, and full IP address of internal servers and network nodes
- Insert an IWF (InterWorking Function) in signaling route when required
- Screening SIP signaling information based on source or destination
- Screening information according to operator policy (e.g., remove information that is regarded as confidential by the operator)
- Generation of CDRs, logging what is crossing the border (addressed in Chapter 8).

7.2.2.3 Routing to External IP Networks

The IP Border servers must cope with any type of IP connection. This may be between:

- Two IMS CNs, but different administration
- IMS to SIP-based non-IMS multimedia network (typically a specialized service provider)
- IMS to non-SIP network (e.g., H.323 networks; typically Enterprise)
- IMS to Internet

7.2.2.4 The Border with Access Networks

The border with the AN (Access Network) varies greatly between technology domains, that is, NGN, Mobile, and WLAN. Mobile networks are deemed secure enough between the handset and the access point over the RAN because the spectrum is restricted and the transmission is encrypted. However, there is now growing awareness that added security is needed for IP and SIP signaling between the mobile GPRS/UMTS nodes and the Core IMS network.

In NGN, the protection from unscrupulous use of the network by the clients has been paramount since the concept of VoIP was established over the Internet. This endpoint can be a soft client or a customer-premise gateway. The endpoint is considered "untrusted," and the connection with it must be monitored for signaling as well as for media flows. In all cases, the likelihood of difficulties with clients' software versions is very high and should be resolved at the gate into the operator's network.

The need to protect the identity of internal servers from the access devices is common to Mobile 3G and NGN. The servers that can be discovered by the user device must allow no further penetration and restrict the user's ability to bypass the network admission procedures.

7.3 Security Aspects

7.3.1 Security Threats

The border management and security measures must combat many different threats and vulnerabilities, including future ones yet to emerge. The list below is, therefore, not exhaustive.

Threats to *confidentiality*, where the user's confidential information is compromised, include:

- *Eavesdropping.* An intruder intercepts messages without detection.
- *Masquerading as a user.* An intruder pretends to be an authorized user to receive confidential information.
- *Masquerading as a network server.* Intruders can impersonate a serving network. An authorized user connecting to it, believing it to be the *bona fide* network server, can proceed to disclose private data.
- *Traffic analysis.* An intruder observes the time, rate, length, source, and destination of messages to determine a user's location or to learn whether an important business transaction is taking place. This can be achieved by passive means (by observing traffic) or by active means (by sending a query or signal to the system to obtain access to information).

Threats to *data integrity*, where corruption of the data inflicts network disruption, include:

- *Manipulation of messages.* Messages can be deliberately modified, inserted, replayed, or deleted by an intruder.
- *Corruption of data.* Illegal access to a database enables an intruder to "poison" the data and change details, for malicious purposes or for fraud. This includes DNS cache poisoning that corrupts the DNS database.

Abuse of *network resources*, where an attacker can manipulate resource usage, includes:

- *Intervention.* An intruder can prevent an authorized user from using a service by jamming the user's traffic, signaling, or control data.
- *Stolen ID.* An intruder can pretend to be another user toward the network, obtain authentication, and then hijack this connection to gain access to network facilities that are charged to the stolen ID.
- *Misuse of privileges.* A user or a serving network can exploit their privileges to obtain unauthorized services or information.

DoS (denial-of-service) attacks that engage the recipient servers with excessive and unnecessary processing, thus denying access to *bona fide* users include:

- *Abuse of services.* An intruder can abuse some special service or facility to gain an advantage or to cause disruption to the network.
- *Resource exhaustion.* An intruder can prevent an authorized user from using a service by overloading the service, leading to denial of service or reduced availability.
- *Protocol intervention.* Intruders can prevent signaling from being transmitted by inducing protocol failures.

- *Abuse of emergency services.* Intruders can cause serious disruption to emergency services facilities by abusing the ability to make free calls to emergency services.

Repudiation of commitment, where user transactions are disputed after the effect, includes:

- *Repudiation of charge.* A user could deny having incurred charges by denying that the service was actually provided.
- *Repudiation of user traffic origin.* A user could deny that he or she sent user traffic.
- *Repudiation of user traffic delivery.* A user could deny that he or she received user traffic.

Unauthorized access to services, where attackers "spoof" their identity to obtain unpaid services, includes:

- *Masquerading users.* Intruders can access services by masquerading as *bona fide* users or network entities, and gain access to services.
- *Exceeding access rights.* Users or network entities can get unauthorized access to services by misusing their access rights.

7.3.2 Data Security

The network functions are driven by control data, both static and dynamic, which is stored in various servers. Tampering with this data can amount to a serious attack, defrauding the operator, and compromising user privacy. Data that needs protection can be either dynamic or static permanent data. The following table lists data items, dynamic and static, that should be protected:

Data Type	Type of Contents	Storage	Description
User data	Profiles	HSS/UPS	User personal details records
	Identification	HSS/UPS	The associated public and private identities, implicit registration group of IDs. GRUU data (associating device number with logical name)
	Services profile	HSS/UPS	User services subscription records, including service triggers and specific parameters
	Location data	HLR/CLF	Dynamic mobility information, including current user or device location
Admission control	Credentials	HSS and NASS/FPDB	Username, password, shared secret, encryption keys
	Access data	UE, APN, AMF	Physical connection, ports, well-known addresses for network entry points

Data Type	Type of Contents	Storage	Description
	Roaming data	UE, APN,	Roaming agreements, access nodes contacts
Session control	Session data	S-CSCF	Dynamic session data, dialogues, originating and terminating parties
	Presence	Stored on CSCF and Presence Servers	Dynamic status of a subscriber, including user connectivity preferences, generated by session controllers
Charging and QoS data	Charging records	CCF (CDF + CGF)	Charging information per session or usage
	Charging rules	PCRF/SPR	Rules stored by the operator to be applied to specific services and particular users, including nonchargeable items
	QoS profiles	HSS/UPS, also GPRS nodes and NASS	Level of QoS allocated per service and per user
	Billing	BD (Billing Domain)	Billing data and invoices from the billing system
Traffic and routing	Routing Data	CSCF	Session control data that includes path for routing
	User Plane Media	Media gateway and border gateways	Payload content in RTP or SCTP
	Addressing	DNS/ENUM server	Data in DNS/ENUM for domain name resolution and TDM numbers association

7.3.3 Privacy

The term "privacy" refers to information that users choose to keep confidential and not disclosed other parties in the communication.

7.3.3.1 Anonymity

Anonymity in IMS is a service that users can subscribe to, as they did in legacy networks, when they choose to be "ex-directory." Keeping anonymity also means that the called party is prevented from finding out the calling number. This is performed via a SIP privacy vector that is inserted into the message header according to user preference, and carried over through the various processing nodes. Anonymity can be permanently turned on or temporarily, session by session.

During session initiation, the UE can request privacy for the P-Asserted-Identity, to prevent revealing it to the other party. The network can override the subscriber's privacy instructions for Emergency services, to allow locating the caller by the Emergency Agency.

Anonymous Communications Rejection is an associated service that allows the recipient of the call to refuse to answer an unnamed caller, in case it is a nuisance call.

7.3.3.2 Privacy for Presence

Presence and user status in NGN IMS are subject to special privacy preferences. The Presence Principal owns a "presentity" that contains configurable preferences, including privacy parameters. These parameters include controls of who is allowed to "watch" the presence and what part of the presence information can be disclosed to that watcher. The parameters can define set times and special dates for the presence discloser (e.g., allow disclosure during working hours).

The Presence Principal can also grant or withhold consent for watchers to use the Presence information in some applications (e.g., for smart diversion). For further details on Presence management, see Chapter 3.

7.3.4 Zone Security Principles

7.3.4.1 Security Zones

As the trend toward implementing global IP networks grows, there is a need to create segmentations with regional variations within the network. Global operators now maintain IP networks over several countries. Large IP network management is made easier by segmentation. In addition, identifying the "Trust Zone" within the IP network is instrumental in controlling security.

To that end, the standards define the NDS area with two levels of secure connection: (1) across the border (Za) and (2) within the Trust Zone (Zb). The more secure connection (Za) is managed by the SEG (Security Gateway).

Figure 7.2 illustrates a security model that shows security service connection between two Security Domains using the Za interface, which is untrusted link and between internal Network Elements (NEs) over the Zb interface, which is a trusted link.

In IMS, a user in a Visited Network reaches the Visited P-CSCF and requests registration or session initiation. The signaling between the P-CSCF and the SEG is carried over the Zb (internal security), shown as Zb-1 in Figure 7.2. The signaling is then transported from the Visited SEG to the Home SEG over the Za interface (external security). Within the Home Network, the signaling is supported by the Zb-2, as defined by the Home Network operator.

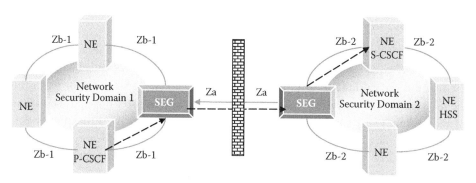

Figure 7.2 Secure connection between IMS security domains.

7.3.4.2   The Network Domain Security (NDS)

The NDS is a network zone that belongs to a single administrative entity, where all the internal servers are entirely "trusted." Within the zone, lower security mechanisms can be applied, thus saving unnecessary processing. However, when exiting the NDS toward another NDS, a higher level of security must be used over the inter-connection.

The concept of NDS allows for the implementation of different security regimes within each zone, where end-to-end security is managed on a hop-by-hop basis. This concept mandates that no direct communication between the NEs of one NDS and another NDS can be made unless the signaling traverses the SEGs.

7.3.4.3   The Security Gateway (SEG)

An SEG is a network function on the border between different Network Domain Security (NDS) zones. The SEG handles communication across the border over reference point Za. It is responsible for enforcing security policies between network segments and for interworking between them. The SEG is responsible for enforcing the security policy of each IP security domain, and over the link with other SEGs.

The SEG uses IKE for its security key management. IKE is defined by RFC 2407, 2408, and 2409. It enables the parties to negotiate, establish, and maintain the SAs that are necessary to establish secure connections. SA definition is provided in RFC 2401.

All IP traffic must pass through a SEG before entering or leaving a security domain. The SEG performs security policy based filtering and firewall functionality. It maintains a secure database of long-term persisting keys used for IKE authentication.

The number of SEGs in a security domain depends on the need to differentiate between the externally reachable destinations, the need to balance the traffic load, and the need to avoid single points of failure.

7.3.4.4   IP Security (IPSec) and Encapsulation Security Payload (ESP)

Security methods are applied differently according to the mode of the traffic. In Transport Mode, the data in the payload is protected by encryption but the headers are not encrypted so that they can be examined and routed by. In Tunnel Mode, the whole IP packet is protected. The encrypted packets are transported across nodes that have no knowledge of their contents. Therefore, Tunnel Mode is more secure but requires both ends to collaborate in creating the tunnel.

IPSec provides the means for creating such tunnels. IPSec is a set of security facilities and set parameters that are combined in an SA (Security Association). The IKE Protocol is used for negotiation of IPSec SAs for the inter-security domain over the Za-interface.

IPSec SA defines the security protocol to be used, the mode of the communication security, and the types of endpoints. IPSec provides authentication of the data originating source, protects data integrity and data confidentiality, and provides anti-replay protection (recording a dialog and replaying it back to gain admission by deception).

The ESP Protocol is generally used for IPSec. The ESP header provides parameters that define a combination of security services for both IPv4 and IPv6. These security services are provided between a pair of security gateways or a gateway and a host.

ESP is capable of protecting confidentiality, data origin authentication, and connectionless integrity. It provides an anti-replay service (protection against blind replay of authentication for masquerading attacks) because the dialog cannot be distinguished. ESP also provides some traffic flow confidentiality, which is effective when the parties' addresses are concealed in Tunnel Mode, or when ESP generates dummy traffic to conceal the relevant packets.

It is possible to use lower level ESP that allows an encryption-only SA, which provides lower security but better performance because generated dummy traffic increases the network loading.

These security functions and their particular details are defined when the SA is determined for the type of connection, and are dependent on the network topology and the parties' relative locations.

The SEGs establish and maintain IPSec and ESP and with SA in tunnel mode between security domains. SEGs will normally maintain at least one IPSec tunnel available at all times to a particular peer SEG. The NE (Network Element) may be able to establish and maintain ESP SAs toward a SEG or other NEs within the same security domain.

As an option, operators can establish only one ESP SA between two security domains. This simplifies the security interaction and also helps protect against the threat of Traffic Analysis because it masks a particular session. However, the drawback is that in doing this, it is not impossible to provide fine-tuned security protection for different types of communication.

The NDS for 3GPP supports the use of the IETF IPSec protocol suite, but certain options are regarded as unnecessary for GPRS (due to better air-interface security) while others are deemed as mandatory, thus simplifying the implementation.

7.4 Security Mechanisms

7.4.1 Firewalls

7.4.1.1 The Functions of the Firewall

Firewalls are designed to protect the network and the user from malicious attacks or fraud. The first packet filters were designed for data services and only recently needed to be applied to real-time interactive communication services. Firewalls essentially perform packet filtering according to set rules. Only packets that pass all the filters can proceed. Otherwise, they are discarded (rejected with an error response) or dropped (rejected without error messages).

There are two different strands to security strategy:

1. Default = Permit: permit whatever is not prohibited by the rules.
2. Default = Deny: deny whatever is not permitted by the rules.

In practice, both these strategies are employed, but at different scenarios.

The firewall type and the level of security it needs to provide depend on the type of the connection:

- Connection between a single node and the network, or between two or more networks
- Connection that requires session tracking or just stateless filtering
- Connection intercepted at the Application layer or at the Network layer

7.4.1.2 The Evolution of Firewalls

Early "stateless" firewalls were not service-aware, therefore, they handled the packets without reference to the media stream to which these packets belong. The filters, naturally, did not contain conditions that depended on the type of service. Filtering was based on information in each packet's header, including the IP address of the originator and the destination, the protocol, and (for TCP and UDP) the port number. The use of "well-known ports" meant that the port number identified the type of data transmission, such as Web browsing, remote printing, e-mail transmission, and file transfer.

The next generation of firewalls became "stateful" (i.e., connection-aware), where the state itself can trigger a particular filter. Being able to determine which packet belongs to which connection provides a tool to prevent DoS attacks, such as SYN (synchronized) flood and TCP RST (reset) flood attacks.

Third-generation firewalls became application-aware, also described as "proxy firewalls." Application layer firewalls do not route traffic at the Network layer. Traffic is terminated at the firewall and only valid packets are forwarded to the new connection made by the firewall. This type of firewall contains filters based on knowledge of which application is used (i.e., FTP, DNS, or Web browsing), recognized by their protocols. The firewall can detect whether another communication dialog or rogue protocol is initiated on a non-standard port, or a known protocol is misused for another purpose.

7.4.2 Intrusion Detection and Prevention Systems

Firewalls protecting the network against unwanted intrusion, such as worms, first appeared as separate types of security servers, especially sought after by enterprises. These functions require monitoring packets and shaping the flow to prevent and detect malicious attacks.

Intrusion Detection Systems (IDSs) became popular initially, but were soon superseded by Intrusion Prevention Systems (IPSs). IDSs and IPSs work by matching patterns ("signatures") that identify viruses, worms, Trojans, etc. In addition, these solutions have the ability to decode Layer 7 (Application layer) protocols such as HTTP, FTP, and SMTP (Simple Mail Transfer Protocol), and base many more rules and conditions on protocol-specific parameters.

IDS solutions fell out of favor when it became evident that they produced too many alerts for legitimate traffic ("false positive") and did not actually act upon intrusive traffic to stop the attack. The IPS, in contrast, aims to prevent the intrusion and deal with the rogue packets promptly. However, an IPS cannot operate efficiently unless it also has a network detection system that is fast and reliable.

For data applications that tolerate a time delay, a firewall can handle traffic that is duplicated offline. For real-time protection, an online security mechanism is required so that action can be taken at the time of forwarding packets, thus preventing the intrusion there and then.

The IPS mechanism differs from IDS by placing the detection system in-line. To do this, it is necessary to ensure that the overall performance is not degraded and the quality of the communication is not affected. Such real-time response is essential for Voice and Conversational Video applications, which need to block traffic as the attack occurs without an undesirable impact on the perceived session quality.

7.4.3 Deep Packet Inspection (DPI) Techniques

7.4.3.1 DPI Principles and Uses

Deep Packet Inspection (DPI) is a method of packet analysis that examines the whole packet, not merely the header. This enables performing packet filtering that considers more information about the originating service than what is conveyed by header fields. DPI filtering can distinguish between media flows, regardless of protocols. DPI inspects data that is related to multiple layers in the OSI layer model, up to the Application level.

Using DPI, packets are classified according to preconfigured rules and conditions, and the packets are marked or tagged so that these rules can be enforced by network nodes along the transport path. This capability can be used to:

- Manage QoS according to media flows
- Prioritize according to service
- Block unwanted traffic
- Limit the flow rate, to control network congestion
- Create a mirrored flow data reporting for Lawful Interception

7.4.3.2 DPI for Firewall Security Function

DPI is an important tool for security monitoring, and is now integrated into firewalls or security servers within access or network nodes. DPI need not reside only in such firewalls, but also can be integrated into existing network nodes such as SGSN or GGSN, thus providing security across the network rather than merely at the border. The same applies to private networks and LANs, where DPI-based firewalls can be utilized to shape internal traffic.

DPI firewalls can be seen as the amalgamation of the IDS and IPS capabilities with traditional stateful firewall technology. Combined with DPI, the stateful inspection extends the packet analysis to all relevant OSI network layers, from Layer 2 to Layer 7. In this way, events occurring at the Transport Level can be interpreted against information from the Applications Layer.

DPI must ensure that performance is not degraded to enable it to operate in real-time interactive services. Real-time applications cannot tolerate the additional performance overhead imposed by the complex pattern (signature) analysis and state tracking. Therefore, DPI provides a new mechanism that executes these procedures in parallel, and avoids impact on the performance.

DPI firewall also must provide wire-speed inspection and filtering for Secure Sockets Layer (SSL) sessions. This entails the decryption of the packets within the SSL session and then the reencryption of them when they have been cleared to proceed.

7.4.4 Application-Level (ALG) Firewalls

7.4.4.1 Requirements for the Application Layer Gateway (ALG)

Applications in the IP communication network are very powerful because they can use the Control layer without the session controller's awareness of how the session is constructed. For example:

- Web applications, using XML, SMTP, or SOAP, can modify the session by carrying signaling instructions in the payload, thereby avoiding network session control.
- Applications that modify the communication port during the session aim to bypass filtering that is configured to watch a particular port.
- With general-purpose port usage, when a tunnel is created and data is encrypted, the application can avoid scrutiny and evade network policies.
- Handset applications can attempt to modify a session after it has been authorized and connected.

Simple pattern recognition does not stand up well against proliferation of attacks because it may require matching numerous variant patterns. Protocol-based analysis, in contrast, can provide protection with a single signature (sequence of values). Protocol analysis includes decoding protocol messages and evaluating them for anomalous behavior or unexpected values. For example, a large binary file in the User-Agent field of an HTTP request could be an intrusion.

7.4.4.2   The Generic Concept of ALG

An Application-level stateful firewall aims to prevent disruptive behavior per application. It supports packet analysis per session and is protocol-aware. It examines packet flow up the protocol stack, including application settings, to gain knowledge of the type of session and the conditions that it can fulfill.

Where certain applications can carry network addresses in the payloads (e.g., SIP/SDP), an application-aware method is required, that is, an ALG. This function is protocol dependent. It requires decoding more than just packet headers but the whole message and may also require knowledge of the application, the protocol, and the dialog sequences. It can then apply customized security filters that interpret the protocol messages.

An example of using the ALG is when it allows an application host to communicate transparently with another host running the same application but in a different IP version. By comparison, the NA(P)T-PT (Network Address and Port Translation—Protocol Translation) is a translation method that is not aware of the application protocols and cannot cope with additional address information carried in the SDP.

7.4.4.3   SIP Screening

The versatility of SIP is also what makes it vulnerable to malicious attacks and fraud. Screening SIP messages for undesirable commands or hidden addresses is required for fraud prevention and security enforcement. This necessitates evaluation of the whole SIP message, the SDP as well as the headers, and a stateful understanding of the dialog. The exact procedures depend on the operators' policies.

IMS is treated as an application in the context of ALG. The IMS ALG is therefore SIP based. It needs the ALG to provide the necessary awareness of the contents in SIP and SDP, to enable interworking and efficient security filtering.

7.5 Network Address Translation (NAT)

7.5.1 NAT Requirements

7.5.1.1 Private and Public IP Addresses

Because IPv4 addressing capability cannot satisfy the demand for unique addresses for global IP routing, the common practice to extend the addressing capability by creating islands of private IP addressing became widespread. Private addressing is now deployed not only for IP-VPN in the Enterprise space, but also in the growing number of Home LANs for residential use. Private addressing also brings additional benefits in terms of hiding internal topology, which helps prevent some hacking and malicious attacks.

Private addressing works on the principle that only a fraction of the total endpoints with the potential to connect to the external networks are actively connected at any one time or seeking connection outside the private domain. Therefore, fewer globally routable IP addresses are needed than the total endpoint devices within the private domain. Private addressing allows the members to communicate internally at any time and request binding with an external IP address only when accessing an external service.

The process of binding an internal private IP address to an external public address is NAT — Network Address Translation.

A similar process also occurs for allocating a port number. Ports are virtual data connections on a physical channel that allow dynamic data exchange for a particular protocol transmission. A finite pool of such ports is shared among sessions, and a port is allocated only as and when it is needed for setting up a session.

7.5.1.2 IPv4 and IPv6 Interworking

IMS was originally designed only for IPv6. In reality, the migration from IPv4 is not progressing as fast as was initially expected and can hold back IMS implementation. To enable IPv4 to be used with IMS as well as IPv6, a number of functions need to be added or expanded. For example, the TrGW (Transition Gateway) was introduced to translate between IPv4 and IPv6. Another solution for the IP versions issue is implementing a dual stack that can cope with both IPv6 and IPv4 on all servers and all handsets.

Figure 7.3 shows an example of routing from an IPv6 domain to an IPv4 domain. In this example, both a mobile handset and a soft client are based on IPv6. Their attachment to the network is performed via an IPv6 DHCP that allocates IPv6 addresses. In routing to a destination in the IPv4 domain, the TrGW must translate on-the-fly addresses between IPv4 and IPv6, and vice versa.

The transition to IPv6 encompasses all elements in the network, including the UE's own software, the DHCP that allocates the addresses, the servers that interpret and route by the addresses across borders that deploy different versions and networks that are gradually migrating to IPv6. In the interim, IP version translation must be available at every step, in both directions, even if it means translating from one version to another and back again in one session leg.

UE supporting IPv6 must find DHCPv6 to allocate it an IPv6 address. On the way into a network, perhaps a Visited Network, the UE's IPv6-based messages may need to be translated to IPv4. DNS servers must also interwork with the prevailing IP version of the network. As the session request is forwarded toward the terminating network, it can pass through a transit IMS network, with yet another translation, and the target network also can be using another IP version. The TrGW is installed at the NNI border to facilitate IP version translation for these cases.

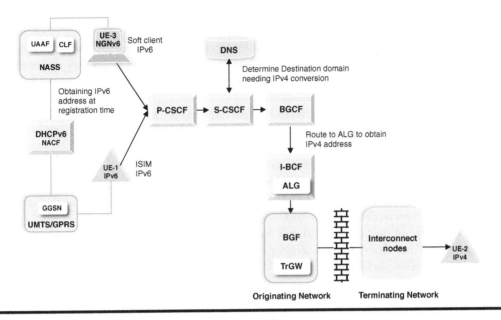

Figure 7.3   IP-to-IP routing with IP version translation.

Not only signaling packets need translation, but all data packets are also affected. Such TrGWs are, therefore, in the path of all media payload packets and must perform this function without any noticeable impact on session quality.

7.5.2   NAT Methods

7.5.2.1   NA(P)T for IPv4 Private and Public Address Translation

NAT and PAT (Port Address Translation), NA(P)T, and NAPT all refer to Network Address and Port Translation.

NAPT or NA(P)T (Network Address (Port-Multiplexing) Translation) is used for translating between private IP addresses that are recognizable only within a VPN or a closed private network, and a public network addresses that can be globally routed. This is performed at the edge, on the border of the private addressing "island" with the public network.

As traffic is routed from the local network, the source address in each packet is translated, on-the-fly, from the private IP address to the public IP address. The local server or router tracks each active connection details in terms of source and destination addresses and ports, to enable all subsequent packets to be routed with the same translated address.

NAT or NAPT methods are described in RFC 2663. Two functions are distinguished:

1. NAT converts private addresses to public ones, and vice versa. It temporarily associates a global IP address, not currently in use, to the required private address, which is only valid within the private domain.
2. PAT is required to address different ports flexibly. Ports are assigned to specific protocols; therefore, a session may need to address different ports. Port mapping allows several terminals to use one port. Port mapping may or may not be used at the same time as NAT.

7.5.2.2 NA(P)T-PT for IPv4 and IPv6 Address Translation

NA(P)T-PT is Network Address (Port-Multiplexing) Translation–Protocol Translation. This solution aims to provide transparent routing to end nodes in a IPv6 realm when they communicate with end nodes in IPv4 realms, and vice versa. This is achieved using a combination of Network Address Translation and Protocol Translation.

The NA(P)T-PT method does not call for supporting dual-stack IPv4/IPv6 protocol or special-purpose routing requirements (such as requiring tunneling support) on end nodes. This scheme dynamically assigns globally reserved unique IPv4 addresses to IPv6 nodes on a temporary basis, for sessions that require cross-IP version connectivity.

NAT-PT therefore dynamically binds addresses in the IPv6 network with addresses in the IPv4 network for the session duration, and frees the IPv4 addresses for reuse after session termination. It also provides translation of transport identifiers such as port numbers.

The advantage of this method is that it does not require any change in the endpoint UE, but it does mandate that all communications for the specific session traverse the *same* NAT-PT server that holds the binding information.

The NA(P)T-PT method is described in RFC 2766. There are two methods:

1. *NAT-PT.* This method uses a pool of globally unique IPv4 addresses for assignment to IPv6 nodes on a dynamic basis when sessions are initiated across the IP version boundaries. NAT-PT binds addresses in the IPv6 network with addresses in the IPv4 network, and vice versa. This provides transparent routing between the two IP domains without requiring any changes to endpoints, UE, or CPE. NAT-PT needs to track the sessions it supports. It mandates that inbound and outbound data for a specific session traverse the same NAT-PT router.
2. *NAPT-PT.* This method provides additional translation of transport identifiers, for example, TCP and UDP port numbers and ICMP (e.g., Ping) query identifiers. This allows the transport identifiers of a number of IPv6 hosts to be multiplexed into the transport identifiers of a single assigned IPv4 address.

7.5.3 NAT Scenarios for the Access Boundary

7.5.3.1 Positioning the NAT at the Edge

There are several methods of serving NAT to support different scenarios. In general, addresses can be translated in a NAT gateway transparently to the end device, or the payload is analyzed at the Application level, with the UE collaboration, in the ALG. This translation is not necessary where media is flowing from the UE within its Home Network or inside an Enterprise VPN, where the internal addresses need no translation, and where only IPv4 or only IPv6 is used.

The scenarios shown in Figure 7.4 suggest different realizations of the ALG and TrGW functions in nodes located either at the edge or at the core:

■ Scenario A shows NAT performed at the edge, then splitting session control signaling and the media payload packets. The signaling is forwarded to the P-CSCF ALG function for application-aware analysis, and the media packets are processed by the Border Gateway without any further NAT processing. This model is suitable for traffic coming from another business entity at the access side, such as an Enterprise. It is aimed to avoid unnecessary NAT processing.

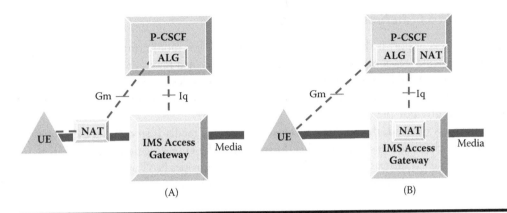

Figure 7.4 Remote or integrated scenarios of NA(P)T translation.

■ Scenario B shows NAT performed by both the P-CSCF (Signaling Proxy) and the Access Gateway (Media Proxy). The P-CSCF also contains the ALG that controls the filtering of SIP messages. NAT is performed separately for signaling and media.

Figure 7.5 shows other implementation options where NAT is performed in the IMS Access Gateway for incoming traffic:

■ In Scenario C, the local signaling requires no translation but the media may be crossing IP-type boundaries, and therefore does. To decide whether unnecessary NAT can be avoided, the SIP-aware ALG analyzes the messages to determine originating and terminating addresses, thereby discover the need to perform NAT or IP version conversion. It also determines whether it is safe and acceptable by operator's policy to do so.
■ In Scenario D, both signaling and media traverse the IMS Access Gateway, which contains both NAT and ALG, before reaching the P-CSCF. This scenario is typical of implementations of the SBC, which acts as both Signaling Proxy and Media Proxy, and also performs various security filtering functions. This approach increases the load on the NAT gateway and is less efficient.

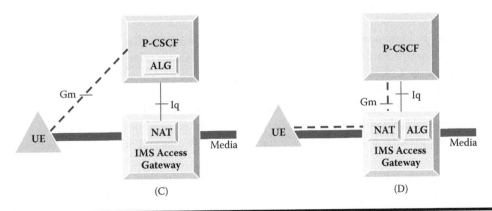

Figure 7.5 Media and signaling traversing NAT.

7.5.3.2 Address Latching at the Access Border

The latched address or port is the address or port that the BGF is monitoring for received media packets for a particular session. If the NAT procedure is performed for an endpoint transmission before reaching the A-BGF, this BGF cannot use the modified destination information carried in the SIP SDP to send the media stream to it. Instead, it must use the reserved IP address of the NAT gateway, as given by the endpoint.

When media is received, the remote address or port of this media is stored as the destination address. This "latched" address or port is then forwarded to the UE as the destination address. In fact, the remote latched address or port can be the entity that provides NAT at the destination end, rather than the recipient endpoint itself, operating the same principle at the other end.

7.6 The Converged Network Border

7.6.1 The IMS Border Network Elements

7.6.1.1 VoIP Border Issues

Many of the concepts of borders and network security came from investigations in the NGN world, where implementations of VoIP have highlighted many of the issues for the IP Border, while the GPRS/UMTS IP service remained protected by the radio interface and the approach of the "walled garden."

Such issues included aspects of NAT such as traversal of signaling across NAT gateways where the original address is no longer known. Other examples include the treatment of encryptions and protecting authentication signaling, and in general the requirement to increase the level of security. In particular, NGN thinking has contributed to the awareness of security toward access and the untrusted user endpoint.

7.6.1.2 Border Control for Signaling and Media

The Border Gateway is a gateway between an IP CN and an external IP network, that is, another operator's packet network. The external IP network can be another IMS network (with GPRS/UMTS, WLAN, or xDSL access) or non-IMS networks (e.g., VPNs running H.323 protocols over packet).

The border control functionality resides in several network elements. The IMS session controllers need to define which border point to select, in the BGCF. This is often a module within the Session Controller, the CSCF.

The signaling border control and the media border gateway are usually implemented in separate devices. The signaling IBCF (Interconnection Border Control Function) coordinates the various processes that take place in crossing the border. This is often combined with the ALG and NAT for signaling packets, as well as the IWF for non-SIP networks and other SIP implementation differences.

The border gateway for the User Plane (i.e., the media flow) is positioned closer to the edges of the network, as it handles large volumes of packet traffic. This gateway enforces the security measures dictated by the signaling servers and performs the packet filtering as instructed by the PDF, SPDF, or PCRF.

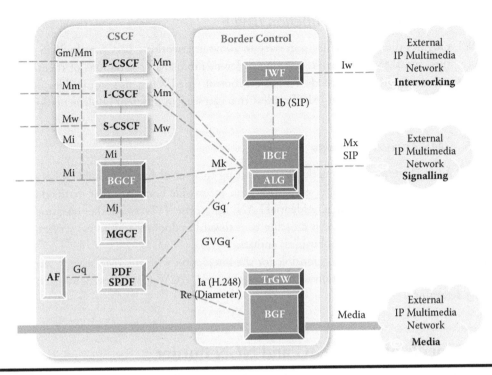

Figure 7.6 **The interconnecting border controllers and gateways in NGN.**

This server also contains the Network Address Translation Gateway, performed for the media packets. The I-BGF (Interconnection Border-Gateway Function) is situated between networks while the C-BGF is notionally situated between the AN and CN, performing similar functions.

Border Control must provide interoperability across the connection while ensuring that the appropriate levels of security procedures are engaged to protect the network and its subscribers, and facilitate two-way authentication between the networks.

Figure 7.6 combines the TISPAN and 3GPP models to show how the border control function connects to other SIP or IMS networks and non-SIP networks that need some interworking functions and protocol conversion to smooth the differences.

7.6.1.3 *Topology Hiding IP Gateway (THIG)*

The Topology Hiding function is deployed to avoid revealing to external entities the names, addresses, and capabilities of internal servers. This also helps in network management. For example, when scheduled maintenance is performed, traffic should be temporarily diverted to alternative instances of the same function by an administrative action, without any outward change. Similarly Topology Hiding is necessary for scaling up the network, enabling the addition of more replica servers (instances) to allow better load-balancing of traffic among the added servers.

The THIG was originally included in the I-CSCF in its role as a Redirect Server, internally and externally. However, the THIG function is now considered part of the IBCF, while the I-CSCF is used mostly for interrogating SLF and HSS and for internal routing only.

7.6.1.4 The InterWorking Function (IWF)

The IWF provides the functionality necessary to allow interworking between the IMS network and any other non-IMS IP network, running different protocols or variations of them. In particular, it includes interworking to H.323 session control. The IWF is also required for inter-IMS connections, to interpret different "SIP Profiles" that can vary in different implementations.

7.6.1.5 IMS Access Gateway

The IMS Access Gateway is a generic name referring to a set of functions applied between the Access Network and the Home Network. This is an access-side boundary that marks the ingress into the operator's network, from a residential LAN, Enterprise VPN, or another operator's Access Network, that is, when the traffic is crossing over toward endpoints and access servers that are possibly administered by different business entities.

The IMS Access Gateway is responsible for the interaction with the IMS servers, namely the P-CSCF. It supports the connection to the IMS-ALG and TrGW for address conversion and for security screening.

7.7 Aspects of Mobile 3G and NGN Border

7.7.1 General Security Principle for Mobile 3G

In 3G Mobile, five secure associations are defined:

1. *The interface between the EU (ISIM) and the HSS for mutual authentication.* The HSS generates keys and random challenge for security dialogs performed by the S-CSCF. The long-term key in the ISIM and the HSS is associated with the IMPI. The subscriber must have one (or more) IMPI and at least one external public user identity (IMPU).

2. *The interface between the UE and the P-CSCF.* This secure connection is for protection of the Gm reference point that transports SIP messages for session control. This security association must provide corroboration that the source of data received is as claimed.

3. *All interfaces to HSS over the Cx-interface.* This protects the privacy and integrity of the central data repository.

4. *The interface between different networks for SIP communication across the border.* This security association is applicable when the P-CSCF resides in the Visited Network and is interfacing to the Home Network servers. This requires network-to-network mutual authentication.

5. *The interface between internal SIP servers within the trusted network.* These interfaces, which include the P-CSCF interface to other IMS nodes within the same network, may need some security as may be suitable for internal or inter-segment communications for a global IP network.

7.7.2 The 3G Border Elements

The border elements specified by 3GPP have undergone an evolution. The various aspects of security and interworking are merged into a number of functions that are now consolidated with

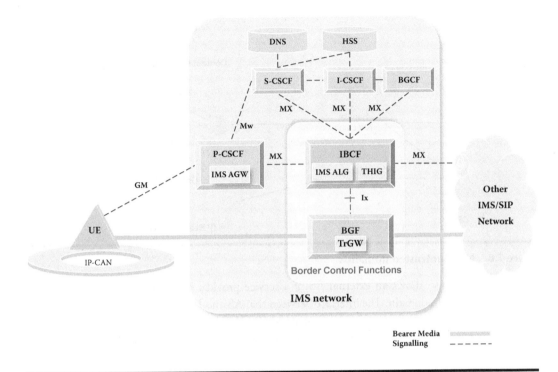

Figure 7.7 IBCF, IMS Access Gateway, and TrGW in 3GPP.

the TISPAN definitions. The positioning of these functions within network nodes is a matter of implementation.

Figure 7.7 shows an example of a border node, where the IBCF contains both THIG and ALG functionality, and the BGF contains the TrGW. The border elements between the Access Network and Core IMS vary between Mobile and Fixed Networks. Equally, controlling the bearer through the border requires bearer-specific management.

7.7.3 The NGN Border

7.7.3.1 Trusted and Untrusted Domains

In Fixed Networks, the AN can be provided by many types of technology (WiFi, Cable, WiMax, Ethernet), which often are offered by another business entity. Hence, TISPAN developed the notion of an Untrusted Domain versus the IMS Trusted Domain, where the user CPE is regarded as untrusted and the AN may or may not be trusted.

In Figure 7.8, the IMS CN is "trusted"; that is, the internal IMS elements need not apply special security measures to interact between them. However, the CPE is regarded as untrusted, and all communication with it is subject to various means that verify the source and authorize any requests.

In this example, the IP-CAN may be untrusted if the device is nomadic and is accessing through a Visited AN, for example, a WiFi hotspot. In NGN, even the regular subscriber connection may be using an untrusted access network, for example, an xDSL broadband connection from a wholesale carrier.

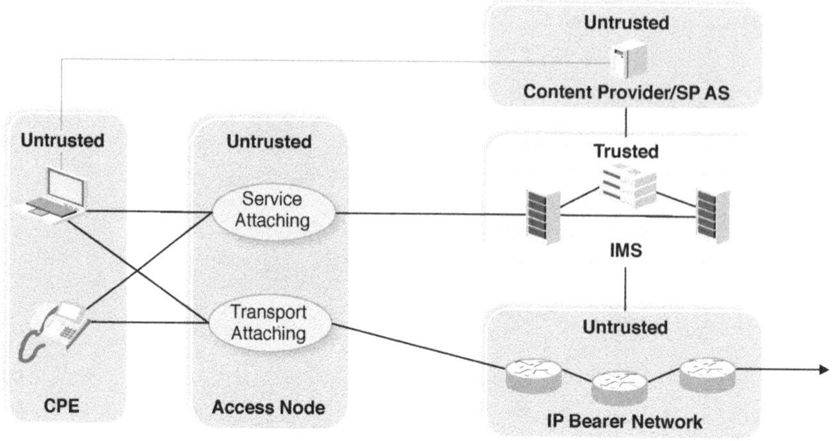

Figure 7.8 NGN untrusted domains.

This example also shows an external AS of a service provider or content provider regarded as untrusted by the IMS domain. The interface between that AS and IMS should also be protected. The media in the User Plane must traverse the untrusted IP-CAN to reach the UE. It may well traverse several other bearer networks to connect to the destination, which are all untrusted.

7.7.3.2 IPSec for Nomadic Users

Where it is possible to create SAs for IPSec, it can deliver secure access for nomadic users. This is already utilized for secure e-mail access for remote Enterprise workers.

The model can also be extended to service providers' applications that should be delivered reliably, maintaining confidentiality and data integrity.

Figure 7.9 demonstrates a scenario where the UE is entering the network via an AN, perhaps an Interworking WLAN, traversing the Visited Network onto the Home Network, where services can be delivered through third parties from either the Visited Network or the Home Network.

TLS (Transport Layer Security) is mandatory for SIP proxies according to RFC 3261. It can be used for added security, to provide confidentiality and integrity inside the networks instead of IPSec. TLS also can be used over IPSec, especially for the intra-domain Za interface, between separate IMS networks.

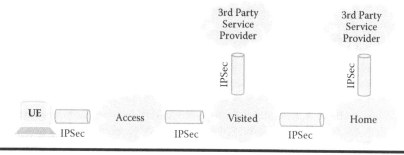

Figure 7.9 Secure tunneling between multiple transit networks.

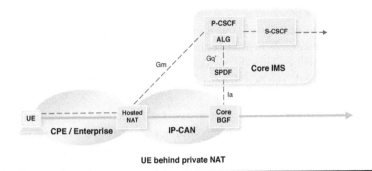

Figure 7.10   Enterprise-hosted NAT.

7.7.3.3   Hosted NAT

In NGN solutions, the Enterprise can undertake performing the network translation for traffic in and out of the IP-VPN. This enables the Enterprise to control better what the employees can or cannot do because this gateway is also a security firewall. Because NAT has already been performed, the Operator may want to avoid unnecessary processing. However, further SIP screening and enforcing operator policies may still be required.

Figure 7.10 shows hosted NAT performed remotely by the Enterprise at the boundary with the IP-CAN. Signaling is processed by the IMS-ALG at the P-CSCF. Media packets traverse the C-BGF that may apply media-based filtering. Furthermore, NAT may be required again when proceeding to a destination within a different IP addressing scheme, a private plan, or a different IP version.

7.8   Control Border Elements

7.8.1   Border Signaling

7.8.1.1   The BGCF: Border Session Control

The BGCF is the signaling server that determines the next hop for a session exiting the current network or incoming into the network. For incoming messages, the BGCF receives forwarded signaling bound for the S-CSCF, though other IMS network nodes can take this role. Therefore, the BGCF main role is directing sessions that need to exit the network.

When the IMS session control determines that a session is not bound for internal destinations, it is up to the BGCF to select the path for the session exit point. The BGCF selects the next-hop exit server, based on the information received in the SIP message from the S-CSCF, possibly in conjunction with an AS. It can also utilize internal configuration data, information from other protocols or other databases that may be at its disposal, or any source of administrative information (e.g., network management data).

The originating BGCF can forward the request to:

- MGCF within its own network
- Another network containing BGCF or I-CSCF
- IBCF in its own network, if deployed

For sessions bound for breakout to a circuit network, the BGCF may determine that the breakout point is within the same network and route to it. If the destination is in another network, it will route to a transit network. When the breakout point is in the same network, the BGCF selects the MGCP server that will execute the breakout and connect to the Media Gateway. When the exit is in another IP network, the BGCF must establish the optimal exit point for a transit network or other operator's contact point.

When the BGCF has decided that exit to another IP network is required, it can route to a selected IBCF server, which handles topology hiding and other functions for an IP-to-IP connection. If IBCF is not implemented, the I-CSCF can be used to route to external nodes.

The BGCF ensures that the messages contain information required for inter-operator charging. When forwarding the request to the next hop, the BGCF inserts previously saved values into the P-Charging-Vector and P-Charging-Function-Addresses headers to ensure that these sessions can be charged properly. It will also insert the value of the ICID (IMS Charging Identifier) parameter received in the P-Charging-Vector header to identify the chargeable session in each message.

Because the BGCF function is completed once the next hop is defined, the BGCF does not insert Record-Route in the INVITE request, so it will not stay in the signaling message path. The BGCF is therefore "stateless."

7.8.1.2 The IBCF: Signaling Border Controller

The IBCF (Interconnection Border Control Function) acts as an entry point or an exit point for the network signaling when interfacing to other IP networks. The IBCF forwards the session signaling to the destination network, applying the operator's security and network policies and rules, as received from the PDF/SPDF or PCRF. The IBCF provides application-specific functions at the SIP/SDP protocol layer to perform interconnection between two operator domains.

The IBCF provides:

■ Interworking of addresses across IPv6 and IPv4
■ Network topology hiding
■ Controlling Transport Plane functions
■ Screening of SIP signaling information
■ Selecting the appropriate signaling interconnect
■ Generating charging data records for interconnect accounting

For incoming calls from other networks, the IBCF is the known address that can be ascertained when resolving the domain names in DNS search. The IBCF can then forward the signaling to the appropriate internal CSCF server, thus ensuring that the number of CSCF servers and their names and assignations are not visible to external entities.

The entry point for NNI when there is no border control in place is the I-CSCF, as previously defined by 3GPP. The I-CSCF is capable of performing the IP-to-IP routing and the topology hiding, masking the identity of other network servers, but it is not designed to perform the remaining functions of the IBCF.

On the access side, when a user is entering via a Visited Network, it reaches the Visited Network P-CSCF. The P-CSCF forwards requests to the Home Network "well-known" entry point. This will be the Home I-CSCF if there is no border control in that network, or to the IBCF if there is.

The IBCF provides control functions to the BGF in the Transport Plane. For example, if it receives an error state in the media flow for the session from the involved BGF, the IBCF will generate a BYE message and terminate either the originating leg or the terminating leg that it is monitoring.

Like other session-aware functions, the IBCF requires periodic refreshment of the session to avoid hung states. This means that at least one UE must be involved in the session.

7.8.1.3   IBCF Role as THIG

In performing the THIG (Topology Hiding IP Gateway) Function, the IBCF applies the hiding procedures to all headers that reveal topology information, such as Via-Route, Record-Route, Service-Route, and Path.

The IBCF processing of a REGISTER request includes modifying the PATH header. The IBCF replaces the address of the originating server (S-CSCF or P-CSCF) with its own address, and saves the originator server addresses for subsequent communications. The IBCF can also insert an indicator of the direction of the session, that is, incoming to the network or outgoing session. This can be encoded by different means, such as a unique parameter in the URI, a character string in the username part of the URI, or a dedicated port number in the URI.

7.8.1.4   The IBCF Role as ALG

The ALG function can reside with the P-CSCF or the IBCF, or both. It is required whenever an untrusted element is communicating into the network. It provides protection (both by screening SIP messages, headers and SDP) to ascertain the actions taken and prevent misuse of network resources. The ALG supports scenarios where the UE is behind a NAT gateway or operates in a different IP version (IPv4 or IPv6) than the network servers.

When the P-CSCF recognizes that the UE requires NAT, it calls upon the ALG. The IMS-ALG requests a routable IP transport address and assigned port for this communication from an Access Point (i.e. IMS Access Gateway) to perform attachment of the UE to the network. The IMS-ALG also communicates with the TrGW that performs network and port translation, as required.

The ALG acts as a SIP B2BUA. It terminates incoming messages and creates new ones with different parameters. It changes SDP information to enable the use of IPv4 address in one direction and IPv6 in the other. Where there are no IP version differences, the ALG can manage in the same manner the conversion between private IP addresses and public addresses. The ALG also applies various security checks based on its understanding of the dialog sequence and the IMS application. In all other respects, the ALG forwards messages transparently.

The IBCF ensures that all messages must pass through the ALG. It also filters out erroneous messages, inconsistent with the forwarding procedures, to prevent them from entering the network, and can perform any available remedial error recovery procedures.

In its role as an ALG, the IBCF saves header field values such as the Contact, CSeq, and Record-Route received in the INVITE request so that it can track the session and release it when necessary. When receiving a session termination message (BYE request, CANCEL request, or non-200 final response), the IMS-ALG releases the session and requests the TrGW to release the address bindings that had been established for the session.

7.8.1.5   IBCF Role in SIP Screening

In its role of performing SIP Screening, the IBCF filters out undesirable SIP messages or harmful information carried within SIP messages. The IBCF performs SIP Screening by evaluating the headers and the SDP payload. According to the network policy, the IBCF can, for example, identify a signaling message of excessive length and remove the body of the message. It should, however, preserve header information, such as Service Route, that is essential to the correct delivery of the session.

7.8.1.6   IBCF Encryption

The IBCF also encrypts and decrypts certain information to cross the border. The IBCF encrypts all the headers in one string. For ease in decrypting, it does not modify the order of the headers. However, where the username is included in the encrypted part, the IBCF reconstructs the NAI in the format of "username@realm". The encrypting network identity is added as an appended "tokenized-by=" indicator.

If the IBCF encrypts an entry in the Route header, then it also inserts its own URI before the top-most encrypted entry to make sure that return messages pass through the same place. When receiving an encrypted message, the IBCF follows the same rules to decrypt the message and reinserts stored information for the session to route internally to the serving session controller or the P-CSCF.

7.8.2   Border Transport Enforcement

7.8.2.1   The BGF

At the border, the BGF has multiple roles applied to the media flow packets:

- Acts as a security firewall to ensure valid communication
- Protects the network from malicious attacks
- Translates addresses from private ones to public routable addresses, and vice versa
- Converts IPv4 to IPv6, and vice versa
- Enforces QoS parameters

Chapter 8 describes the QoS functions. The security functions and address translation for media flow packets are the same as described for signaling.

7.8.2.2   Transition Gateway (TrGW)

The TrGW operates at both the signaling level and in the media path. It is called upon to perform the translation by the IBCF. The TrGW function is to provide network address or port translation and IPv4/IPv6 protocol translation for media packets as well as signaling.

Where IPv4 is used, even where there is no IPv6 translation, local private addresses must be translated to public addresses. This enables local networks (VPNs) to keep their servers' addresses anonymous for added security, and provides greater flexibility in utilizing their own range of addresses.

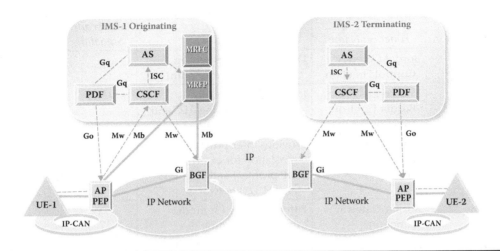

Figure 7.11   Media from MRF passing through BGF, for 3G Mobile networks.

The IP address or port translation methods used are the NAT or NAPT method for private or public IP address translation, and the NA(P)T-PT method for IPv4/IPv6 translation.

7.8.3   Media Server with Border Crossing

All media packets must pass through the BGF. This includes media stream injected by a media server. Normally, the application, via the session control, will instruct the media server controller (MRFC) to send the media contents to the handset or the local gateway. Where the media is to be played to an IP remote end, it is directed toward the border gateway.

Figure 7.11 shows an originating 3G Mobile IMS network that is delivering a service with the help of an MRF. The media must reach the border gateway first, before crossing over to its destination. Because some of these MRFs are used frequently (e.g., system announcements and playing DTMF tones), there is good reason for placing MRF functionality within border or access nodes, as is often done in practice.

7.8.4   Summary of IP Border Control Elements

Border facilities are presented in a multitude of terms and functions, and are yet to be streamlined and consolidated for the converged network. Furthermore, the realization of these functions into network nodes varies in different implementations.

There are no set rules, for example, where the ALG should be — within the P-CSCF or the IBCF. Another bone of contention is whether the border signaling control should be together with the media filtering function of the border, as a combined media and signaling firewall that is known as the session border controller (see description at the end of this chapter).

Certain elements can have the same functions, expressed differently in 3GPP and in TISPAN, or may be split in a different way (e.g., the TrGW and the IWF). In some cases, the apparent overlap of functionality occurs because the functions should be located at different network points (e.g., C-BGF and A-BGF).

To help understand the relative positioning of these elements, a summary is provided here. The table below lists the combined border elements as defined in both 3GPP and TISPAN.

Abbreviation	Function Name	Description	Media and Signaling	Signaling Only
BGCF	Border Gateway Control Function	Directs messages toward the appropriate border point		Y
IBCF	Interconnection Border Control Function	Interfaces to external networks, applying ALG and TrGW as required		Y
IWF	Interworking Function	Provides inter-protocol translation for external IMS (SIP profiles) and Non-IMS (e.g., H.323) networks		Y
BGF	Border Gateway Function	Filters packets for media streams according to security rules	Y	
C-BGF	Core Border Gateway Function	As above for connection into CNs (network side)	Y	
A-BGF	Access Border Gateway Function	As above for connection with the AN (user side)	y	
IMS ALG	IMS Application Level Gateway	Filters signaling messages according to rules that are application-aware and SIP protocol-aware		Y
TrGW	Transition Gateway	Translates network addresses for IPv4 private/public domains and for IPv4/IPv6 conversions, for both signaling and media paths		Y
L2TF	Level 2 Traffic Function	Filters packets at the bearer level for media and signaling	Y	
RCEF	Resource Control Enforcement Function	Opens and closes the gate according to instructions from resource and security controllers	Y	
IMS AGW	IMS Access Gateway	Hosts IMS traffic filtering in the access network	Y	
SEG	Security Gateway	Provides IPSec tunneling between network segments	Y	
SBC	Session Border Controller	Provides all-purpose border functions, with security filtering, ALG, and NAT	y	

Further descriptions of the functions are provided below.

7.8.5 IP Border Reference Points

The following reference points are involved in IP Border activities:

- Mk: the reference point between a BGCF or IBCF and a BGCF. The Mk reference point allows the BGCF and the IBCF to forward session messages to other BGCFs. The Mk interface is based on SIP.
- Mx: the reference point between a CSCF or BGCF and an IBCF. The Mx reference point carries the communication between a CSCF or BGCF and an IBCF when a session needs to connect to another IP network. The Mx is based on SIP.
- Ix: the reference point between an IBCF and a TrGW. The Ix interface is used by the IBCF to control the TrGW (e.g., to request network address translation binding).
- Mm: the reference point between a CSCF or IBCF and Multimedia IP networks. This is an IP interface between the CSCF or IBCF and IP networks. This interface is used, for example, to receive a session request from another SIP server or terminal.
- Za: the reference point across a border. The Za interface is between SEGs of different security domains (NDS), belonging to different administrative entities or segments within one network. For this interface, the protection of authentication and data integrity is mandatory and encryption is recommended. ESP is used for encapsulating information securely over this interface. The SEGs use IKE to negotiate, establish, and maintain a secure ESP tunnel between them. The SA for Za can be subject to roaming agreements.
- Zb: the reference point between internal entities. The Zb interface is the security service between SEGs and a Network Element (NE) or between two NEs in the same security domain. The Zb interface is optional, according to the operator's needs for internal security, and generally provides a lower level of security than the Za. If implemented, it is based on ESP+IKE.

7.9 The Session Border Controller (SBC)

7.9.1 The Scope for the SBC

SBCs appeared on the market ahead of standards, to support Fixed-line IP-based communications. Such communications included not only SIP, but also other protocols, predominantly MGCP for softswitches and H.323 for the Enterprise VPN edge. As the name implies, the SBC is designed for sessions that consist of one or more signaling streams and one or more media streams.

The SBC for SIP operates as a B2BUA for both signaling and media packets. This means that the SBC terminates incoming flows and initiates new ones, having checked or modified the data.

The SBC provided packet filtering security gateways positioned between the user device and the network server or between two IP networks. The main purpose of the SBC is to protect the network from hackers, intruders, and fraudsters, and to traverse NAT borders (private/public or IPv4/IPv6) with Enterprise and Residential VPNs.

However, the strategic positioning at the border meant that the SBC could also provide session tracking for billing purposes, measuring volumes of data that crosses over the border, which can be utilized for network management as well as for inter-administration accounting. In addition, the border functions can also include some admission control, both network level and access level. This entails enforcing policies for admission and QoS profiles.

Some data conversion facilities to enable full interworking are also included in some SBC implementations. Where the SBC is the only server on the media path between two end users, some products evolved further to provide media functions such as tone detection (DTMF), media replication (e.g., for Lawful Interception), codecs conversion, and signaling conversion (e.g., H.323/SIP, MGCP/SIP).

7.9.2 SBC in the IMS Context

The SBC functionality has evolved into a number of separate functions in IMS, and therefore does not exist as such. These functions are now specified by the standards, within the P-CSCF, IBCF, C-BGF /I-BGF, ALG, TrGW, and IWF, and even some RACS and media server functions (e.g., DTMF tones). While combining all these functions may be useful at an early stage, for small deployments or for IMS trials, putting all these functions in a single physical "box" is not suitable for most network topologies in the long run.

Most designers want to place the media-bearing BGF in different physical locations than the signaling to keep the main volume of data packets from traversing the length of the network. It also is considered risky to maintain the publicly known point of entry (P-CSCF) with the security function of media filtering.

If data from the endpoint arrives in encrypted form, the SBC must have the key to decrypt the data packet to check it and possibly modify it. Otherwise, it can only address the unencrypted part of the packet. Peer-to-peer communications deploy end-to-end encryption, which makes it impossible for an SBC to decode and check the contents. This can be overcome by the SBC "devolving" this responsibility to other trusted network elements, using SIP-TLS, IPSec, or SRTP (Secure RTP). NAT can be performed remotely, hosted by the recipient, when using methods such as STUN (Simple Traversal of UDP through NAT), TURN (Traversal Using Relay NAT), ICE (Interactive Connectivity Establishment), or UPnP (Universal Plug-and-Play) to enable traversing the network transparently when both ends can collaborate. However, in doing so, some of the main functions of the SBC are not required.

7.10 State Security

7.10.1 Lawful Intercept Requirements

All communications, Fixed or Mobile, Voice or Multimedia, are subject to LI (Lawful Interception), according to the security regulations within each country. LI standards and terms are different for CALEA in the United States and LI in various countries in Europe. The description here is giving the ETSI generic views for IMS.

A Law Enforcement Agency (LEA) can request to report on real-time communication activities. In fact, several LEAs may request the LI data input at the same time. This means that the LI system must create and manage multiple concurrent streams, copying the data to multiple addresses.

The LI service is invoked for target users that are "marked" by the LEA to be monitored. This can be the session originator, the destination user, or any other session participants (e.g., in a conference call). The delivery of the ID for all involved parties in a dialog with the monitored party is also a mandatory requirement.

The LI service can be invoked for a particular access node or a defined area, for example, where a certain location requires higher security scrutiny. In the case of a moving target, in a car or on a train, for example, the LI service can be handed over from one authority to another, when crossing LEAs' areas of jurisdiction. This causes the LI session to send data to another server address.

To be effective, interception must be unobtrusive, without the knowledge of any of the session participants, and with minimum impact on the session performance.

Where the network deploys encryption, the LI information must be decrypted or the LEA provided with the decryption key. Decryption also must take place without either party being aware that it is happening.

7.10.2   *Generic Lawful Intercept (LI) Architecture*

7.10.2.1   *Lawful Intercept Component Functions*

The standards define a generic architecture, although there may be many ways to implement it. The architecture is designed to provide a single interface to all the network elements, while the data replication and management of transmission of the LI information are executed on separate LI servers. In this manner, the network elements are shielded from possible performance impact created by initiating the monitoring service.

The architecture defines a Delivery Function between the LEMF (Law Enforcement Monitoring Function) and the Intercepting Control Element (ICE). This Delivery Function (DF) performs the following:

1. Distributing the Intercept Related Information to the requesting Agency via HI2 (signaling)
2. Distribute the Content of Communication via HI3 (media).

Independently, the INEs (Intercepting Network Elements) also receive requests to commence sending their media payload to their replicating server, the Delivery Function 3 (DF3). The addresses of the LEMF servers, details of required interception, and target identities are sent over the X1_3 to DF3. The DF2 receives signaling data (interface X2) from both ICE and INE. The DF3 receives media data from the INE over the X3 interface. Both signaling (DF2) and media (DF3) are forwarded to the LEMF.

This generic architecture defines the network elements that contribute LI data. The ICEs include IMS control functions: P-CSCF and S-CSCF. Access control elements (i.e., INE) are also required to report LI data, for example, SGSN, GGSN, HLR, AAA Server, and PDG.

The targets to monitor can be identified by normal identifiers, as appropriate to the network element. For 3G IMS elements, both the SIP URI and Tel URI can be used. MSISDN can be used if available, but the IMEI is not applicable for 3G gateways. For target entities in the WLAN (the AAA Server and the PDG), the NAI can be utilized.

To manage the relationship of IMS elements with the various LEMF agencies, the Administration Function (ADMF) was defined to:

■ Interface between the ICE and multiple LEAs
■ Provision monitoring requirements and target information
■ Convey LI instructions (activate, deactivate, etc.)

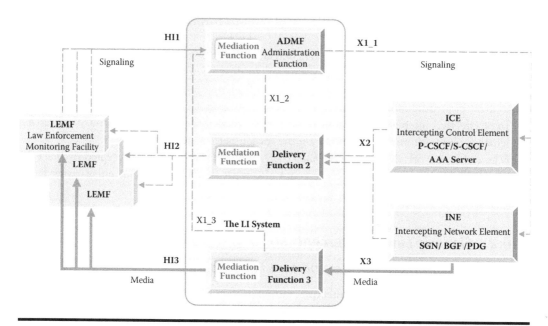

Figure 7.12   LI generic architecture.

■ Maintain state of each LI reporting for individual LEA
■ Maintain state of each ICE reporting

The ADMF can be partitioned to ensure confidentiality and security even between different agencies.

Figure 7.12 depicts the LI delivery scenario for the IMS session control elements and for media-bearing elements. The ICE receives instructions to commence an LI session and send signaling messages to the DF2. The DF2 gets instructions (via the X1-2) where to send the replicated data, which is the addresses of the LEMFs.

7.10.2.2   LI System Interfaces

■ The HI2 interface is the LI reporting of signaling data, and is used by control ICEs.
■ The HI3 interface is the LI reporting of media data packets, for example, the RTP, which is used by media-bearing elements in the access network.
■ The HI1 interface carries LI service instructions that enable the ADMF to activate and deactivate LI activities and also interrogate the network element about the monitored sessions.
■ The X1-1 connects the ADMF with ICE. Every physical ICE is linked to the ADMF by its own X1-1 interface. This means that the LI reporting from each ICE can be performed independently.
■ The X1-2 and X1-3 interfaces enable the Administration Function to communicate with the Delivery Functions (DF2 for signaling, DF3 for media) and control their sessions.

Chapter 8

Charging

8.1 This Chapter

This chapter explains the charging functions and processes from an IMS point of view. This includes details of charging procedures for IMS that are shared with non-IMS functions, such as FBC, but does not detail bearer service functions such as MBMS.

This chapter describes both online and offline charging subsystems, as well as the contributing factors that trigger charging in IMS elements and in various access networks. It also covers the PCC function and procedures to charge for content.

Note that the charging architecture and participating elements have changed considerably between 3GPP Releases 5/6 and 7/8, as the concepts of Content Charging, Event Charging, MBMS, and FBC have been evolved.

8.2 Basic Principles

8.2.1 The Evolution of Charging

8.2.1.1 Positioning of Charging

The charging system is central to any commercially operated network. At any point in the network transformation, the charging system must continue to function and ensure the collection of revenues.

The charging system must be enhanced to support the new revenue streams, while retaining the ability to support existing and aging services for a long time. Therefore, the charging system should be seen as an independent system that evolves separately. While it is not an integral part of Core IMS, it morphs to support the demands placed on it by multimedia, flow-based services, content delivery, and combinations of IMS services.

8.2.1.2   Multimedia Charging

Because IMS is access-agnostic, charging for IMS must support sessions to and from different bearer networks, for fixed or mobile users at their own networks or while roaming. IMS charging can be performed in stateless mode (event-based charging) or in stateful mode (session-based charging), and can involve both at the same time.

The evolving requirements for charging also bring about far-reaching changes in the stored information, the sources of the information, and the triggers to produce it. In particular, new data is needed for real-time event-triggered charging and FBC, which meters volumes and characteristics of media streams. The type of service affects how it is charged, with information extracted from different bearer nodes, session controllers, media bearing gateways, and application servers.

With such a wide range of information contributors, the means of correlating charging information from different network elements must be standardized and simplified, and the facility to exchange unambiguous information across business borders must be facilitated. Billing systems are increasingly required to deal with greater complexity, as more technology-driven capabilities become available — and chargeable.

8.2.1.3   Widening the Sources of Charging Data

Previously, charging methods were tied to the means of access to the network, based on geographic location and duration. Because traditional charging methods are yielding less and less, profitability must now depend on new ways of charging according to what is valued by the user. In the new packet-based network, there are also new methods of metering usage.

The requirement to streamline billing becomes more important when organizations with Fixed Networks and Mobile Networks merge. In addition, traditional telcos get involved in providing Contents and Internet access at the same time as providing voice communication.

Many services require combining Voice and Data services in "Rich Calling" and various multimedia services. Therefore, there is a need to correlate session charging data that is generated in different parts of the network. The inability to charge appropriately can often hinder the introduction of a new service.

Figure 8.1 shows the numerous sources of charging information on the left and the various operator departments that make use of the information to produce bills, business decision, marketing, network analysis, and statistical reports. The figure also shows many (but not all) of the types of charging required by the converged billing system.

8.2.1.4   Converging Online and Offline

Online and offline systems as defined by the standards are not the same as traditional prepaid and post-paid services. For example, the operator can have post-paid subscribers on credit control by using online charging mechanisms, and certain real-time event charging generating offline charging records.

There is now a trend to combine the OCS (Online Charging System) and OFCS (Offline Charging Sysem). When a Charging Record is closed on the OFCS, it is immediately transferred to the Charging Gateway to be processed by the Billing Domain (BD). The timing of transfer depends on the operator's configuration but there is now a requirement to do so in real-time or near-real-time.

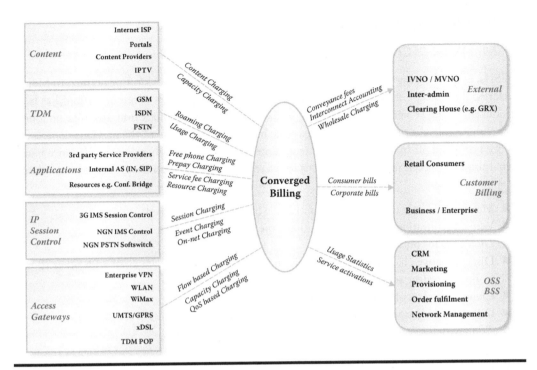

Figure 8.1 Converged billing.

8.2.1.5 *Charging Changes in 3GPP Releases*

The concepts and capabilities of charging have undergone major changes. These changes enhance the system's ability to provide media-oriented and flexible charging. These changes are, in brief:

- *Policy decision.* The Charging Rule Function (CRF) was defined in 3GPP Release 6, alongside the Policy Decision Function (PDF). In later releases, the PCRF (Policy and Charging Rule Function), as part of the PCC (Policy and Changing Control), combines the roles of both the PDF and CRF, but is designed to interface to the AF (Application Function) and to interwork with existing resource management and charging.
- *Flow-based charging.* This charging mechanism is defined to enable tariff differentiation according to individual media flow. This requires inspection of the service data flow to determine whether it requires separate charging rules.
- *Data management.* Release 6 defines the OFCS containing the CCF (Charging Collection Function). Release 7 splits this into two functions: (1) the CDF (Charging Data Function) and (2) the CGF (Charging Gateway Function). This allows network elements to address each function independently.
- *Bearer detection and enforcement.* The PCEF (Policy and Charging Enforcement Function) was defined as part of the PCC in Release 7. This function is a more advanced form of the TPF (Traffic Plane Function) and is a specific case of the generic PEF (Policy Enforcement Function).

8.2.2 Charging Principles

8.2.2.1 What Is Charged For?

In CSNs, usage charging was based on duration, distance, and often peak/off-peak time. In Mobile networks, the concept of roaming (i.e., differentiating rates for off-net communication) has been introduced. In the IP network, with strong influence from the Internet world, distance no longer plays a major role. As the price for Voice service has decreased, much of the previous complexity has been dropped off from billing tariffs (e.g., time-of-day bands of charging and distance bands of charging). However, new considerations for charging differentiation have emerged, including:

- Measure not only the duration, but also the bandwidth consumed
- Very long duration, where calls maintain the line "open"
- Discounts for on-net sessions
- Discriminative charges for off-net calls and to PSTN or GSM compared with other IP networks, according to levels of cross-charging
- Chargeable class of QoS, with guaranteed bandwidth
- Differential charging per media type (e.g., video streaming)
- Content delivery via Content and Digital Rights Management
- Large file transfers as a class of communication
- Higher levels of security and privacy
- Chargeable events and new types of supplementary services
- Sessions with multiple concurrent media streams
- Flow-based charging, enabling one to distinguish between media flows
- Charging for access to special resources, such as conferencing bridge
- Charging service providers through metering usage according to SLA and QoS

Broadband enables more video-based services and broadcast and interactive video. Charging for an individual media stream is now required, as is charging for the combination of media types in a single service. This means that media flows must be distinguished and metered. It should be possible to apply differentiated charging for the traffic flows belonging to different services carried by any IP-CANs.

8.2.2.2 Sources of Charging Information

Charging triggering can occur from incoming requests from external domains, from internal network elements, and from within the bearer. Initiation of the charging procedure is permitted from any network element configured to report charging events or session data.

In Bearer and Access Networks, access nodes provide charging information. Such nodes can be SGSN and GGSN in UMTS, or BRAS, WLAN AP, and other APs in Fixed Networks. In the Core, session controllers contribute charging information. The softswitch and the AGCF also produce charging records. Applications may manipulate charging, change the charged party, or split charges, and produce charging records accordingly. When they drive a Media Server, records will be produced optionally by the MRF.

For inter-operator accounting, network elements responsible for border control, such as BGCF, IBCF, and other Border Gateways and Media Gateways, all generate charging information records. All the contributing Network Elements generate the charging details in CDRs (Charging Data Records) that are stored on the CDF (Charging Data Function).

8.2.2.3   Charging Events and Sessions

Essentially, charging occurs for an event, for a session, or for a media flow. Both online and offline charging systems can process event charging as well as session charging. Charging for an event involves a single user-to-network transaction, for example, the sending of a multimedia message, where the charge is known and the service is prequantified. Charging for a session is usually a user-to-user interaction, which involves recording start and end of the session.

In the case of event charging, resource usage can be one action (e.g., sending a short message) with a one-time charge that debits the subscriber's account as soon as permission for the resource usage is given. This is used where the performance of the service is certain or can be guaranteed by the network because it provides no way of refunding if the service is not subsequently executed.

Where performance is not guaranteed or there is no knowledge at the beginning of the process of the full amount to be debited, the online charging service reserves units of charge. These units may be monetary or nonmonetary, as defined for the type of service.

When reserved, these units cannot be spent by other services. However, if the units are not all consumed at the end of the session, the remaining reserved units are "released" and are made available again as credit. The process of reserving units when the previous ones have been exhausted can be repeated several times during the session.

Three types of credit control procedures can be categorized, depending on the type of session or event and on the operator's network policies:

1. *Immediate Event Charging.* This means an immediate granting of accounting units to the requesting IMS network element is performed in a single operation that also includes the deduction of the corresponding monetary units from the subscriber's account. Such event charging can also produce a CDR that is opened and completed in one operation.
2. *Event Charging with Unit Reservation.* This is the online charging process for requesting, reserving, releasing, and returning unused units for events. The amount due, in the appropriate monetary units, is deducted when the transaction is finalized.
3. *Session Charging with Unit Reservation.* This is used for credit control of ongoing sessions in an online charging scenario. It includes the process of requesting, reserving, releasing, and returning unused units at the end, at which point it deducts the calculated charge from the account credit amount. This process continues throughout the session. There can be multiple unit reservations and debit operations within one session.

8.2.2.4   Charging Measurement Units

Charging attributes are assigned for each activation of the charging facility prior to starting the service delivery. The chargeable unit attributes affect how the charging amount is calculated. These attributes include the definition of the measurement unit and the rate of consuming units. The measuring unit can be a defined service data unit, which could be data volume blocks, time duration periods, or specific events. The service data unit has particular associated attributes that are used, for example, to calculate the number of units required to be in credit before service delivery can commence.

The measuring unit attributes are determined by:

- The charging system, in a network that centralizes charging control, determines what charging attributes to use, including the chargeable data unit or event type. The decision is made on the basis of triggering information, service identifier, etc.
- The PCRF, where implemented, provides network-wide policies and rules that determine how to charge the particular session or event.
- By an operator's choice, determination of the charging unit attributes can be devolved to the local nodes. In this case, the charging triggering function makes the determination of the unit on the basis of given session information and pre-embedded rules.

8.2.3 Flow-Based Charging (FBC)

Charging for many new services no longer fits the old model of charging for a voice call. Call duration may not be the chargeable factor, for example, when users download media or view streamed video. Usage is not always ideally measured by call minutes, but could be better measured by types of IP flows or by volumes of packets. To charge for user-perceived value, the network must have the means to measure flow of media and charge accordingly. FBC provides such a mechanism.

FBC can operate in online mode or offline mode, linking to the OCS or the OFCS systems, respectively. FBC can be triggered by an event or controlled by a session. Events must be detected to start charging for the data and to stop the process, or as one-off charge.

To optimize charging processing, the network must identify not only the discrete data flows that have a differential charge attached, but also aggregate data flows that are charged in the same way, to optimize processing and data handling. This entails close interworking with the IP-CAN servers and access nodes. For example, for GPRS, the PDP Context flows must be set up and monitored under the control of the network Charging Rules to report volumes (packet counts) and service context flows (per service).

The FBC service requires the network to provide the following capabilities:

- Distinguish data flows to be charged individually, by rates, measurements, or attributes
- Aggregating data flows that can be charged in the same way to optimize the charging process
- Provision of Charging Rules to be applied to service data flow level
- Applying the Charging Rules to data flows (e.g., per PDP Context in GPRS/UMTS)
- Reporting usage such as byte or packet counts for volume-based charging
- Reporting of session charged duration, for time-based charging
- Indication of events to start and stop online charging procedures
- Indication of events that may trigger a charging action (e.g., stop charging recording if the bearer network has terminated the data flow)
- Activating or deactivating the PCRF according to network policies

8.2.4 Content Charging

Charging for content is becoming a requirement for telecoms, where IMS sessions can include a variety of media delivery services and content can be involved in communication sessions. For example, the Multimedia Ring Back Tone application allows for a subscriber's chosen content to be sent to callers or called parties while they wait to be connected — instead of the traditional ringing sounds.

Charging for content entails:

- Charging the subscriber who requests and receives content (i.e., the consumer)
- Managing the content provider accounts that deliver the content (i.e., the supplier)

More often than not, the source of the content comes from outside the network, mostly from numerous Web sites or corporate storage. The content providers normally seek to recover fees from the subscriber via the network operator. This is performed in the Content Provider Charging Function. This function manages accounts that are maintained for the content providers, based on agreements to supply content to the network subscribers and to be remunerated accordingly.

Upon receipt of a charging request from an application server or a media server that performed the delivery, this function retrieves the content provider's details, including fees and discounts, in order to generate charging requests on the consuming subscribers. The function that initiates such accounting requests can reside together with the external applications and media servers, that is, at the service provider's domain. When such content accounting requests arrive at the operator's charging function, the source must be authenticated before processing.

Content charging requests are processed by the Subscriber Content Charging Function. This function must be native to the operator network in which the account of the subscriber is kept. This function checks the user account, debits it accordingly, and communicates with the subscriber if the need arises (e.g., when there is insufficient credit). This function performs the following:

- Handles charging requests received from the Contact Provider Charging Function
- Determines the subscriber's identity and details of the user's account
- Checks on current credit in the user's account to cover the requested service
- Obtains user confirmation for the charging, if such a confirmation is needed
- Requests to debit the user account by the charged amount
- Requests to credit the user account by a specific amount
- Manages Pre-pay accounts or post-paid accounts

8.2.5 Charging Interfaces Requirements

The protocol used throughout the packet-based charging system over most reference points is Diameter, accompanied by specific AVP extensions. The Diameter client, residing in the contributing network elements that trigger charging, communicates with the Diameter server in the CDF or the online charging function.

The charging interfaces must be resilient and reliable because they transport sensitive, if not critical financial information. Each initiated action (a charging event) is responded to by acknowledgments, and sent back to the network nodes to ensure that action is taken. They must also support high availability mechanisms, such as using alternative CDF or OCS platforms when the primary ones are not available. The data transfer mechanism must support error correction, such as retransmission of data.

8.2.6 The Generic Charging Components

8.2.6.1 Charging Systems

Charging components support two main systems: the OCS and the OFCS.

■ *The OCS.* This system is geared to support real-time billing where immediate results are needed for several reasons, for example, to provide a Pre-pay service, provide advice of charge for calculating available minutes, display call cost immediately after it has terminated, or perform real-time credit checks for transactions.
■ *The OFCS.* This system polls and pulls charging records from disparate network elements, and performs initial processing, such as collating information for the same sessions.

8.2.6.2 Network Functions

The charging mechanisms within the network are based on the following functions:

■ *The CTF (Charging Triggering Function).* This function is the triggering of a charging process that occurs at all network levels. Therefore, the CTF can reside in several nodes, in the IP-CAN servers in particular.
■ *The PCEF or the TPF.* This function supports the FBC which is controlled by the IP bearer, and enforces policies and rules from the service-based Policy Decision function. This function can be implemented in different types of bearers.
■ *The PCRF.* This function is the extended concept of the previous CRF. It interfaces to IP-CAN nodes and provides operator policies on charging for different scenarios.

Bearer and Access charging data can be extracted and sent directly to the BD. The PCEF function resides on traffic bearing nodes. This function consults with the PCRF when setting up media flow sessions that are charged via FBC. The PCEF enforces the charging policies and rules, thus affecting the way the charging is performed.

Figure 8.2 shows the BD with the OFCS and OCS feeding charging information to it. The OCS and OFCS both link to the network nodes that provide the charging information via their CTF. Note that the OCS can be independent of the BD, but the records are normally transferred to it after the effect, for statistics and CRM functions.

8.2.7 The Charging Trigger Function (CTF)

8.2.7.1 The CTF Principles

A chargeable session or event in any domain, service, or subsystem must be recognized within the relevant network element. This means that a signaling message, such as INVITE, is associating with a charging process. The CTF within that network element recognizes that charging action is

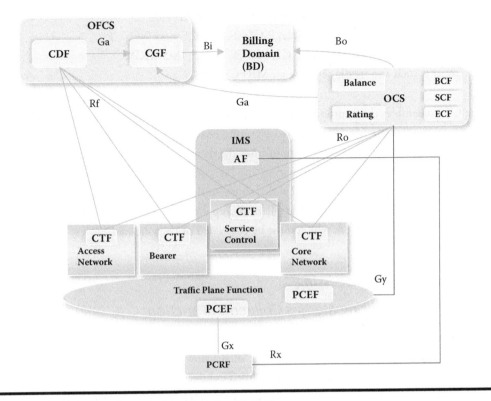

Figure 8.2   Charging elements in network with IMS.

required and creates a trigger (a *charging* event, not to be confused with a *chargeable* event) to be sent to the charging systems (i.e., the OCS or the OFCS).

The charging triggering event contains information sent over from the network element, information that reflects the details of the required service or transaction. This information enables the OCS to calculate the charge, or the CDF to open a CDR for recording charging data.

Note that online and offline charging can occur at the same time for the same transaction, if set up to do so. This can happen if a credit control process is required but also CDR generation on the offline charging. Some online systems create CDRs independently and send them directly to the BD.

8.2.7.2   *Charging Triggering in Network Elements*

The CTF contains a set of triggering conditions that activate charging when certain events are detected and the conditions are met. As a result, the charging process is initiated, modified, or halted. This means, for example, opening a CDR (Charging Data Record) to start recording charging information, modifying existing CDR parameters, or terminating the measurement and closing the CDR.

The CTF must be present in every network and service element that provides charging information. It does not represent a network entity, but an embedded function. The CTF collects information about chargeable events within the NE. The NEs can be:

- IP-CAN access nodes: for example, GGSN, WLAN AP, etc.
- IMS session controllers: for example, P-CSCF, I-CSCF, and S-CSCF.
- Border controllers: for example, BGCF, MGCF, I-BGF, IMS-MGW.
- Resources such as a media server (e.g., MRFC).

The CTF assembles the appropriate data items and compares them to the preconfigured conditions to determine whether to trigger the charging recording. When the received information matches these conditions, the CTF communicates the details to the CDF. The CTF is therefore a mandatory function, and is an integral part of any NE contributing charging information.

Multiple triggering can be associated with a single session, where the access node, for example, is recording usage, as well as session controllers and applications. This means that several CDRs will be produced for one session, which must be correlated or consolidated.

8.2.8 Accounting Processing

8.2.8.1 The Accounting Metrics Collection Function

The function of Accounting Metrics Collection monitors calls, service events, and sessions for triggers of accounting that arise from signaling, media flow, or service delivery to the user. The triggering attributes can be, for example, a fulfilled condition, an activation of a service, or the value chain, that is, the need to charge other networks or external service providers. It collects metrics that identify the user, and it monitors the user's consumption of network resources and non-session-related services in real-time. The type of metrics, the unit of measurement, accuracy level, type of rating required, and other parameters are also determined by the triggering function.

Metrics can be collected in a centralized manner, that is, directly into a central store or in a distributed fashion, to be pulled together later. The functions of Accounting Metrics Collection occur in all NEs and in multiple instances of the same type of NE. As a result, any network with multiple NEs needs a collection mechanism.

8.2.8.2 The Accounting Data Forwarding Function

This function receives the collected accounting metrics and decides on the next action. Based on the received CTF data, it decides whether or not any chargeable events have occurred. If charging is to be triggered, it assembles charging requests that match the detected events, and forwards the charging events information toward the Charging Data Function (CDF). The information is sent over the Rf reference point to the CDF. The information includes data about the source of the chargeable trigger, the network resource usage, and the identification of the involved parties.

The Account Data Forwarding function is performed in real-time for both online and offline charging, as well as for FBC. The same process takes place if there are mid-session changes that might affect the charging information. This process is also performed on events that signal session termination.

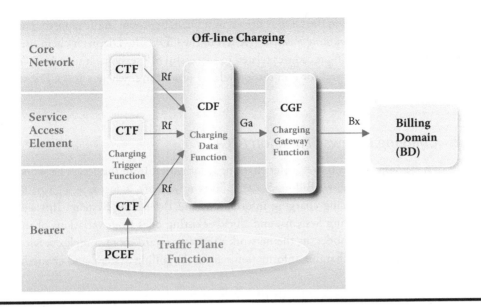

Figure 8.3 The OFCS overview.

8.3 The Offline Charging Subsystem (OFCS)

8.3.1 *Offline System Architecture*

The OFCS relies on clients in the network nodes (CTF) that initiate, modify, and terminate charging reporting according to a set of parameters, relevant to each NE.

Figure 8.3 shows the OFCS components and their links to elements in the network. The CTF in the control nodes sends triggering information via the Rf interface to the CDF or multiple CDFs. The CDF collects data and sends it to the CGF via the Ga interface. Finally, the CGF sends collated data to the billing system.

The two components in OFCS are:

1. The CDF, which creates the CDRs, and records charging duration and timing and other session details.
2. The CGF, which centralizes the collection of CDRs and performs some basic preprocessing and transformation to provide a consistent interface to the BD.

In Release 6, the CDF and CGF were combined in the CCF. Release 7 has separated these functions. This allows one CGF to service multiple CDFs. The CDF can reside on different nodes, making it easier for local databases to store the data in real-time.

8.3.2 *The OFCS Components*

8.3.2.1 *Charging Data Function (CDF)*

The CDF is the logic that performs the creation of the CDR, logging of metrics, modifying parameters according to triggers data from the reporting elements, and closing the CDR when the session terminates. The network nodes communicate with the CDF as a result of various SIP messages, and draw the relevant information. For example, the signaling messages inform the CDF about the parties' identities and types, about significant events, etc.

The accounting reporting on the CDF is driven by a Diameter client in the reporting network nodes, sending Diameter *Accounting Requests* (ACRs) to the CDF. The Diameter client uses ACR "start," "interim," and "stop" in procedures related to successful SIP sessions. It uses ACR "events" for unsuccessful SIP sessions and for non-session-related procedures, where a full CDR will not be generated.

8.3.2.2 Charging Gateway Function (CGF)

The CGF provides a mechanism to transfer charging information from the access network nodes to the network operator's chosen Billing Systems. This way, there is just one logical interface between the elements producing charging information and the Billing System. This simplifies the interface to any new Billing Systems, and shields existing ones from variations of information received from different nodes, as the network migrates toward more complex services. The preprocessing optimizes the charging information sent to the Billing System, and can reduce the load on it.

The CGF can have a diverse set of activities, including:

■ Act as the storage buffer for near-real-time CDR collection.
■ Provide the CDR data to the billing system in the expected format.
■ Perform consolidation of CDRs.
■ Preprocess certain CDR fields, filtering unnecessary CDR items.
■ Add predefined fields to suit the requirements of a particular billing system.

As the demand for timely billing information increases, the CGF should be able to receive CDRs from the network nodes in near-real-time mode. It should have enough storage to enable it to collect charging data for a period of time before transmitting it to the billing system, plus additional space, in case the data transfer cannot be performed as planned.

The CGF can reside on a central, stand-alone node, or as a distributed function built into NEs such as SGSN, GGSN, AAA Server, and other network nodes. The CGF can be distributed to specialized CGF nodes that keep only certain types of CDRs to streamline processing, or collect CDRs within a certain region, or other criteria. This enables both scalability and aggregation of CDR data.

The CGF should be implemented on more than one node to enhance reliability. If the Primary Charging Gateway entity does not respond to communication originating from the reporting nodes, these nodes will try to send the CDR data to a Secondary Charging Gateway entity (in 1+1 redundancy scheme) or another node out of a pool of servers.

The CGF opens a file to store the received CDRs. When a file is closed, the CGF ensures that another opens and is ready to receive the CDRs. CDR files are closed according to preconfigured parameters, for example, file size limit, time limit, time of day, and maximum number of CDRs.

When CDR files are closed, they can be preprocessed and then sent to the BD. The transfer to the BD can be in either pull mode or push mode. In pull mode, the BD initiates the transfer from each CGF instance. In push mode, the CGF platform imitates the transfer. The synchronization of file transfers can be configured to occur on particular time slots to avoid overloading of the BD. Usually, the requirements are for timely transfer, as close as possible to real-time.

8.3.3   The Offline Charging Links to IMS Elements

Most IMS network elements and application servers are required to be involved in triggering charging. An IMS element, such as P-CSCF, I-CSCF, S-CSCF, MGCF, BGCF, or AS (with or without MRF), can initiate charging recording by sending a Diameter ACR message.

The operator can configure the IMS NEs to enable or disable the generation of an ACR message in response to a particular triggering SIP message. It is also possible to enable or disable them, depending on the response to the SIP INVITE message, when it carries the appropriate user data.

IMS session-based charging is initiated on each of the IMS NEs at a particular point in their dialog. As an example, when the parties agree on the session particulars and are ready to start the media flow, the S-CSCF sends a Diameter ACR to the CDF, with the *Accounting-Record-Type* parameter indicating "START_RECORD." This will create an S-CSCF CDR on the CDF and will start recording charging information for this session. The same process occurs with the P-CSCF, creating a P-CSCF CDR when it forwards the appropriate SIP message that contains the triggers for charging.

When the UE is in another network (i.e., roaming), the Home I-CSCF interrogates the HSS to find the S-CSCF address. Having received it, the I-CSCF requests the Home CDF to create an I-CSCF CDR via an ACR with Accounting-Record-Type set to EVENT. This CDR will contain roaming information, and will record the start and end of the session and other details pertaining to the Visited Network. It can be used for an audit trail of roaming.

Connecting to PSTN, the MGCF requests the CDF to create an MGCF CDR when a session bound for PSTN is established. The BGCF does the same, triggered by its own set of conditions. CDRs from the MGCF and BGCF are used for inter-administration accounting, for billing other operators and service providers.

Applications also can initiate CDR creation on the CDF. For example, a Number Translation service details (i.e., the generated number to be redirected to) will be recorded when the AS sends an ACR with Accounting-Record-Type indicating EVENT_RECORD to start an AS CDR.

Figure 8.4 details the IMS network elements that play a role in charging procedures. It also shows the access nodes in the IP-CAN that get involved, including WLAN, xDSL and GPRS/UMTS AP nodes, which may be also contributing CDRs. Above the IMS nodes, the diagram shows the CS network elements and CS service elements that interwork with IMS.

8.3.4   Offline Charging Reference Points

The offline charging reference points are those connected to the OFCS system, that is, the CDF and the CGF.

Rf, the reference point toward the OFCS: In a convergent charging architecture, all communications for offline charging purposes between the CTF and CDF use the Rf interface. This interface carries trigger data from the CTF within various network elements to the CDF. This is an interface between a Diameter client on the CTF and a Diameter server on the CDF. Although the Rf connects to the offline system, it must be capable of supporting real-time transactions, due to the convergence of the online and offline operations.

The data through this interface identifies the event, the user, the session, and the service. It contains, for example, operator identifier, access-based media flow identifier, etc. The CDF uses this information to construct and format CDRs.

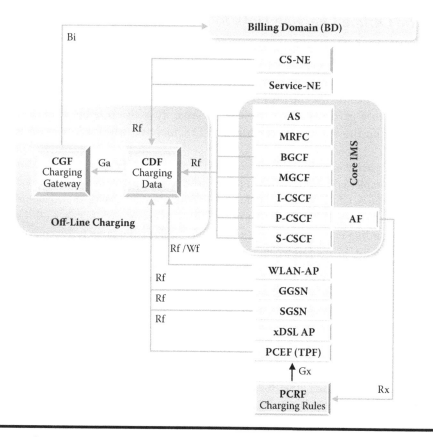

Figure 8.4 The OFCS contributing IMS elements.

Ga, the reference point between the CDF and GDF: This interface is used for interaction between a CDF and a CGF. The CDF sends CDRs onto the CGF, and the CGF acknowledges receipt of CDRs. This interface works in near-real-time, as the data is used by applications that require fast turnaround of charging information. For greater efficiency and speed, it should be possible to send more than a single CDR.

Gz, the reference point between PCEF and CGF: The Gz reference point between the PCEF/TPF and the OFCS is functionally equivalent to the Ga. The Gz reference point enables transport of offline charging information of service data for FBC, for which there is no session control in core IMS.

8.3.5 The CDR

8.3.5.1 CDR Triggers

The CDF is triggered from the CTF (i.e., various IMS network elements) by an ACR. This trigger identifies sessions, including an IMS SIP session. The session CDR reflects the specific media component details that are held in the media "container" within the CDR.

The triggering ACR can be ACR [Start] (starting the session); ACR [Interim] (mid-session changes); or ACR [Stop] (terminating the session). Unsuccessful attempts to connect are also recorded in CDRs. These are CDRs triggered by an event message, i.e., ACR [Event], where the event represents the failure to connect, recording the reason for the failure.

Other non-session-related events are also handled as Event CDRs. For example, registration and deregistration can trigger an Event CDR by sending an ACR [Event]. The Event CDR is created and closed immediately, and there is no possibility of follow-on partial CDRs.

8.3.5.2 CDR Creation and Closing

When the CDF in the OFCS receives a charging trigger (also called Charging Events), it activates recording of charging information. This means that the CDF uses the received information to open a CDR, noting the session start time. Subsequent charging events can occur during the session. These are processed and added to the CDR. Finally, the CDR is closed as a result of call termination or other preset conditions, such as time limit, volume limit, or memory constraints.

If an ongoing session exceeds limits set for the CDR, the CDR can be closed and a follow-up associated CDR can be started. This mechanism ensures that the system will not carry many open-ended CDRs that cannot be processed. However, this means that several CDRs can exist for a single session. In this way, CDR production is independent of the session itself. To reconstruct the session information, all related CDRs must be correlated eventually.

8.3.5.3 Partial CDR

Partial CDRs are CDRs that provide information on just a part of the subscriber session. A long session can be covered by several Partial CDRs. This means that a series of CDRs must be consolidated for a single session. After the first Partial CDR has been closed, the follow-up CDRs need not carry the full information again, thus saving on CDR storage space. Such subsequent CDRs contain only changes from one CDR to the next, as well as sufficient information to correlate them. For example, the location information is omitted if there has been no change to the positioning of the handset. There are two formats for Partial CDRs:

1. *Fully Qualified Partial CDR (FQPC).* This format contains *all* the CDR fields, although the CDR covers only part of the session. This format is mandatory for the initiating CDR.
2. *Reduced Partial CDR (RPC).* This format contains only the changes from the previous CDR, together with mandatory fields required for correlating CDRs. This format is optional. RPCs can be converted back to FQPCs if they are transferred to systems that do not process PRCs.

8.3.6 Transferring CDR Data

8.3.6.1 Implementation Options

The volume of charging data transferred between network elements and the charging system components is high. This merits close consideration of where the modules are positioned, in order to optimize the length of the transfer and the number of hops. Where modules are co-located on

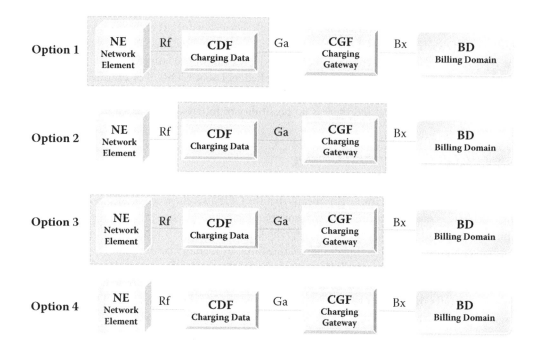

Figure 8.5   Options for OFCS node implementation.

the same node, significant savings are made in terms of network traffic, but at the expense of flexibility, independent scalability, etc.

There are a number of permutations of how the OFCS components can be positioned in the network, as Figure 8.5 shows:

- CDF in the NE, stand-alone CGF node, linking to the BD
- Combined CDF/CGF node between the NE and BD
- Both CDF and CGF in the NE
- Separate CDF and CGF nodes between the NE and BD

Implementation decisions are made according to available vendors' products, migration paths, size of the network, and network topology.

8.3.6.2   CDR Billing Interfaces

Other billing interfaces are defined for special types of CDRs is a converged system, including:

- Bc: reference point for the CDR file transfer from the circuit-switched CGF to the BD.
- Bg: reference point for the CDR file transfer from the GPRS CGF to the BD.
- Bl: reference point for the CDR file transfer from the GMLC CGF to the BD.

- Bm: reference point for the CDR file transfer from the MMS (Multimedia Messaging Service CGF) to the BD.
- Bmb: reference point for the CDR file transfer from the MBMS CGF to the BD.
- Bp: reference point for the CDR file transfer from the PoC (Push-to-Talk over Cellular CGF) to the BD.
- Bs: reference point for the CDR file transfer to the BD from CAMEL.
- Bw: reference point for the CDR file transfer from the I-WLAN CGF to the BD.

8.4   The Online Charging Subsystem (OCS)

8.4.1   OCS Principles

8.4.1.1   Uses of Online Charging

There are several reasons for needing real-time or near-real-time calculation of charging rates and session charges. For example:

- Calculation of rates and before the start of the session to inform the caller of the status of the account in terms of a call to the indicated destination
- Calculation in mid-call to determine when the user's credit has run out
- Calculation immediately after call termination to display the costs of the terminated call

The OCS is not entirely synonymous with pre-paid, although pre-paid service must always work through the OCS. Real-time charging information is useful for a host of other purposes, including internal monitoring and supplementary services such as "Advice of Charge" (AoC).

8.4.1.2   The OCS Architecture and Principles

Like the OFCS, the OCS is triggered from all the network layers, from bearer nodes as well as application servers. But unlike offline charging, the OCS has mechanisms that can influence the flow of the service and cause call termination (e.g., when the credit runs out). Therefore, a direct interaction of the charging mechanism with session or service control is essential for the OCS.

The OCS provides both bearer charging and session/service-based charging. It also provides a correlation of bearer charging with IMS (service-based) charging.

Figure 8.6 illustrates the OCS triggered via multiple CTFs in various network nodes. The controlling function, usually an application, monitors the service and can terminate it when the credit is exhausted. The main functions of the OCS are: Charging, Rating, Account Balancing, and Recharging.

The primary role of the OCS is to manage user accounts that are depleting through usage and replenished through payment. Unlike the OFCS, which merely records charging functions, the OCS contains elaborate logic to calculate monetized charges for the consumed services.

OCS processing enables the following:

- Calculate charging for sessions, events, and bearers services.
- Calculate whether there is enough credit to start the session and advise the user of talk-time.

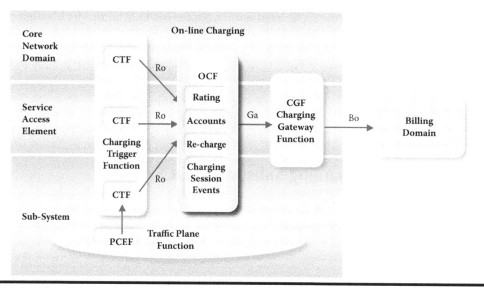

Figure 8.6   The OCS architecture.

■ Balance accounts with debit (when services consume credit units) and credit (when more funds are added to replenish the credit).
■ Recharging mechanisms, including a portal, voice procedure, prepay cards or vouchers, etc. that allow replenishing the credit amount in the account.
■ Report to post-processing system, like OFCS and statistics reporting, via CDR.

Although OCS is self-contained, with its own rating engine, the CDRs are normally required to be sent to the BD so that statements can be produced for the pre-paid accounts and statistical records can be maintained.

8.4.1.3   Reservation of Units

Except for *immediate event charging*, where the charging is performed as a one-off charging together with the transaction, any other online charging process requires a reservation of charging units. This is necessary because the OCS cannot predict the total charge, but is required to make credit control decisions. The reservation of units therefore allows approving credit in steps, where each step confirms availability of funds against a reserved quota.

The OCS, having reserved the first bundle of resources, can then start the charging process and allow the session controller to connect the session. The reserved charging unit bundles in each step are determined by the OCS according to operator policies applied to session details (i.e., type of service, roaming, user ID).

When the reservation quota is exhausted, the process of reservation of another amount is repeated, and the session continues. The process of reservation also can be triggered by mid-session changes, for example, where there is a change to the QoS requirement of the PDP Context in a GPRS/UMTS media stream. The change is requested via a charging event sent to the OCS. The

OCS reevaluates the credit and provides unit reservation. If there is no adequate funding cover, the OCS will reject the request.

When the session is terminated, it is likely to have some reserved units remaining. Therefore, on termination, the OCS will release those remaining units back into the available credit for the user account.

8.4.1.4 Advice of Charge (AoC)

The AoC is a supplementary service that provides information about the service charge if requested by an application or by a UE feature.

Two types of AoC are defined:

1. Advice of Charge for Information (AoCI) is used for charging enquiries.
2. Advice of Charge for Charging (AoCC) is used to debit accounts in real-time.

An alternative mode of AoC also can be used to indicate the occurrence of new charges to the user (e.g., when a monthly allowance has been exceeded, or when a service is requested that is not included in the subscription fees).

8.4.1.5 Bearer Charging

The OCS can be called upon for charging non-IMS activities. This involves bearer network nodes interacting with the OCS to allow media streams to be controlled by credit availability in the user account. The measurement of usage is performed at the bearer level and the reservation of units is based on media flow rates.

8.4.2 The OCS Components

8.4.2.1 The OCS Architecture

The OCS contains the following main modules:

- Charging Functions, including Bearer, Session, and Event Charging
- Rating server with tariff data
- Account Balance, containing the account credit and associated procedures
- Recharging server for paying in deposits to increase the credit

Figure 8.7 depicts the OCS with its components — the Charging Function, which includes the session and event-based charging modules; the Rating Function with tariff database; and the Account Balance Manager. There is a possible link to the post-processing billing system via the CGF (Charging Gateway Function), which forwards online CDRs to the offline billing system.

The Rating Function and the Account Balance Manager components can co-locate on one platform with the OCS. Alternatively, they can interoperate from separate platforms. The OCS also connects to the access nodes for GPRS, WLAN, and DSL broadband to provide online charging based on measurements made on these nodes or to process particular events.

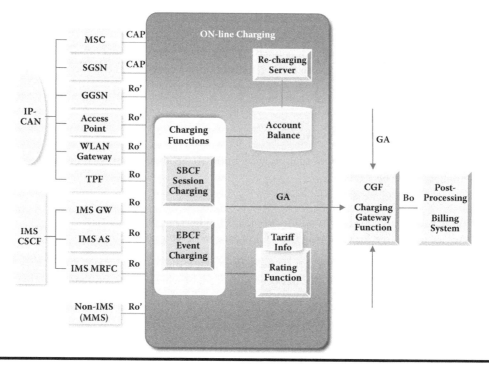

Figure 8.7   OCS components.

IMS network elements and SIP Application Servers connect to the OCS via the Ro interface. The interface to GPRS nodes and existing WLAN solutions is the Ro' interface (to denote some differences in the data structure). The Ro' represents IMS charging data from the bearer nodes or from non-IMS packet-based services, such as MMS. Also available is the legacy connection of CAP (CAMEL Application Part) to the OCS, to support existing online services via the MSC.

Note that the CSCF session controller is not connected directly to OCS, but via applications or through the IMS-GW (IMS Gateway). The IMS-GW provides the mapping of SIP messages to Diameter-based accounting messages.

8.4.2.2   The Rating Function

The Rating Engine calculates the rates for the call charge. It estimates the credit for the session according to the charging rules and session parameters to decide whether the service can start and advise the user about "talk-time" in AoC. The Rating Engine also calculates the actual charge from the session duration and rate of consuming measurement units and the usage of counters after the session termination.

The Rating Function performs both monetary and nonmonetary unit calculations. It performs rating for the on-network services and for external applications (session, service, or event). Rating can be activated before and after service delivery. Rating also can be used for cross-product and cross-channel discounts, benefits, and allowances. As the variety of chargeable services grows, the Rating Function must be enhanced to handle a wide variety of rating types, including:

■ Rating by volume, especially when an access network entity initiates the charging

■ Rating by time duration, normally initiated by a SIP application or service controller
■ Rating of events that signify chargeable activity (e.g., Web content or MMS)

The Rating Function determines the appropriate tariff per unit, for example the price per call minute, data volume unit, multimedia session, or Web content unit. It also determines one-off charges for chargeable events.

The Rating Function maintains tariffs and responds to requests by fetching the appropriate data. Optionally, the Rating Function can be extended to perform the logic and return the calculated results. In such cases, it contains not only tariff data, but also procedures for analyzing the account status and the special accumulation counters. These counters provide historical records of usage, usually accumulating toward a particular target (e.g., for bonus, loyalties, or free minutes). The counter data can be received from the Charging Function or stored locally.

The Rating Function supports two primary functions:

1. *Price Request.* This is the determination of a price for the execution of a service. From the rating perspective, this is the same method when it is run before delivery (e.g., for balance check enquiry or AoC), after delivery (post-rating), or during the service delivery, whenever a new rating is needed. The same method applies for one-time or recurrent charges. The Price Request is used by the EBCF (Event-Based Charging Function).
2. *Tariff Request.* This is the determination of a tariff for a given service. Based on the tariff, the charging function calculates either the amount of units for a given price or the price for a given number of units. The Tariff Request is used by the SBCF (Session-Based Charging Function).

8.4.2.3   The Account Balancing Function

The previously called "Correlation Function" has been renamed the "Account Balance Management Function." This function provides access to accounting and correlation information for the OCS accounts. It performs the following:

■ Provides credit checks to assess available funding for a requested service
■ Updates the available funds according to received payments
■ Decreases the balance according to usage
■ Reserves units for a session
■ Sets and checks on expiry dates for pre-paid accounts

8.4.2.4   The Charging Function

The Charging Function evaluates the credit and reserves units for the type of service requested. It receives requests for charging and connects to the Rating Function to fetch rates (or calculated units) and to the Account Balance to obtain current credit.

8.4.2.4.1   Session Based Charging Function (SBCF)

The Session-Based Charging Function (SBCF) is responsible for Charging for a Session type service, including the session affecting control, such as session termination when the credit runs out.

The SBCF performs the following:

■ At the bearer level, based on bearer requests, it controls the usage per granted time, volumes or bandwidth.
■ At the session control level, based on requests for session resource usage received from the network (e.g., from the IMS CSCF), it controls sessions in the network. For example, it has the ability to grant or deny a session setup request and to terminate an existing session.
■ At the service level, based on requests for services received from the network, it controls service availability in the network. For example, it has the ability to grant or deny using a service or an application.

The SBCF communicates with the Rating Function to determine the value of the requested bearer resources or the requested session. It communicates with the Account Balance Management Function to query and update the subscribers' accounts.

8.4.2.4.2   Event Based Charging Function (EBCF)

The Event Based Charging Function (EBCF) performs charging calculation in response to a single occurrence or fulfilled conditions. The EBCF uses the Ro reference point to interface with IMS Application Servers and Media Server Function (MRF) platforms.

The EBCF makes use of the rating function to determine the value of the service rendered. It can correlate several event-based charging requests. The EBCF provides information that triggers the Correlation Function to debit or credit the subscriber's account. The information sent by the EBCF also can be used in the Account Balance Function to correlate Event Charging with Bearer Charging and Session Charging.

8.4.2.5   The Recharging Function

This module is responsible for user interaction to enable users to authorize payment via various methods and increase their credit balance. The methods may be via a Web portal or voice call commands, or by buying vouchers and activating them. The Recharging Function interacts with the Account Balance Function to update the balance. For recharging during the session, when the application is able to notify the user of low credit, the process of recharging and updating the balance must operate in real-time.

8.4.2.6   Online Charging Reference Points

Ro reference point. This is the reference point for IMS nodes to communicate with the Online Charging. Over the Ro interface, information about chargeable sessions and events flows from the CTF within the IMS components is sent to the OCS. This trigger information is considered

"Charging Events" that stimulate a charging activity (not to be confused with events that are in themselves chargeable). These charging events trigger the process of online charging. The OCS confirms receipts of charging triggers over the Ro.

This interface must perform under time constraints in real-time. Both stateless mode for Event Based Charging and stateful mode for Session Based Charging are supported.

The Bearer Charging Function (Flow Based Charging) needs to support the Ro interface as well as CAP for CS based Pre-pay. Ro variants may be needed for online charging interfaces toward different access network entities.

Ro' reference point. This is an interface similar to the Ro reference point. It is the online charging reference point for non-IMS services (e.g., MMS) interaction with the OCS online charging functions of EBCF, SBCF, and bearer-based FBC. Because this is non-IMS charging, the data details transferred over Ro' are different from the Ro interface.

Gy reference point. The Gy reference point between PCEF/TPF and the OCS is the same as the Ro. The Gy reference point allows performing online credit control for services using Flow Based Charging, (FBC).

8.4.3 OCS CDRs

Although the OCS performs the complete process and the charges are applied to the subscriber's account, a record of the transactions is often required for the production of billing statements and management statistic reports. OCS records can be produced in two ways:

1. The OCS can contain CTF functionality that triggers CDR production on the CDF. This will be an event trigger that forwards the information of the session and its cost to the CDF after the transaction completes.
2. The OCS can contain CDF functionality internally. The completed CDRs are forwarded to the CGF in the same way that they are forwarded by the OFCS.

8.4.4 OCS Design Options

There are some optional designs in the allocation of work between the Rating Function and the Charging Function that affect where the logic resides. These include:

1. *Repository style.* The Charging Function performs the processing, having obtained tariff data from the Rating Function and credit level data from the Account Balance Function, and the logic is performed entirely in the Charging Function.
2. *Rating engine calculation.* The Charging Function asks the Rating Function to perform the rating and return the final result (a price or AoC to advise the user). Then the Charging Function instructs the Access Balance Function to update the account.
3. *Extended rating with counters.* Either the Rating Function or the Account Balance Function can keep the counters that affect the calculation (e.g., bonus counters or loyalty accumulations).

While option 1 can provide high performance, options 2 and 3 are less complex and more flexible because the data formats are internal.

8.5 Integrating IMS Charging

8.5.1 IMS Charging

The charging system, by its nature, must interwork with all domains and all types of charging. That means:

- The OCS and OFCS are able to work together.
- Packet network as well as circuit-based network charging can be supported.
- Mobile, Fixed, or Wireless LANs can all share the charging facilities.
- Multiple access nodes and CN nodes can contribute charging information.
- Any kind of Application Server can participate in the chargeable service.

Although the OCS and OFCS are considered independent systems, they are often used at the same time. Capabilities developed in the OCS, such as content charging and event charging, are generalized for use in both Pre-pay and post-pay modes.

The collection of usage data from the access nodes for FBC applies for both. Further integration emerges in the relationship of resource allocation rules with the charging rules. Often, services delivery depends on both sets of rules being harmonized.

The charging system also integrates CSN elements with packet network. The CS NE can use the OCS/OFCS as well as IMS. Formats of CSN charging records must be acceptable to the IMS charging systems for the purpose of inter-charging for sessions to and from CSN. IN based Pre-pay services can continue to operate alongside new IMS prepaid services.

Figure 8.8 illustrates the complete charging system, including both the OFCS and OCS, the IMS entities, CSN network entities, and some of the IP-CAN nodes (GSN and WLAN). This diagram shows the AS as well as other IMS entities communicating directly with the OCS over the Ro interface and to the CDF over the Rf interface. The bearer is represented by the PCEF, also contributing charging triggers and data to both the OCS and OFCS.

8.5.1.1 Billing Domain Reference Points

- Bo: reference point between the OCS and BD. The Bo reference point is the interface from the BD toward the OCS. This interface transfers online charging records from the OCS to the BD.
- Bi: reference point between the OFCS and BD. The Bi reference point is the interface between the BD and the OFCS, that is, the CGF. This interface transfers offline charging records from the OFCS to the BD.
- Bx: reference point toward the BD. It is a generic reference for the transfer of all types of CDRs, for online and offline systems alike.

8.5.1.2 The IMS Gateway Function (GWF)

The S-CSCF does not contain the CTF within its scope and does not create any online charging events, but does respond to chargeable actions. The S-CSCF handles SIP messages to and from all

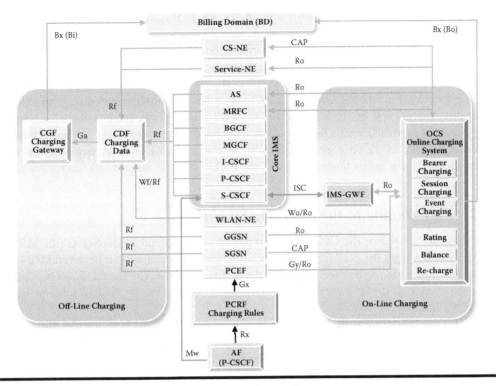

Figure 8.8 Overall network charging system and contributors.

application servers via the standard ISC, relaying charging triggering transparently, as it does for any other service. In this way, the S-CSCF regards the OCS as an application server.

The introduction of the IMS Charging Gateway between the S-CSCF and the OCS enables the CTF function to be performed for the S-CSCF in a separate entity, thus maintaining the S-CSCF functionality as it is, with the same ISC interface to all application servers.

The charging IMS GWF translates between the SIP service control toward the S-CSCF (ISC) and the Ro credit control toward the OCS. The charging IMS GWF module can co-reside on the S-CSCF, on the OCS, or provided as a separate component.

8.5.2 *Flow Based Charging (FBC) and IMS*

FBC is designed to charge for media delivery services that are not governed by a session control server (i.e., non-IMS). However, with the growth of "blended services" such as video streaming combination services, there is a need to have a single umbrella service control that can share the IMS-defined authentication, resource management (QoS), and charging.

FBC for IMS session charging is controlled via a Charging Key that indicates a set of rules on the PCRF. If the session is charged via an IMS service, the bearer flow of media packets must be correlated to avoid duplicate charging. This can be done simply by zero rating the session CDR.

Previously, the OCS and OFCS determined such record correlation, and therefore had to contain the logic and rules to do so while differentiating the rates. Where PCRF is employed, a Charging Key can be assigned to the IMS traffic. This Charging Key is associated with a set of charging rules provisioned independently onto the PCRF. In this way, the central PCRF can provide a consistent approach and data integrity.

In IMS, the AF provides the means to correlate flow based charging information with service-level charging records. The AF is a function residing within the P-CSCF or other proxies. It interacts with the PCRF to activate charging rules. The AF contains the logic to determine which Charging Key to use for the IMS session. The Charging Key can be set to zero rating, for example, so that FBC charges will not be levied where the IMS session is charged at service level, or vice versa, where the particular media flow should not be charged (e.g., for Presence or status information messages). The AF also triggers a CDR to be written on the CDF. This charging information contains details of the media flows along with the ICID.

8.5.3 IMS Routing to the Charging Systems

The address of the OCS or the OFCS servers is passed to the AS and Media Server (the MRF) in the SIP signaling from the S-CSCF via the ISC interface. This address can be overridden by the operator's configuration or by preconfiguring the addresses on the AS. The CDF or OCS address can be retrieved from the HSS via the Cx interface and passed by the S-CSCF to subsequent entities, the AS and MRF.

The AS can also retrieve the charging function address from the HSS via the Sh interface. CDF or OCF addresses can be allocated as locally preconfigured addresses. These addresses are not passed to the Visited Network entities or the UE.

8.5.4 IMS Charging When Roaming

Charging information is produced for the benefit of the native network that generates it. There is no arrangement for charging triggers to cross business borders. Charging rules and policies are valid only within the native network. Billing information is therefore exchanged only at the BD levels, where inter-administration accounting data is produced from the consolidated CDRs.

The user in a Visited Network obtains authorized access only when there is a mechanism of operators' settlement in place. A session based charging agreement is confirmed at the time of Home Network authentication and subsequently starts a charging process there. Information from the local access nodes also triggers charging processes in the Visited Network. CDRs are produced within each network, to be processed by the respective internal billing systems.

Each network interprets the charging data and consolidates its own CDRs. Inter-operator settlement occurs on exchange of processed information between the billing systems, which usually aggregate the information and expose only what is needed to be disclosed.

At the Visited Network, the IP-CAN can monitor the media flow if such charging is activated for a particular service. The CDRs are produced at the Visited Network by the access nodes, and the P-CSCF. At the Home Network, CDRs are produced by the CSCF and AS.

In scenarios where services are provided from the Visited Network, the local Proxy contacts Visited Network session controllers that can invoke an application. The Visited Network's session controllers and ASs produce CDRs for the service. These CDRs will be transferred to the Visited Network billing system to be processed further and used for inter-operator settlement or charged at zero rate.

Figure 8.9 shows an example of two IMS mobile phones with offline charging for both parties while they are both roaming.

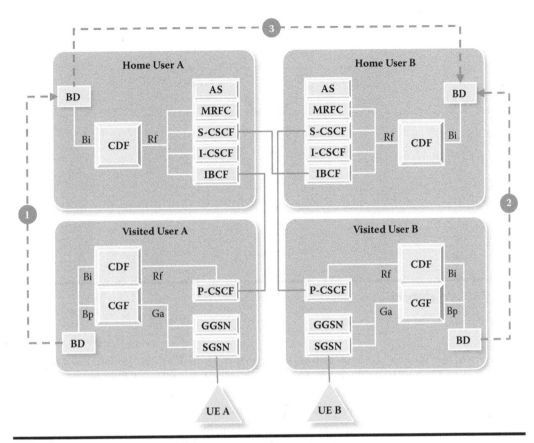

Figure 8.9   Exchange of charging information when roaming.

User A is in a Visited Network that forwards the session request via a local P-CSCF to Home Network A. Visited Network A has an agreement with Home Network A, and will start charging metering when the session is connected. Home Network A authenticates the user, enabling IMS servers to produce charging information via the Rf interface to the Home A CDF.

Having been authorized, the Home A CSCF contacts the Home B CSCF through border control (IBCF). The CSCF ascertains that User B is roaming and initiates a session toward Visited Network B. When Visited Network B accepts the session, it starts a charging process there, and the Home B charging process is also engaged when the media flow is connected successfully between User A and User B.

Finally, the BD in Visited Network A sends roaming settlement fees to the Home Network A. Visited Network B sends roaming charges to the Home Network B. The Home Network B passes these charges (and its own) to the Home Network A, where the calling User A will be billed.

8.5.5   IMS AS Charging

Pre-pay service, as a rule, has been based on an application. In CSN, this is an IN/CAMEL-based application on an SCP. In such cases, the OCS may be an integral part of the application on the AS, or utilized as a separate "rating engine." Other IN applications can determine the charging modes, for example, Free-Phone, Local-Rate, Split-Rate, Reverse-Charge, Charge-Account, and

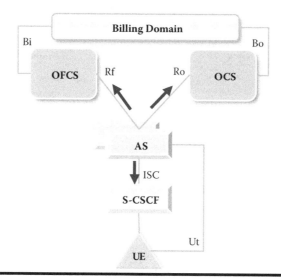

Figure 8.10   AS charging.

others. These applications continue to fulfill their role in IMS. They can initiate the creation of CDRs on the CDF and affect the values inserted into session controllers' CDRs (e.g., modify the billed party).

Figure 8.10 shows the means of connecting an AS to the charging systems. The AS can initiate charging procedures on behalf of the user, as indicated by the arrows in bold.

The separation of the OCS from the application allows many more applications to use the OCS, from other application servers, via a standard interface. The AS with its media server can establish whether the service requires online or offline charging, and act accordingly. This means that the AS application will send charging information over the Rf interface to the CDF in the OFCS, or over the Ro interface to the OCS, or can use both if the service requires it.

The AS receives information that starts the charging process via the SIP ISC standard interface to the S-CSCF. The AS can start both OCS and OFCS processes at the same time. For example, an AoC using the OCS may be needed for interaction with the user at the start of a session, but the following session is charged via the OFCS from offline CDRs. For the offline charging, the AS starts and closes AS-CDRs on the OFCS. These records will be reconciled with CDRs from the S-CSCF and maybe the media server (MRF) if used in the service.

The AS starts a charging process when it acts as a User Agent (UA) and initiates a call using a third-party call control mechanism. In this case, the AS is the first network element, so it must generate the session identifier, the ICID (IMS Charging ID). However, in most cases, the AS is called upon by the session controller, passing to it the ICID as well as other session details. The AS will build these details and identifiers into the AS CDRs and send them to the CDF. The AS CDR can also contain service-specific parameters and media attributes descriptions.

8.5.6 IMS Charging Correlation

8.5.6.1 Correlation Principles

Correlation of charging information is essential for managing numerous CDRs produced by numerous network entities. Correlation is needed from several points of view:

1. *Within sessions:* a number of media components that are involved in a single session.
2. *Within media streams:* related media streams that make up a single service (e.g., synchronized audio and video).
3. *Between network elements:* consolidation of CDRs from different IMS components, such as P-CSCF, BGCF, I-CSCF, S-CSCF, etc.
4. *Between the IMS and the PS Domain:* when the service-level charging takes into account flow-based charging recorded at IP-CAN level.
5. *Between OCS and OFCS:* where both are used for a single session or event.
6. *Between several partial CDRs:* for long sessions that generate several follow-on records after the initial CDR.

The correlation of charging information for an IMS session is based on the use of ICIDs and IP-CAN identifiers (e.g., GCID for GPRS Charging ID) for FBC services. These identifiers are generated at the initiation of a charging action and are passed between all the participating network elements, so that all relating records will contain these identifiers. Where the session is delivered in several networks (Visited, Home, intermediate), then the involved Operator ID is also added.

Charging information, including correlation identifiers, are encoded in the P-Charging-Vector header in the SIP message. This P-Charging-Vector contains:

- ICID
- IOI
- IP-CAN information and addresses

To prevent fraud, these charging correlation identifiers must not be revealed to the user. Usually, they are used only within the same network but may be shared with trusted servers, such as a content server that is connected as an endpoint.

8.5.6.2 IMS Charging ID (ICID)

The ICID is an IMS Charging Identifier for a particular session that is shared between the various IMS elements that participate in the delivery of the service. The same ICID is also shared between the networks that are involved in the session.

The ICID is produced by any IMS network entity that starts the charging process, usually the P-CSCF or the AS. Thereafter, the ICID propagates to other network elements in the SIP signaling related to the same session, and is sent across to the terminating network.

The ICID is generated by:

- The P-CSCF for UE-originated calls
- The AS application when acting as an originating UA
- The I-CSCF or IBCF for UE-terminated calls

If there is no ICID in a connection request, the network entity dealing with it will generate it. This can happen if the originating network does not support ICID. Similarly, the MGCF will generate an ICID for PSTN-originated calls that arrive without an ICID. Therefore, at each SIP session establishment, a new session-specific and unique ICID is generated. This ICID is then used in all subsequent SIP messages for that session (e.g., 200 OK, Re-INVITE, BYE) until the session is terminated. The ICID is encoded in the P-Charging-Vector header of the SIP message, when it is present. The ICID is a temporary identifier that remains valid only for the duration of the session or the event it represents. Thus, a registration ICID can be valid for the initial registration and any subsequent reregistrations. It expires after the final deregistration.

The IP-CAN Charging ID is also generated and conveyed. For example, to correlate between IMS and GPRS, one or more GPRS Charging Identities (GCIDs) are used to identify the PDP Contexts of the session. These GCIDs are generated at the IP-CAN and passed from the PS bearer to the IMS entities, along with the node addresses.

The ICID is used for session-unrelated messages too, for example, SUBSCRIBE, NOTIFY, and MESSAGE. These session-unrelated services should be correlated in the same way as user-to-user sessions.

To correlate historical data, the ICID value must be globally unique, at least for a period of time (e.g., one month). This can be achieved using node ID and time or date stamps, or other methods according to specific implementations.

8.5.6.3   Inter-Operator Identifier (IOI)

Any session that requires a connection across two or more carrier networks must have network identifiers in the charging information. These are the IOIs, which are globally recognizable and are shared between sending and receiving networks.

Each network generates its own identifier for its own leg of the session. The IOI for the originating network identification is inserted in the "orig-ioi" parameter in the P-Charging-Vector, and the terminating network identification is inserted in "term-ioi" parameter.

The IOI is used to identify operators in the following interaction scenarios:

1. *Between the Home Network and the Visited Network when the user is roaming and gains access via P-CSCF in the Visited Network.* This Visited P-CSCF will generate the originating IOI as the Visited Network Operator identifier, and the Home Network S-CSCF generates the terminating IOI.

2. *Between the Home Network of the originating caller and the terminating network that owns the subscription of the called party.* The S-CSCF in the originating party's Home Network or the originating MGCF is responsible for generating the originating IOI, and the S-CSCF in the terminating party's Home Network or the terminating MGCF is responsible for generating the terminating IOI.

3. *Between the terminating network which is the Home Network of the called party and the redirected network where the called party is receiving the call in this instance (i.e., a Visited Network).* The IOI is produced as for case (2) but the terminating Home Network is regarded as an originator IOI, and the Visited Terminating Network as the terminating IOI.

4. *Between the Home Network and a Service Provider.* When the S-CSCF invokes an external AS, the Home S-CSCF generates the originating IOI, and the target external AS generates the terminating IOI. When the AS initiates the session, the AS generates the originating IOI and the recipient node (I-CSCF, IBCF, or S-CSCF) will generate the terminating IOI.

Both originating and terminating IOIs are extracted from the SIP messages and inserted into CDRs that are produced for the IMS elements on the CDF. These identifiers are used, in particular, for inter-operating accounting purposes.

8.5.6.4 Access Network Charging Identifier

To correlate service data flow (i.e., a media stream that is identified within the IP-CAN) with a session charged at the IMS service level, the particular media flow identifier must be conveyed to the IMS elements preparing charging information. This can be, for example, the GPRS identifier and the PDP Context information for it. To enable the correlation of media streams with a session, the Access Network Charging Identifier is inserted into the P-Charging-Vector and sent in the SIP message to the IMS session controllers.

The details of the access network charging information are passed from IP-CAN to the P-CSCF via the PDF, SPDF, or the PCRF. This data is updated during the session when media flows are added or removed. The Access Network Charging Identifier is passed in the P-Charging-Vector header from the P-CSCF to the S-CSCF and other IMS servers, and is also shared with an AS that might need it for charging-affecting services.

The Access Network Charging Identifier is not shared with other networks and is solely used within the same network. This means that there is no exchange of IP-CAN identifiers between Visited and Home Networks, and external ASs do not receive such information.

8.6 Policy and Charging Control (PCC)

The PCC system is common for both charging functions and resourcing functions, (QoS). This section discusses the charging aspects of the PCC. For resource-oriented PCC, see Chapter 6.

Note that the previous concept of CRF is replaced by the PCRF, and that the enforcement function is now referenced by PCEF.

8.6.1 Charging Rules Principles

8.6.1.1 Applying Charging Rules

Charging rules are applied to session-based charging as well as FBC, events as well as sessions, and offline as well as online charging. Previously, charging rules were built into services that affected charging, such as Freephone or Premium Rate. Now, charging rules have far more scope and variations, and can affect any session.

The introduction of media-flow-based services, not just voice calls, means that many more decisions must be taken when setting up the session. Some such sessions, such as games or music downloads, are not necessarily governed by the session controllers but must have rules for the behavior of the media bearing nodes. For all these reasons, there is a need to formalize and streamline charging rules for all types of services.

The Charging Rules function can:

- Distinguish between service data flows that are charged individually
- Provide specific rules per service data flow

- Provide charging rules per PDP Context containing several commonly charged flows
- Maintain counters measuring volumes or elapsed time per session or service data flow
- Report events to create charging triggers
- Activate network-based policies
- Decide which policy prevails in case of conflict

8.6.1.2 Types of Rules-Based Charging

Flow-based bearer level charging can support dynamic selection of charging rules. The PCRF can examine several factors of the bearer and the service request before making a decision on the specific charging rules to apply.

Charging rules can enforce limits on usage of certain service data flow, according to the service type, for either online or offline services. The online charging systems can provide both volume credit and time indication.

Charging rules can distinguish between different levels of QoS that are charged different rates. The charging rate can be set when a bearer media flow is created for a given QoS level. They can be changed as a result of an event, for example, inserting a paid advertisement in the media stream that should not carry any charge.

The detected event is sent to the AF within the appropriate network element, and the AF notifies the PCRF. Other changes can occur as a result of an event in the OCS, for example, reaching thresholds in bonus counts that alter the charging rate.

Charging rules can overlap each other. Rules in the Traffic Plane layer can clash with rules from the Service layer. In such cases, the PCRF uses the associated precedence order given to each rule to resolve the conflicts. When overlaps occur between a dynamically allocated charging rule and a predefined charging rule at the IP-CAN bearer, and they both share the same precedence, then the dynamically allocated charging rules are applied first.

The charging rules can support several charging models for different types of data flow charging:

- Volume-based
- Time-based
- Volume- and time-based
- No charging (periodic fee only)
- Prepaid and post-pay
- Per service type or class
- Per media type
- Per level of QoS
- Roaming add-on fee (rather than distance)
- Online or offline
- Credit limited

Different charging rules can be applied for downlink and for uplink. Charging rules can be configured for both user-initiated and network-initiated flows. Charging rules may be different for each user or may be shared by multiple (or all) users. Charging rules usually vary for roaming users and may depend on the agreements between operators.

8.6.1.3   Charging Rules Data

Charging rules contain information that the OCS or OFCS will need to execute the appropriate charging. In the first place, the charging rules define whether the service needs online rating, offline charging, zero rating, or no rating at all. The rules indicate what charging unit to use. This is a matter for policy, depending on the type of service. For offline charging, the rules can identify whether to apply volume-based accumulators or time-based counters.

Charging rules data also contains:

- Charging key for PCRF to identify a package of policies
- Service data flow filter
- Service identifier
- Precedence (used at the PCEF to determine the order in which charging rules are applied to a packet flow)
- Charging rule identifier (used between PCRF and PCEF for referencing charging rules)
- AF record information
- Service identifier level reporting: mandated or not required

8.6.1.4   IMS Media

IMS media flows can be one of two categories:

1. *Peer-to-peer IMS media flows.* In this case, filters for the flows may need to be dynamically defined according to the session control signaling for that particular peer-to-peer session. The details of the filters (e.g., destination address) may need to be dynamically provided to the PCEF via the PCRF, using the Gx interface. The PCRF selects the Charging Key according to information forwarded from the bearer via the AF (Application Function in the P-CSCF or similar function), using the Rx interface.
2. *Client/server IMS media flows.* In this case, filters for these flows should be identified dynamically when the session control is established with the IMS server, but can reference existing settings for the client/server services used. The PCRF builds the information from what it receives from the AF to select the appropriate Charging Key.

8.6.2   Policy and Charging Rules Components

8.6.2.1   The Role of Charging Rules in the Network

Just as rules for allocation of resource and a level of QoS are defined in the Policy Decision Function, the network's policies for charging are defined and stored on the PCRF. Instead of allowing for charging policies to be determined by each application, a network-wide charging policy decision function is introduced to select charging rules for all applications.

FBC depends on the ability to control the gating of packets at the bearer. This necessitates the correlation of the charging and the resourcing for particular media flows, hence the convergence of charging rules with resource policy decision. The PCRF instructs the PCEF to open the gate to traffic when it finds appropriate filters for a particular service data flow. If none are found, the packets will be discarded and the gate is closed. It is possible to implicitly block a specific service data flow when there are no other charging rules matching the service data flow for any bearer.

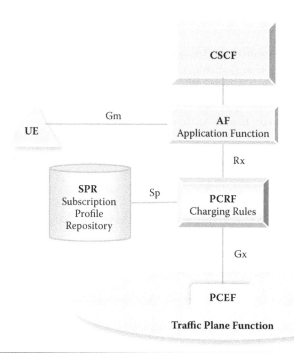

Figure 8.11   PCRF connecting to PCEF.

Figure 8.11 shows the components of the PCC. This centers on the PCRF, connecting to the Traffic Plane PCEF via the Gx interface and to the AF (Application Function) requesting policy and charging rules via the Rx interface. The PCRF contains a repository of policies and rules in the SPR.

8.6.2.2   The Policy and Charging Rules Function (PCRF)

The PCRF (previously CRF) supports both activation of predefined charging rules already installed in the PCEF (previously TPF) and dynamic charging rules that are downloaded to the PCEF when the media stream is created. These rules are installed for both offline and online charging. The PCRF, as a central function, takes responsibility for resolving policies conflicts and contradictions between installed and downloaded instructions by determining the precedence of these rules.

The PCRF gets involved in both service-based and flow-based charging:

■ For service-based charging, the PCRF selects the charging rules and their priorities for each session request or session modification request via the IMS session control. The PCRF decision is based on information made available to the PCRF by the UE, the AF (from the Service layer), and the bearer nodes (i.e., information related to the user details, the bearer characteristics, and the network routing).

■ FBC rules can be optionally built into the bearer nodes but ideally an external function can provide uniform handling of rules and links to the service type. Certain charging rules are defined within the bearer (e.g., on the GGSN, SGSN, or DSL access nodes) but the PCRF can still address them by reference. These rules relate to media flow type rather than to sessions and user particulars.

Multiple charging rules can be supported simultaneously per user and bearer, and differential charging regime for different media flows can be maintained, even within the same media flow.

However, note that the PCRF does not operate on a per-MBMS bearer context basis for Multicast Broadcast service.

The PCRF can dynamically generate and install charging rules to support scenarios where filtering information is dynamically negotiated at the application level. Such filters can detect packets associated with particular application protocols, for example. An application can provide the PCRF with input to the filter definition, for example, the scope of the subscriber's access to the service or any restrictions to it. Using these filters in the charging rules can refine further the way services are charged, and therefore help to fine-tune the rules.

To apply the PCRF, the underlying network must be able to distinguish between different flows that need to be handled according to different rules and charged at different rates. Those media flows that do not require special treatment can be aggregated together for more efficient processing. Charging rules can apply to service data flows by their duration or by byte counts for volume-based charging rules, in which case the network must be able to meter such volumes. The network must also be able to detect and monitor events, such as Accounting Start or Stop requests.

The PCRF needs an appropriate provisioning mechanism to enter the charging rules into a database (SPR) with their charging keys.

8.6.2.3 Policy and Charging Enforcement Function (PCEF) or the Traffic Plane Function (TPF)

The PCEF or TPF is the function within the transport plane that can interwork with the charging systems. This function must be present in the bearer nodes to trigger and manage charging processes. The charging activities in the PCEF are initiated by IMS at session establishment, but the PCEF itself triggers charging for FBC.

The PCEF must be able to differentiate between data flows that belong to different services. It should be able to apply rules to these media flows and activate pre-defined filters. In the case of OCS, for both flow-based and session-based charging, the PCEF or TPF must be able to prevent starting the session until credit has been authorized, and to terminate ongoing media flow when this credit runs out.

For IMS sessions, the PCEF receives details of the user session at the time of establishing a connection. This includes the user ID and terminal identifier, location (country code and network code), local access node details, bearer characteristics, and QoS capabilities. The information is incorporated into the charging triggering event data that is sent to the PCRF.

The PCEF charge triggering events can provide the following information:

- Subscriber identifier
- IP address of the UE and location of the subscriber
- IP-CAN bearer attributes
- Request type (initial, modification, etc.)
- Type of IP-CAN (e.g., GPRS, I-WLAN, etc.)
- A PDN identifier
- IP-CAN bearer establishment mode

At session initiation, the bearer node requests to receive instructions (i.e., charging rules) from an available PCRF instance. The selection of the PCRF address may be made depending on the user's identity or on preconfigured node addresses. Similarly, for resilience purposes, more than one bearer node should provide the PCEF facilities.

The received rules are used alongside any predefined local rules that may have been configured on the PCEF. The PCRF can activate or deactivate preloaded rules by their references. This can be performed dynamically during session establishment. If there were no rules installed, neither from the PCRF nor predefined within the Traffic Plane, then the PCEF will reject the bearer connection.

The activation of a list of charging rules is performed by communicating a Charging Key that is assigned to a bundle of rules. The Charging Key is decided by the PCRF for each instance of service data flow. The Charging Key can refer to locally installed rules within the PCEF. The PCEF applies the gating decisions to all media flows that share the same Charging Key.

8.6.2.4 Subscription Profile Repository (SPR)

The SPR is the subscriber data that is specific to the usage of the IP-CAN services. This data can exist within the Access Nodes for non-IMS services, for example, for a user of data packet network services in GPRS or WLAN data services.

The SPR provides information for a subscriber, including a user category and charging parameters, the subscribed services IDs, and for each allowed service, a preemption priority level. The SPR also maintains the subscribed guaranteed bandwidth and QoS class identifiers that determine charging bands.

8.6.2.5 Application Function (AF)

The AF is the element in the proxy node that requests resources for session initiation. It can reside in P-CSCF or an AGCF and gets involved in setting up the session according to the network's rules for resource allocation and charging, in consultation with the PDF/SPDF and RACF as well as the PCRF for charging. For resilience purposes, the AF can contact one of several PCRF instances. The AF will select PCRF according to the user's IP address or other UE identifiers.

The AF is responsible for the negotiation of dynamic policy and charging rules with the IP-CAN nodes, which may have their own set of rules. The AF communicates to the PCRF dynamic session information obtained from the UE. The PCRF can reject the particular requested settings but will offer alternatives. The AF proceeds to negotiate with the UE until agreement is reached and resources can be allocated.

The AF sends charging events to the PCRF to obtain charging rules. These charging events contain the session request information that should enable the PCRF to make the appropriate decision. This information identifies the service data flow but it may include "wildcard" to denote groups of flows. The AF provides information sufficient to identify the service data flow with a packet stream. This information includes:

- Application or service and event identifier
- Type of stream (e.g., audio, video) (optional)
- Data rate of stream (optional)
- User information (e.g., user identity)

To be aware of PCRF activities related to the PCEF, the AF can request the PCRF to report on events. These events may indicate a change that invokes a reassessment of the resourcing and charging details. For certain events related to policy control, the AF may be able to give instructions to the PCRF to act on its own, that is, based on the service information currently available without input from the UE.

8.6.2.6   Relationship between Functional Entities

The AF can act as an external proxy function in roaming when it is situated in another network. It can communicate using the Rx interface even for connections that are inter-domain or across operator networks. However, the PCRF operates only within its own network and works closely with the policy enforcement rules and procedures in the same network. The PCRF and the PCEF must belong to the same network.

8.6.2.7   Charging PCC Rules

Charging rules, like the resourcing rules (described in Chapter 6), flow between the PCRF in the Control layer and the PCEF in the Traffic Plane. PCC handles both resourcing and charging aspects but the table below details charging rules only.

Charging Control Rules		
Information Name	*Description*	*Modifiable by PCRF*
Charging key	The charging systems (OCS or OFCS) use the charging key to determine which tariff to apply for a service data flow.	Yes
Service identifier	This is the identity of the service data flow to which a charging rule can be applied.	Yes
Charging method	This indicates the required charging method for the PCC rule. Values: online, offline, or neither. It is mandatory only if there is no default charging method for the IP-CAN session.	No
Measurement method	This indicates what service data is measured: volume, duration, combined volume/duration, or event. This is applicable for either online or offline, but event-based charging can only use predefined PCC rules.	Yes
Application function record information	An identifier provided by the AF, correlating the measurement for the charging key or service identifier values in this PCC rule with Application-level reports.	No
Service identifier level reporting	This indicates that separate usage reports will be generated for this service identifier. Values: mandated or not required.	Yes

Source: Based on 3GPP TS 23203.

PCC rules are referred to by a unique Rule Identifier that is assigned by the PCRF. Both PCRF and PCEF must recognize these identifiers.

8.6.2.8   PCC Rules for the IP-CAN

Implementations can vary according to characteristics of IP-CANs and local operator's preferences. However, some rules must not be modified to make sure of interworking with external networks. The attributes of PCC rules that are related to IP-CAN bearer and IP-CAN session are listed in the table below:

Parameter	Description	PCRF Can Modify the Attribute	Scope
Charging information	Contains the OFCS and OCS addresses	No	IP-CAN session
Default charging method	Contains the default charging method for the IP-CAN session	No	IP-CAN session
Event trigger	Contains the events that trigger a re-request of PCC rules for the IP-CAN bearer	Yes	IP-CAN session
Authorized QoS	Contains the maximum authorized QoS for the IP-CAN bearer	Yes	IP-CAN bearer

Source: Based on 3GPP TS 23203.

8.6.3   PCRF Charging When Roaming

In the case of roaming users, the Home Network provides subscription charging rules. In such cases, there is an assumption that Charging Keys will not be recognized across business borders. The use of PCC in the Visited Network is therefore limited to dynamic PCC rules. The Visited Network may well have its own set of rules, which it enforces within the part of the session under its control.

Figure 8.12 illustrates the multi-faceted interfaces across a business border when a user is roaming. The Visited PCRF (V-PCRF) interacts with the Home PCRF (H-PCRF) to establish agreed-upon rules for a chargeable service and resourcing levels. The Online Charging service relays queries regarding credit control. Finally, the billing systems exchange records to allow for inter-administration cross-charging.

The Home Network supports an H-PCRF (Home PCRF) that is acting as a Charging Rules server, and the Visited Network supports a V-PCRF, acting as a proxy. The interface between the V-PCRF and the H-PCRF is the reference point Gx. The rules to be activated must be previously agreed to in the SLA between the network operators.

8.6.3.1   The PCRF Reference Points

The Rx reference point. The Rx reference point is between the AF and the PCRF. This interface, which is based on Diameter, enables the transfer of subscription information related to the user's

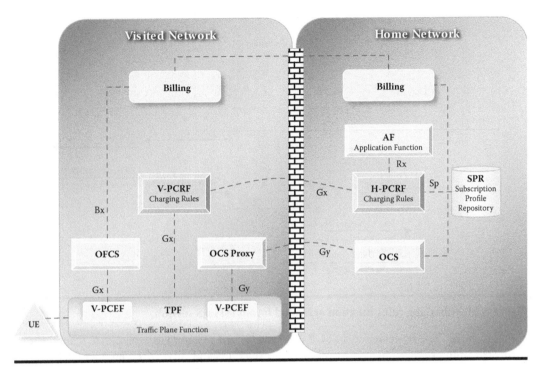

Figure 8.12   Interfaces in a roaming case.

QoS and Charging Profile between the Service layer and the function that makes decisions of resource allocation and charging levels. The information includes IP filters to identify the service data flow in an effort to differentiate the charging and determine policies. The Rx interface should be compatible with previous definitions of both Rx and Gq to allow for interaction with earlier implementations of the PDF and AF.

The Gx reference point. The Gx reference point resides between the PCEF, the Traffic Plane, and the PCRF and it is based on Diameter. The Gx communicates policy decisions for both SBLP and FBC to the media bearing devices. It conveys requests and responses of the PCC information. The enforcement of QoS and charging procedures for the session are determined via this reference point, through negotiations with the UE and possibly other networks.

The Gx also contains the Go interface (COPS) and is backward-compatible with previous implementations.

The Sp reference point. The Sp reference point is defined between the PCRF and the data repository for it, the SPR. The Sp reference point allows the PCRF to request subscription information that is associated with the subscriber's usage of a particular IP-CAN, such as an identifier for a PDN. The SPR is intended as an Access agnostic data repository.

The Sp interface delivers information about the user's subscription QoS profile. It also allows the SPR to notify the PCRF when the subscription information has changed. The SPR will stop sending the updated subscription information when the PCRF cancels the notification request.

8.6.4   Charging Procedures within the Network

Charging rules are normally activated by the IP-CAN nodes (FBC) or by session controllers (the AF in IMS). However, they also can be determined by applications. A commonplace example of

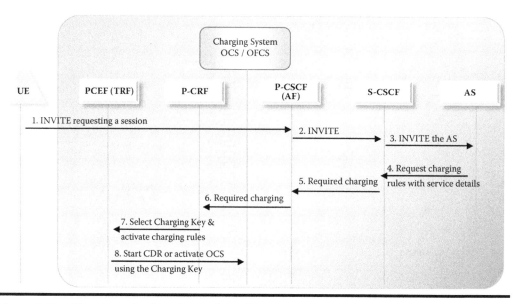

Figure 8.13 Charging rules function activated for an IMS AS session.

the latter is the Pre-pay service on an SCP platform. When a Credit Control service is applied to a media streaming application, credit is checked via the OCS and rules are set by the PCRF to control the flow. The applications furnish the charging requirements via the S-CSCF and the AF.

Figure 8.13 gives an example of an IMS session, with an AS determining which Charging Key should be used, and activating charging rules, enforced by the PCEF. The sequence of signaling is performed as follows:

1. UE requests an IMS session from a P-CSCF.
2. P-CSCF forwards to S-CSCF.
3. S-CSCF selects an AS according to filter criteria.
4. The AS sends input through CSCF via the Rx interface to identify the charging rules.
5. S-CSCF forwards it to the AF in the P-CSCF.
6. The AF requests charging rules and provides the service information to the PCRF.
7. PCRF uses the AS input to select the Charging Key and sends it to PCEF.
8. PCEF activates the charging rules according to the Charging Key, which is input to the rating logic in the offline or online charging system.

8.7 Charging for Different Access Networks

8.7.1 Access Network Charging Data

Charging is not entirely access-network-agnostic. Different IP-CANs must be able to communicate with the charging systems for:

■ Triggering charging events
■ Reporting
■ Enforcing charging policies

When implementing FBC, data from the IP-CAN on the media flows must be correlated with service-level data from the session controllers and applications. Both bearer and IMS produce CDRs, which the Charging System must consolidate or discard, as indicated by the network's charging rules and policies. In particular, such correlation is performed to prevent double charging if both FBC and session records produce bills.

Charging data is generated in the Access Network when resources are committed in the IP-CAN. This data is conveyed to IMS for communications with the UE. The involved IMS network nodes also produce data in P-CSCF CDRs, S-CSCF CDRs, I-CSCF CDRs, and MRF CDRs. Applications, when invoked, also produce the relevant CDRs.

Additional data is generated at the AN as a result of some changes, modifying the media flow, and adding or terminating existing flows. Such mid-session changes also act as triggers for reevaluating the charging rules and procedures. Session terminations by UE actions are also conveyed as charging triggers. In such cases, they cause CDRs to be closed or OCS to complete the process and release unused reserved units.

Any generated charging data is used only within the native network. The originating network will use its own data for the subscriber's charging information, while the terminating network will record the information for the called party. The access network charging information is not forwarded to the external application servers either. This means that each network remains responsible for its own data integrity and security.

8.7.2 Charging in GPRS and UMTS

Charging in GPRS ANs may be using CSN protocols such as IN/CAMEL to provide Pre-pay services, but use the Ro interface to the OCS to provide any credit control services online.

For offline charging, the OFCS CDRs are produced by the MSC in CSNs, or by IMS network elements for SIP-based communications.

In terms of session-based charging, the GPRS-based charging can address:

■ Charging data related to PDP Contexts
■ Charging data related to service data flows within PDP Contexts

To perform the charging procedures using this data, the session-based online charging capability is required in the GGSN.

Changes to charging information are provided by charging events that trigger mid-session charging activities. Management intervention of any kind can also force triggering a charging event.

GPRS nodes accumulate details of sessions, per user and per PDP Context, including the duration of the PDP Context and data volumes transferred during the lifetime of that PDP Context. Quantitative data (volumes of packets sent and received) is collected per QoS level that is applied to the PDP Context. This information is sent to the charging system.

The charged party, the User, is identified in GPRS/UMTS by either MSISDN or IMSI and the IMS User is identified by the IMPI/IMPU. The PDP Context Identifier is used in GPRS to relate records belonging to the same session. The PDP Context is recognized by a unique identifier generated by the GGSN when creating a new PDP Context. This identifier is also forwarded to the SGSN so as to allow correlation of SGSN PDP Context CDRs with the matching GGSN CDRs in the BD.

For OCS, the GGSN collects charging data for a user, per PDP Context. The GGSN is able to count volumes of data packets transferred or received by the UE. When an online chargeable service begins, a PDP Context is created and, at the same time, a Credit Control Request is launched toward the OCS. Events that signify session termination, or reaching volume limits, or time limits, are also conveyed to the OCS. Other changes, such as QoS change or tariff time change, can cause a new charging condition that needs to be captured on the OCS.

8.7.3 Charging in xDSL

NGN Charging follows the same principles and the same procedures. xDSL parameters are used in the IP-CAN-related identifiers, while the IMS charging procedures remain the same as for UMTS service based charging.

The xDSL access node maintains xDSL information in a set of parameters within the access-network-charging-info, in the P-Charging-Vector header in SIP messages. The access-network-charging-info parameter is an instance of "generic-param" from the current "charge-params" component of P-Charging-Vector header. The access-network-charging-info parameter includes alternative definitions for different types of access networks. The xDSL charging data contains:

■ BRAS address (bras parameter)
■ Media authorization token (auth-token parameter)
■ Parameters "dsl-bearer-info," containing information for one or more xDSL bearers

The "dsl-bearer-info" has:

■ One or more dsl-bearer-item values with dsl-bearer-sig, dslcid, and flow-id
■ The value of the dsl-bearer-item is a unique number that identifies each of the dsl-bearer-related charging information within the P-Charging-Vector header

The dsl-bearer-info contains:

■ An indicator for an IMS dsl-bearer (dsl-bearer-sig parameter)
■ An associated DSL Charging Identifier (dslcid parameter)
■ An identifier (flow-id parameter)

The flow-id parameter contains identifiers of associated m-lines and the relative order of port numbers in an m-line within the SDP from the SIP signaling to which the dsl-bearer charging information applies. The Charging ID "dslcid" has the same structure as that of the GGSN Charging ID.

8.7.4 Charging in WLAN Networks

WLAN is another access network that supports services in Core IMS. This means that services delivered through a WLAN are charged for in the same way by the OCS and OFCS charging mechanism. WLAN users often encounter a WLAN AN of another Service Provider, and therefore charging while roaming will occur frequently. When roaming, WLAN-enabled UE discovers a local WLAN node of a network that has interconnection agreements.

WLAN charging can be performed within the WLAN network or as combination charging with service-level charging. WLAN-enabled UE detects the local WLAN Access Point Name (W-APN) and gains access to it via local authentication.

The UE connects to the 3GPP AAA Server via the WAG to be authenticated for IMS services. The AAA Server retrieves user profiles from the HSS, and authorizes the user for all subscribed services in IMS. The 3GPP AAA Server starts the charging process on the CDF.

A session charging identifier, like the ICID, is generated in the PDG. This charging identifier is propagated in messages toward the AAA Server to enable correlation of charging data from the PDG and CDRs from the AAA Server.

The WLAN session is charged according to a Charging Profile, that is, a set of rules and policies, as defined for the PCRF. The PDG enforces the Charging Rules, acting as a PCEF. The PDG may also contain the PCRF or board, selecting and enforcing charging rules.

If the W-APN contains preconfigured charging rules, the PDG uses them by preference. If no special charging rules are chosen, the default charging profile is used.

Figure 8.14 demonstrates the relationships of Visited Network charging and Home Network charging. The Visited Network obtains charging information from the local WAG, and the Proxy AAA Server generates CDRs onto the Visited Network's own CDF. Later, the BD systems will exchange inter-operator settlement information based on this data. The Home Network produces its own CDRs from the Home PDG and from the Home AAA Server, using both OCS and OFCS, as required.

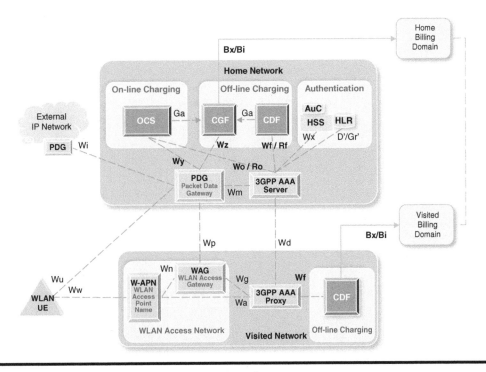

Figure 8.14 Charging with WLAN access network.

8.7.4.1 WLAN Charging Reference Points

This is a brief description of the reference points for the charging functions for WLAN. Note that the reference points for authentication and other functions are not shown here. For a full description of WLAN charging interfaces, see Chapter 10.

- *Wf reference point.* The Wf is the same as the Rf reference point. It connects an OFCS and a 3GPP AAA Server or Proxy. It forwards charging information toward the OFCS offline charging system, located in the Visited or Home Network.
- *Wo reference point.* The Wo reference point is the same as the Ro. This is an interface for the 3GPP AAA Server to communicate with 3GPP OCS. It transports online charging-related information toward the OCS for any credit control or online charged service.
- *Wy reference point.* The Wy reference point is used by the PDG to communicate with the OCS. This interface allows transporting online charging-related information so as to perform any real-time charging (e.g., credit control for the online charged subscriber).
- *Wz reference point.* The Wz reference point is used by the PDG for the transfer of offline charging records from the PDG CDF (if it is integrated with the PDG) to the CGF. As such, it is the same as the Ga interface.

Chapter 9

Circuit-Switched Network Interworking

9.1 This Chapter

This chapter considers aspects of interworking of IMS with CSN/TDM, where these terms refer to the combination of PSTN, ISDN, and PLMN legacy networks.

Implementations of softswitches, IP-MSC, and IP-VPN are not described here. This chapter concentrates on the IMS specific features that interface with CSN.

The key IMS element for interworking with CSNs is the MGCF, which is responsible for managing the gateways toward CSNs and performs much of the feature translation. Aspects of PSTN Emulation and Simulation are explained here, along with the treatment of supplementary services, though not full service emulation details. The focus here is on the interaction with IMS, and in particular—the solution of connecting POTS to IMS via an AGCF utilizing the IMS session control is detailed. Finally, this chapter also describes the IMS Transit facility to transport transparently both CSN communications as well as IP across other networks.

9.2 IMS Connectivity with CSN

9.2.1 The IMS Control of PSTN Breakout

In the earlier phases of packet network implementations, the bulk of the traffic is likely to flow to or from CSNs. Therefore, the network exit points, where the CSN breakout functions are located, must be capable of handling large volumes, and their positioning must be carefully designed to take advantage of the packet network attributes and cost savings.

Network breakout procedures deliver the following functions:

- Discover next hop contact point in external networks, for both CSNs and PSNs.
- Identify how to route to the exit point, by a transit network or by an optimal route to the target network.
- Establish the optimal exit point, between alternative instances of the signaling controllers (the BGCF, the MGCF, and the SGW).
- Select the egress BGF and the MGW that carry the bulk of the media flow.

When it is established that the destination is in the CSN domain, the S-CSCF forwards the SIP request/response to a BGCF for routing to the PSTN or other CS Domain. The BGCF selects the MGCF to engage, if there are multiple servers. The MGCF can be selected according to proximity to the preferred exit point to the CSN network. The MGCF is responsible for selecting the media path and media conversion gateways.

9.2.2   The Media Gateway Control Function (MGCF)

9.2.2.1   The IMS MGCF Functions

The MGCF is the server that enables the flow of IMS communication to and from PSTN/ISDN by determining the media path and media management requirements. The MGCP is responsible for the following functions:

- Control the media channels in an IMS-MGW and the call state that relates to the media connection.
- Receive requests from the CSCF for media connection to PSTN/ISDN, and communicate with the CSCF to control the media part of the session.
- Receive requests from BGCF for incoming sessions from PSTN entities, and communicate via the BGCF to set up the media part.
- Determine the next hop for the media path, depending on the routing number for incoming calls from legacy networks. This includes selecting an MGW to establish the media flow.
- Perform protocol conversion between ISUP or TCAP and the IMS call control protocols.
- Forward out-of-band information received by the MGCF to the CSCF for incoming calls and to the IMS-MGW for calls to the PSTN.

The MGCF is the session controller that performs the task of interpreting the session requirements, while the MGWs and SGWs transform the data units from packet to circuit, and in reverse. The MGCF must translate not only the session signaling, but also the associated services, which are delivered in a different manner in their respective domains. It also needs to provide suitable information to identify the route for the call and the identity of the destination network.

9.2.2.2   Procedures at the MGCF

The MGCF acts as a user agent for PSTN bound communication. It receives SIP messages, interprets them, and initiates H.248 instructions toward MGWs and CSN nodes. The MGCF address is known to peering network elements by preconfiguration or by setting up a GRUU for its address (see GRUU definition in Chapter 5). The MGCF inserts its GRUU address in the P-Charging-Vector and the P-Charging-Function-Addresses headers before sending out messages related to the session. This GRUU remains in force for the duration of the session and ensures that all subsequent dialogs with PSTN will be proceed by the same instance of MGCF server.

The MGCF generates a SIP INVITE when it receives communication from the CSN that is bound for an IMS network element. To do that, it sets the URI to the "Tel" format using an E.164 address or to the "sip" format using an E164 address in the user portion with the "user=phone" parameter. It also furnishes details for the P-Asserted-Identity header, including the display name, if it is present, in the received information from the CSN.

To identify the charging entity for each call, the MGCF inserts an ICID into the P-Charging-Vector header. To identify the network, it inserts an IOI in the P-Charging-Vector header. To indicate the direction of the communication, an IOI type indicator shows whether this is an originating SIP message (orig-ioi) or a terminating (term-ioi) one.

The MGCF does not support PATH and Service-Route headers because the message route information is not passed on. The MGCF determines new routing.

The MGCF is responsible for negotiating a codec (coder/decoder) that both ends can support for the session, which is included in the SDP (Session Description Protocol). It also sends an indication of its DTMF capabilities that may interwork with the CSN party. Acting as a user agent, when the MGCF receives notification of a call release from the CSN, it generates a BYE message. It can also send error information for a premature bearer release, when it occurs.

9.2.2.3 The 3G Mobile MGCF

The MGCF allows interworking between 3G handsets IMS clients and devices on CSNs of any kind — Mobile or Fixed. Typically, a Mobile MGCF needs to support the MSC set of services while Fixed MGCF supports PSTN/ISDN services.

Figure 9.1 shows the main components involved in exit from the network: the CSCF components and the BGCF, which decides which exit (IP or CSN) the session will take. The MGCF is

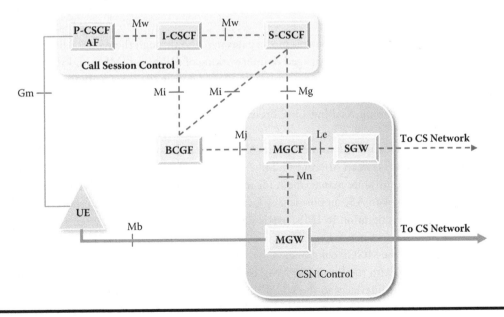

Figure 9.1 Network breakout to CS networks.

shown with its interfaces to the MGW and the SGW that translate signaling and media codecs, and transform packets to circuit and vice versa.

9.2.2.4 The MGCF Reference Points

- *The Mn reference point between the MGCF and the IMS-MGW.* The Mn reference point contains the interfaces between the MGCF and the IMS-MGW, which is complex and feature-rich. It includes the set of commands specified by the H.248 standard functions for interworking between IMS and PSTN. This interface should be flexible enough to deal with the CSN style call model as well as H.323 and others. It must manage the processing of a variety of media types.
- *The Mg reference point between the MGCF and the CSCF.* The Mg reference point allows the MGCF to forward incoming session signaling from the CSNs to the CSCF, and vice versa. This means that the MGCF can process SIP messages and convert the meaning to TDM signaling and the reverse.
- *The Mj reference point between the BGCF and the MGCF.* This reference point allows the Border Gateway Control Function to forward the session signaling to the MGCF when communication with the PSTN or PLMN is required. The Mj reference point is based on SIP.
- *The Le reference point between the MGCF and the SGW.* This reference point converts signaling for communication between IMS and PSTN or PLMN. The SGW can be incorporated with the MGCF or with the MGW.

9.2.2.5 The TISPAN MGCF

The MGCF has a significant role in PSTN emulation solutions, as described in the PES in TISPAN 283 022. The MGCF controls session establishment when it involves IMS and a legacy network.

As in 3G Mobile, the MGCF controls media gateways, and through them — the flow of media to and from PSTN. Similarly, it is engaged in interworking of SIP-based IMS with ISUP features. PSTN networks have a great number of features developed over many years, with varying degrees of popularity. The MGCF provides feature-level interworking of supplementary services such as Call Transfer, Call Waiting, Calling Line Presentation, Calling Line Restriction and more, or variations of them.

The MGCF provides more than mere protocol conversion. It contains logic that translates session control functions between SIP and H.248. It also provides call flow logic for a sequence of multiple features and call state management for interworking with legacy Fixed networks. It can, usually with the help of an AS, support legacy Centrex and Centrex features. In short, it plays an important role in the migration to IMS, particularly where legacy network systems are replaced gradually.

Figure 9.2 shows the MGCF components that include the PES Media Gateway Controller and the Feature Manager, with the Event and Notification module that triggers the activities between them. It also shows the interfaces to the NGN components involved in routing, the Connectivity session Location and repository Function (CLF), the resource allocation, Resource and Admission Control Subsystem (RACS), and the access media gateways.

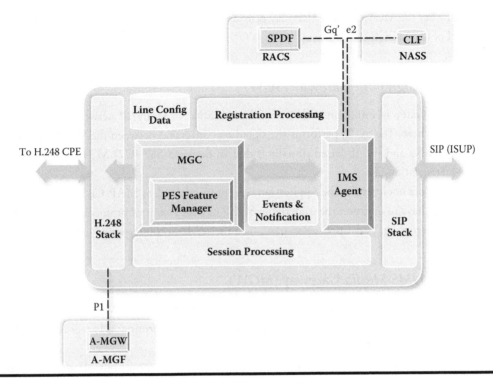

Figure 9.2 ES-based MGCF with extended PES Feature Server role.

When the MGCF receives an incoming call from PSTN, it must negotiate Packet Network resources via the RACS. The SPDF may contain rules and policies for such traffic, and the RACF negotiates the resources alongside other PSN multimedia sessions.

The MGCF also provides SIP-based transit through the IP network, where the transit is performed through encapsulation of ISUP in SIP, and unpacking it for local H.248-based CPE.

9.2.3 Packet and Circuit Conversion

9.2.3.1 Bearer Transcoding

IMS functions facilitate voice calls signaling between IMS and CSN, both Mobile and PSTN, to be set up via the MGCF and the SGW. The voice media is connected between the IP bearer and the CSN via the MGWs. The intervening gateway performs the conversion between packets of media and circuit-based transmission.

This media conversion is also needed to support transcoding between different codecs that can be used by the parties involved in the call. IMS defines default codecs to enable IMS to interwork end-to-end with other networks; but if that is not possible, then interworking is achieved via transcoding. IMS should be able to interwork with the CSNs by supporting, for example, AMR-to-G.711 transcoding in the IMS MGW.

An important feature for service interworking is the support for DTMF, which can be controlled or interpreted by the IMS UE, MGCF, or MRF. IMS, therefore, must be able to send or receive DTMF tone indications using the bearer in-band signaling as well as out-of-band DTMF signaling. Either the MGW or the MRF may provide tone generation and DTMF detection under the control of the MGCF.

9.2.3.2 The Signaling Gateway (SGW)

The Signaling Gateway Function (SGF) is well established in many IP networks and does not merit detailing here. The SGW performs the signaling conversion between the SS7 -based signaling used in CSNs and the IP-based signaling, that is, between SS7 (C7) MTP and SIGTRAN (Signaling Translation).

The SGW does not interpret the Application layer protocols such as MAP, CAP, BICC, ISUP, or INAP but may have to interpret the underlying SCCP or SCTP control protocols to ensure proper routing of the signaling. The "application part" (i.e., the services) is interpreted in the MGCF.

9.2.3.3 The IMS Media Gateway (MGW)

The MGW is already well defined and operational in many networks. The MGW is responsible for converting CS communication to PS networks and vice versa. It terminates bearer channels from CSNs and media streams from a packet network.

From the CS point of view, the MGW is a transport termination point. It supports media conversion, bearer control, and payload processing (e.g., codec, echo canceler, even conferencing bridge).

The MGW roles are:

■ Resource control interaction with MGCF, MSC server, GMSC server, and PSTN switches
■ Handling resources such as echo cancelers, etc.
■ Supporting various codecs

Various standards distinguish between different MGWs. This is usually due to identifying specific codecs or parameters carried through specific interfaces, such as AMR for Mobile networks. Therefore, CS MGWs must have the correct codecs for interfacing to PLMN and PSTN or ISDN, while IMS-MGWs must comply with the IMS-specific requirements and media types for a multimedia packet network. When acting as a gateway between two different domains, the MGW must comply with both.

An MGW also can be referred to as an Access-MGW (A-MGW) when residing on the operator's premises, connecting a number of external analog lines, and as a Residential-MGW (R-MGW) when residing on the customer's premises, connecting a small number of lines.

The MGCF should be able to address multiple IMS-MGWs. For resilience purposes, each IMS-MGW should be able to connect to more than one session controller. The dynamic sharing of IMS-MGW physical node resources entails partitioning it into logical entities (i.e., multiple virtual MGWs). The assignment of endpoints or CPE usually is achieved by static configuration or local configuration data.

9.3   PSTN Emulation

9.3.1   IMS-PES Principles

PSTN Emulation that is provided by non-IMS solutions (i.e., softswitches) is not described here. The IMS-based PES (PSTN Emulation System) supports emulation of PSTN services for analog terminals connected through Residential Gateways or Access Gateways (A-MGW), which are controlled via an Access Gateway Controller Function (AGCF).

IMS-based PES is a mechanism that allows non-IMS devices to appear to IMS as normal SIP users. The communication from these devices pass through the Access Gateway Controller to present them to IMS as SIP messages, allowing these devices (H.248 based or POTS) to register in IMS and make or receive calls.

The IMS-based PES differs from softswitch solutions in that it implements the principles of layer separation, where the session control is served by Core IMS servers (CSCF) and the services are delivered from one or more SIP ASs. This means that these functions are not embedded within a single node that also controls the MGW. This approach streamlines the control of sessions and provides greater flexibility in introducing and changing services, without disturbing existing functions.

9.3.2   IMS-Based PES Architecture

IMS-based PES provides IP network services to analog devices. Analog terminals using standard analog interfaces can connect to IMS-based PES in two ways:

1. Via an A-MGW (Access Media Gateway) that links to an AGCF, using H.248.1 over the P1 reference point. The AGCF is placed within the Operator's network and can control multiple A-MGWs.
2. Via a VGW (VoIP GW) on customer premises. The VoIP Gateway links to IMS using SIP (RFC 3261) over the Gm reference point. In this scenario, the POTS phone connects via a VoIP Gateway directly to the P-CSCF. The conversion from the POTS service over the z interface to SIP occurs in the customer premises VoIP Gateway.

Both gateways are stateless and unaware of the services. They merely relay call control signaling to and from the PSTN terminals. Session control and handling of services are provided in the IMS components, via standard IMS session controllers and one or more application servers.

IMS-based PES provides links to other subsystems — NASS and RACS — to enable binding of lines and IP addresses, and to enable allocations of resources. Alternatively, the AGCF can store line configurations with user identities and public user identities in a locally held database. The use of the RACS is also not essential for voice services that predominantly run in the PSTN and have fixed bandwidth and QoS attributes.

In the case of the VoIP Gateway (VGW), the POTS signaling is converted to SIP and passed on to the P-CSCF. The VGW is acting as a SIP user agent and appears to the P-CSCF as a SIP terminal. In this scenario, only very limited support of ISUP supplementary services is possible, but "Basic Calls" are enabled.

In the case of the AGCF, the POTS phone connects to the A-MGW over the z interface. The signaling is converted to H.248 in the A-MGW. The AGCF interprets the H.248 signals and other input from the A-MGW to format the appropriate SIP messages. It forwards the generated SIP signaling messages to the S-CSCF or to the IP border via IBCF (Interconnection Border Control

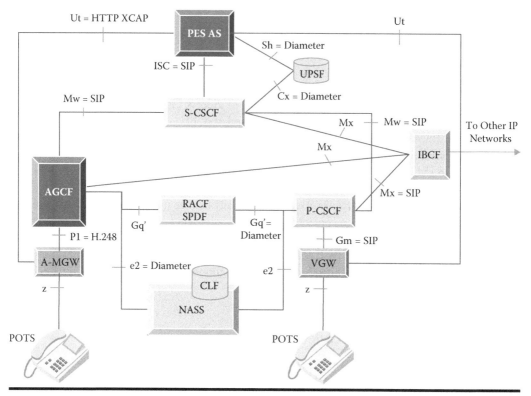

Figure 9.3   AGCF and PES AS links to IMS.

Function). Service features are presented to the S-CSCF through the appropriate SIP messages that trigger the PES AS.

The logic embedded in the AGCF enables it to convert the information contained in SIP message into a service that can be presented to analog lines, and vice versa. It also has certain service-independent logic. For example, on receipt of an off-hook event from a MGW, the AGCF requests the MGW to play a dial tone.

The IMS-based PES architecture, is shown in Figure 9.3. The diagram highlights the AGCF and the PES AS. The PES AS provides emulation services to both VGW and AGCF users.

9.3.2.1   PES Components

The following is a brief description of the network elements that participate in delivering a PSTN Emulation service:

- *CPE Access Gateways.* The user equipment consists of a residential gateway or an access point and one or more analog terminals hanging from it. This gateway can be an H.248-controlled A-MGW or a SIP-based VGW. The SIP-based VGW appears to IMS as an IMS UE connecting to the P-CSCF. The H.248 gateway needs an access gateway controller to manage the conversion to SIP. The gateways can be located in the customer's premises or in the operator's premises.
- *AGCF (Access Gateway Controller Function).* The AGCF takes the role of the PES access point. The AGCF is part of the trust domain. The AGCF combines the functionality of an H.248

MGC and a SIP User Agent, coordinated by the internal Feature Manager, which decides how the services are managed. The AGCF appears to the CSCF as if it were a P-CSCF.

- *PES AS.* The PES AS provides service logic for emulated PSTN. It works with the AGCF in loose coupling mode or tight coupling mode. The AS does not need to be a special platform and requires no specific protocols. It can be a SIP AS that is also involved in non-PES applications.
- *MRFC.* In PES, an MRFC is needed to take the role of the PES Announcement Server.
- *IBCF.* The IBCF connects to both P-CSCF and AGCF, and provides the IP network interconnection function for roaming or for nomadic users.

9.3.2.2   Reference Points

- *The Mw reference point between the AGCF and the CSCF.* This is the same interface as between the P-CSCF and the S-CSCF. The protocol is SIP and it deals with registration and session control. Normally the Mw allows for encapsulated ISUP messages; however, this type of message is not passed between the AGCF and the CSCF.
- *The Ut reference point between the SIP AS and the UE, and between the PES AS and the AGCF.* This is the same interface as between the UE and the SIP AS, and it consists of XCAP (XML Configuration Access Protocol) over secure HTTP. The Ut reference point enables the VGW and the AGCF to transfer service information that otherwise could not be transferred over generic SIP. For example, Ring Tone Management information is passed through the Ut interface.
- *The P1 reference point between the AGCF and the A-MGW.* This interface is H.248 and is used for a variety of call control services.

9.3.3   The AGCF

9.3.3.1   The AGCF Functions

The AGCF generic role is described as a controller of Access Point or Access Gateways that are external to the operator's core network. It is the first point of contact for Residential Gateways (RGWs) and A-MGWs, also referred to as the Access Media Gateway Function (A-MGF). The AGCF is only deployed for emulation of PSTN. It performs the following functions:

- Acts as an MGC (Media Gateway Controller), controlling customer premises MGWs or access point nodes functioning as access-side media gateways
- Obtains resources via the RACS
- Retrieves line profile information from the NASS when line configuration data is not provisioned internally
- Performs signaling interworking between SIP and analog signaling via the P1 reference point, providing some service logic, and also passes on dialed digits for an AS to perform the services
- Acts as a SIP User Agent with regard to IMS SIP functional entities, providing similar functions as the P-CSCF on behalf of legacy terminals that are connected via the A-MGWs (these functions include SIP registration, generating asserted identities, and creating charging identifiers, among others)

The AGCF appears as a P-CSCF to the other IMS network elements. The SIP interface is identical, therefore limiting the scope of SIP signaling to those available at the Mw reference point. For example, flash-hook events may not be explicitly reported to ASs but they trigger appropriate SIP signaling procedures on the AGCF, to initiate calls or mid-call SIP messages.

However, the AGCF performs additional functions that are not within the scope of the P-CSCF. It handles the media functions directly, in particular managing the dial-tone and ring-tone functions and processing mid-call events (flash-hook) functions that are essential for many advanced services.

The AGCF handles the dial tone via the Dial Tone Management system. The UE, the A-MGW, and the AS all need to subscribe to the network's Dial Tone Management. The subscription to Dial Tone Management can be implicit or explicit. If explicit, the subscription identifies the acting AS or, alternatively, the user profile can contain the iFC for the appropriate AS. This PES AS keeps the Dial Tone Management XML documents that define which dial tone is used, at what circumstance, and what service.

The AGCF is instrumental in delivering mid-call services, such as Call Waiting, Inquire/Transfer calls, Call Park, Three-Party call, etc. To do this, the AGCF subscribes to events reported by the A-MGW and informs the requesting AS. The way in which the service is performed depends on network policy of loose coupling or tight coupling (see Section 9.3.3.8 and Section 9.3.3.9).

9.3.3.2   Inter-AGCF Session via IMS

Where IMS-based PES is deployed, routing between networks, even between two AGCF users, is performed in the same way as for standard IMS terminals — that is, via the originating S-CSCF, where the originating AGCF is perceived as a user agent.

The process of finding the destination S-CSCF for another AGCF user is also the same as for the destination IMS user, and the terminating supplementary services are assisted by the terminating PES AS.

Figure 9.4 shows a session between two legacy terminals managed by the AGCF in both originating and terminating networks, controlled by IMS Core components. The figure omits border control servers and gateways between the networks.

9.3.3.3   AGCF Registration

The AGCF registers users to IMS according to their IMS identifiers. The AGCF data contains user identifications, private and public identities. One private user identity can be assigned to a group of lines or subscribers, allowing for multiple Home extensions or IP Centrex groups. But each private user identity is associated with only one Home Network domain name, which is the IMS network that owns the subscription.

The AGCF contains line configuration where a line is represented by a termination identifier on the A-MGW. The AGCF binds a line with one or more public user identities (e.g., published telephone numbers). To do that, the association of public and private user identities from the UPSF is made available to the AGCF via the S-CSCF.

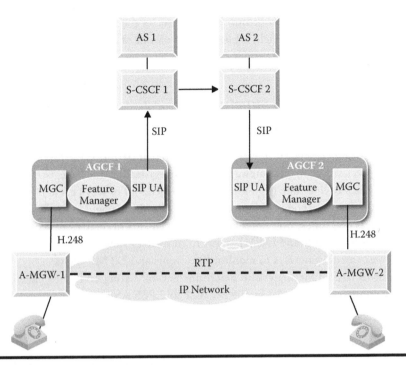

Figure 9.4   Routing between two IMS-based PES users.

The AGCF also must hold information on the MGWs connected to it. For each MGW it can hold a default parameter set, such as the default dial tone identifier (to convey to the Media Server) and a default digit-map, upon which collected dialed digits are analyzed.

The IMS CSCF requires registration of users and also regular re-registration to be performed to confirm the current status. The AGCF initiates standard SIP registrations, deregistrations, and re-registrations on behalf of each A-MGW line or group of lines. The registration takes into account any change notified by the gateways, together with provisioned configuration data.

When one of a group of public user identities is explicitly registered, the rest of the group (sharing the same private user identity in the Home domain) also are implicitly registered. The list of the implicitly registered identities is kept on the UPSF (HSS) and is provided to the AGCF via the S-CSCF, where it can be cached in local memory.

9.3.3.4   SIP Routing to the AGCF

Incoming calls received by a PES access point operate in the same way as UE call termination procedures. The P-Called-Party header value is used to fetch the termination line to which the call will be delivered.

Calls initiated by a PES user via a PES access point operate like UE call origination procedures but the P-Asserted-Identity is populated to ensure keeping the AGCF in the path. The AGCF sets the parameters to point to itself rather than to the UE, so that incoming messages will route via the AGCF to enable interworking.

The AGCF sets the "Contact" header to include SIP URI with its own IP address or the FQDN in the "host-port" parameter. The "via" header also contains the AGCF FQDN in the "sent-by" field. The P-Access-Network-Info header is set as the Fixed access network, e.g., DSL. The P-Visited-Network-ID header field is filled with the value of a preprovisioned string that identifies the network where the AGCF resides.

9.3.3.5 The AGCF Internal Structure

The AGCF contains three logical main components:

1. *The AGCF Media Gateway Controller.* This component performs the following:
   - Keeps track of the media gateway state (e.g., registration and deregistration)
   - Keeps track of line state (e.g., idle, active, parked, out of service, etc.)
   - Controls the connection configuration (media flows topology) in media gateways
   - Connects dial tones, ring tones, and announcements in media gateways
   - Receives line events and DTMF digits from media gateways
   - Requests media gateways to monitor line events and DTMF digits
   - Performs basic digit analysis to determine end of dialing and emergency calls (the full digit analysis is performed by the S-CSCF and PES AS)
   - Provides line control signaling to media gateway lines
   - Downloads "Digit-Map" to media gateways for PSTN number analysis
   - Controls media mapping, transcoding, echo cancellation, etc. in media gateways
2. *The AGCF Feature Manager.* This component coordinates between the IMS Core and the MGWs at a feature level, rather than at a protocol level. Essentially, it translates between the P1 interface to the MGW and the Mw interface to IMS. To do that, it needs to maintain call state and understand the call model that associates lines to IMS sessions.

   The Feature Manager interprets H.248 signaling into service logic with matching SIP message. It has to perform SIP registration procedures on behalf of the lines connected to the A-MGW. It initiates interaction with the CSCF and AS, and ascertains which dial tone to apply according to the subscriber profile.

   When interpreting SIP messges, the Feature Manager needs to apply additional logic to establish what ring patterns are required. In particular, it gets involved with processing mid-call events (i.e., flash-hook). It needs to determine the requirements for digit collection and interpretations of special keys ("#" and "*") and decide, according to network policies and the type of service required, how to involve the PES AS and what to pass on to it.
3. *The AGCF IMS Agent.* This component provides both the functionality of an IMS UE and the P-CSCF. It is responsible for the interaction with IMS entities, the I-CSCF, S-CSCF, and the IBCF, acting as a P-CSCF. The IMS Agent can also interface to the NASS to retrieve information related to the IP connectivity access session (e.g., physical location of the user equipment), when the MGW is located in the customer premises (i.e., R-MGW). Where the RACS is implemented, the IMS Agent in the AGCF can interact with the SPDF, taking the role of the AF that requests resources on behalf of the UE.

As shown in Figure 9.5, the AGCF also contains the internal Line Configuration Database that supports these modules. The platform must have both H.248 and SIP stacks to manage the links to the MGWs and the relationship with IMS. The underlying logic components (Session Processing, Event Notification, and Registration) call upon these modules and protocol stacks as required.

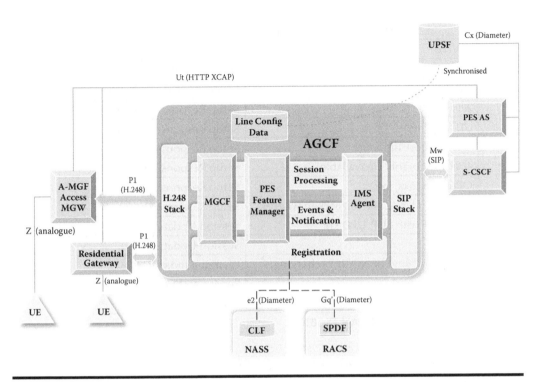

Figure 9.5 AGCF internal structure.

9.3.3.6 MGC H.248 Mapping to the SIP UA

The Feature Manager processes internal events received from the MGC side and requests the SIP UA to generate the appropriate SIP messages. The following table maps MGC commands with the SIP UA messages:

Internal Primitive	SIP Message
Setup Request	INVITE
Session Progress (alerting)	180 (Ringing)
Setup Response (no answer)	480 (Temporarily unavailable)
Setup Response (answer)	200 (OK)
Setup Response (busy)	486 (Busy Here)
Setup Response (reject)	606 (Not Acceptable)
Feature Request	re-INVITE
Release	BYE

Source: Based on TISPAN TS 183043.

The Feature Manager processes SIP messages received from the SIP UA side and transmits appropriate internal primitives (basic commands) to the MGC side, based on the mapping described in the following table:

SIP Message	Internal Primitive
INVITE	Setup Request
183 (session progress)	Session Progress
180 (ringing)	Session Progress (alerting)
200 (OK)	Setup Response (answer)
603 (Decline)	Setup Response (no answer)
408 (Request Timeout)	Setup Response (no answer)
480 (Temporarily Unavailable)	Setup Response (no answer)
Other 4xx, 5xx, 6xx	Setup Response (failure)
486 (Busy Here) , 600 (Busy Everywhere)	Setup Response (busy)
re-INVITE with SDP, UPDATE	Session Update (SDP)
REFER	Session Update (refer)
INFO (charging) or NOTIFY (charging)	Charging Indication
NOTIFY (other)	Service Notification
BYE	Release

Source: Based on TISPAN TS 183043.

9.3.3.7 Events Notifications and Digit Collection

Processing flash-hook event notifications depends on the call type, call parties, and the type of service. On receipt of a flash-hook notification from the AGCF MGC component, the Feature Manager (via the MGC) instructs the MGW to play a dial tone and to collect digits. When the digit collection has completed successfully, the Feature Manager (via the SIP UA) sends an INVITE request to the PES AS, with the collected digits contained in the Request-URI.

9.3.3.8 Loose Coupling Procedures

Processing the feature code depends on whether loose or tight coupling procedures are applied between the AGCF and the PES AS, according to network policy. Loose coupling means that the AGCF provides some local service logic, depending on the service, and the AS gets involved only for certain services and specific conditions, but not every time.

The Feature Manager determines what feature code is required, according to a local mapping table.

For each feature code, there is a service connection description. For example:

- Connect to a held party (call waiting, inquire)
- Connect third party
- Release one call party

The service logic is aware of the session status and can determine the next action. The SIP UA component is instructed to send a specific SIP message, such as INVITE, re-INVITE, OK 200, BYE, etc.

With loose coupling, the AGCF logic is used to manipulate call legs; therefore, session flow is maintained by the AGCF, not the AS.

9.3.3.9 Tight Coupling Procedures

Tight coupling means that the AS provides the *full* service logic and the AGCF is merely reporting to it all the events and the collected digits. With tight coupling, the Feature Manager requests the SIP UA to create a new dialog and send an INVITE request to the CSCF, containing details of the user identity (including P-Asserted-Identity), Home domain, and user input digits, if any have been received. The user identity and domain name are used to ascertain which AS to connect, according to filter criteria in the User Profile.

When the collected digits have been received by the PES AS, it determines the appropriate call processing actions to take. Manipulation of call legs for call waiting and three-party calls is managed entirely by the AS, not the AGCF.

Unlike the loose coupling procedure that dictates that a flash-hook notification must be followed by dialed digits for the AGCF to generate an INVITE request, in tight coupling, the procedure supports the reporting of a flash-hook event even when no additional digits are dialed by the user, and control is passed back to the AS.

The handling of the media is also performed according to instructions from the AS. The Feature Manager requests the MGC component to interact with the MGW to modify the H.248 Remote Descriptor and stream mode according to the SDP information received in re-INVITE messages sent by the AS.

The release of the dialog, which has been opened for the purpose of sending the feature code, is the responsibility of the AS.

9.3.4 Supplementary Services Interworking

9.3.4.1 The Mapping of Supplementary Services

The traditional services are delivered by their equivalents in IMS, with or without some differences (i.e., in Simulation mode). The table below lists the IMS and NGN names and descriptions of the specified services:

Service Abbreviations	Supplementary Service
CLIP	Calling Line Identification Presentation
CLIR	Calling Line Identification Restriction
COLP	Connected Line Identification Presentation
COLR	Connected Line Identification Restriction
TP	Terminal Portability
UUS	User-to-User Signaling
CUG	Closed User Group
SUB	Sub-addressing
MCID	Malicious Call Identification
CONF	Conference Call
ECT	Explicit Call Transfer
CFB	Call Forwarding Busy
CFNR	Call Forwarding No Reply

Service Abbreviations	Supplementary Service
CFU	Call Forwarding Unconditional
CD	Call Deflection
HOLD	Call Hold
CW	Call Waiting
CCBS	Completion of Calls to Busy Subscriber
3PTY	Three-Party
CCNR	Completion of Calls on No Reply
ACR	Anonymous Communication Rejection
MLPP	Multi-Level Precedence and Preemption
GVNS	Global Virtual Network Service
REV	Reverse charging

Source: Based on ETSI TS 183 043.

The MLPP and GVNS services are specified in ITU Recommendation Q.735 and the REV service is in ITU Recommendation Q.736. All others are defined in ETSI EN 300 356.

9.3.4.2 PSS Service Simulation

Only some of the supplementary services are deemed essential and feasible in a Simulation-based solution. The following table provides a listing of IP-equivalent services of the selected PSTN supplementary services that have been defined by TISPAN and endorsed by the 3GPP:

Abbr.	NGN Name	Equivalent PSTN/ISDN Supplementary Service	TISPAN Ref.
AoC	Advice of Charge	AoC	TS 183 012
CCBS	Call Completion Busy Subscriber	Call back when free on Busy	TS 183 015
CCNR	Call Completion No Reply	Call back when free on No Reply	TS 183 015
CDIV	Communication Diversion	Call Diversion	TS 183 004
Conf	Conference Call	Conference Call	TS 183 005
CW	Communication Waiting	Call Waiting	TS 183 009
HOLD	Hold communication	Call Hold	TS 183 010
IBC	Incoming Communication Barring	Anonymous Call Rejection	TS 183 001
MCID	Malicious Communication Identification	Malicious Call Identification	TS 183 016
MWI	Message Waiting Indication	Message Waiting Indication	TS 183 006
OIP	Origination Identity Presentation	Call Identity Presentation CLIP	TS 183 007

Abbr.	NGN Name	Equivalent PSTN/ISDN Supplementary Service	TISPAN Ref.
OIR	Origination Identity Restriction	Call Identity Restriction CLIR	TS 183 007
TIP	Termination Identity Presentation	Call Identity Presentation COLP	TS 183 008
TIR	Termination Identity Restriction	Call Identity Restriction COLR	TS 183 008

Source: Table drawn from information in TISPAN TS 183006 and 3GPP TS 24819.

9.4   Blended Services

9.4.1   CSI Principles

The combination of CS and IMS (CSI) services is also known as CSICS (Circuit Switches and IMS Combinational Services). These terms are used interchangeably. CSI is a mechanism that ties together CSN calls and IMS sessions under a single context, thereby enhancing the user's experience.

CSI enables users to establish combinational service between two users within the same Home PLMN or across the border of different operators. CSI should be available whether the users are in their Home Network or roaming into a Visited Network, as long as roaming is available.

The key requirement for successful CSI services is that a user initiating a CSI session will need to know only one address or number for the destination. The CSI service should be able to retrieve the associated identifier from the HSS.

A CSI session can be initiated from:

- A CS speech call, adding an IMS session
- An IMS session, adding a CS speech call
- A CS Multimedia call, adding an IMS session
- An IMS session, adding a CS Multimedia call

Termination of one of the parallel sessions will not automatically end the other session. Therefore, if the CSN call has ended, the IMS media continues to flow. If the IMS media session is stopped by one party, the conversation between them can continue on the CSN call.

A user should have the power to actively accept or reject media sent by the other user, in an explicit act. Where the mutual capabilities do not match, the service provides prioritization that favors CS speech over IMS audio and IMS video or image over CS video.

9.4.2   The CSI Architecture

9.4.2.1   CSI Network Scenario

The CSI architecture models how CSI sessions can be created from parallel sessions in both domains, while sharing the context of the session. To facilitate CSI capabilities, the UE must be able to support it and an application is needed to coordinate the overall session.

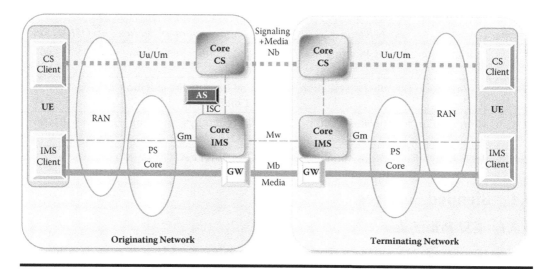

Figure 9.6   The CSI model for a single operator session.

Figure 9.6 shows a scenario of a simultaneous IMS session and CS call between two end users belonging to the same operator. The CS connection of both the signaling and media is shown by dotted lines. Two parallel sessions, one in CS and the other in IMS, can proceed, where the application provides the association between them, creating the user experience of a single session. In this example as shown, the originating network activated the CSI session in the CSI AS, hanging off the Core IMS. This AS service is responsible for the correlation of the two calls.

Alternatively, in CSI phase 2, the sessions are combined to create a single multimedia session. In this scenario, the CS call is converted to an IMS call and the voice is carried over the PS network together with the multimedia.

Figure 9.7 demonstrates the state after the CSI connection has been established as a single multimedia session. In this example, the AS in the terminating network is used to support the combinational session, although this can also be controlled instead from the originating CSI AS. The media flow is now flowing in the PS domain end-to-end.

9.4.2.2   The CSI Active Elements

The architecture elements are:

- *The UE.* The UE should support both CS and PS clients with their respective access procedures. The UE must be able to exchange capability information without which the CSI service cannot run, and it must be able to present the CS call and IMS session within the same context to the user.
- *The RAN.* The RAN requirements for CSI include DTMF for GERAN (GSM EDGE Radio Access Network) and multiRAB (Multi Ratio Access Bearer) for UTRAN (UMTS Terrestrial Radio Access Network).
- *The PS Core.* The PS CN functions as normal, and there is no special media requirement with the prerequisite that it has installed IMS.
- *The CS Core.* The CS CN operates as normal. It should have MSC/VLR, HLR, and other normal CS Mobile network elements.

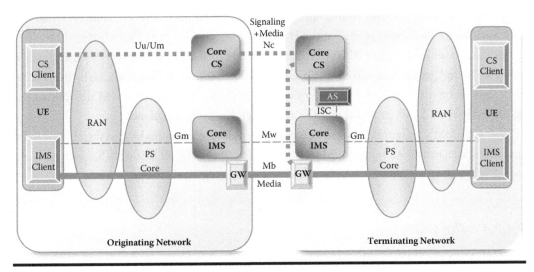

Figure 9.7   CSI origination/IMS termination with CSI-level interworking in the terminating network.

- *The IMS Core.* The IMS session control performs the exchange of UE capability and registers the CSI AS that monitors the session. It also routes the SIP signaling between the parties.
- *The CSI AS.* The AS interacts with IMS in the standard manner as any AS, via the ISC interface. The AS not only controls the logic when 3PCC is required to control both call legs, but also determines service-based charging.

9.4.2.3   The CSI Mechanisms

The process includes:

- Ensuring sufficient resources (e.g., for the air interface for Mobile)
- Negotiation of the respective parties' terminal capabilities to ascertain their ability to support both circuit-based voice and multimedia, as well as the required radio signals
- A mechanism of adding IMS sessions to on-going CSN calls
- A mechanism of adding a CSN call to ongoing IMS sessions

The standards currently support two scenarios:

1. Sessions between two CSI-enabled UEs that can both initiate and terminate (receive) calls
2. Sessions between UEs that use IMS for originating sessions and UEs that use CSI to terminate (receive) calls

The CSI solution should operate without the need for any particular action from the user, having requested the associated session. There should be no need for a user action to create a separate session. To facilitate a transparent service, it must also operate without perceptible delay or degradation of quality, regardless of whether or not the CS call is combined with an IMS session.

The CSI solution must enable the discovery of an associated CS numbers when the CSI service starts with an IMS session using URI in the PS network based, and the discovery of an associated

URI when the call starts with connection to an MSISDN number. This is enabled by inserting such associations in the user profile on the HSS.

At the time of initiating a CSI service, the UE should be registered with its CSN, get attached to the PSN, and get IMS registered. If the UE is not IMS registered, it automatically initiates an IMS registration, using a Public User Identity. This implicitly will register the MSISDN that has been used in the CS call, provided that the association has been provisioned.

The same MSISDN numbers should be used for the users' IMS subscription and their CS subscription. The system behavior is not specified for the case where the MSISDN for the IMS subscription and the CS subscription are different, which makes the automatic association impossible.

9.4.2.4 The CSI AS

The CSI AS in the IMS domain hosts an application that provides session control for the multi-media session leg between UE that uses IMS origination and UE that uses CSI termination.

The functions of this application are to:

■ Perform third-party registration of the CSI AS when the UE registers with IMS.
■ Obtain UE CSI capabilities.
■ Perform 3PCC logic for the CSI termination session (i.e., get the destination connected via a separately controlled leg).
■ Ascertain, according to the media components, how to connect the media flows.
■ Determine the associated SIP URI and Tel URI, in case the call needs to be connected to the CSN client.
■ Handle mid-session changes.

The iFC for the CSI application should be installed on a user's service profile in the HSS, regardless of their other service subscriptions. This is necessary when forwarding the SIP requests for a CSI session to the CSI AS. It also enables IMS registration requests from the CSI-capable UE to initiate third-party registrations for the CSI AS.

The trigger points in the iFC can use the IMS Communication Service ID, or the lack of it. The UE capability of CS voice or video that has been reported during IMS registrations can also be used, and the AS can dynamically activate or deactivate the Service Indication Data in the HSS according to that information.

9.4.3 CSI Interaction with CS Supplementary Services

CSI is a basic connection type that should support supplementary services, if possible. Most CS services are not affected but some will prevent the establishment of a combinational IMS session.

The following table provides the services that interact with CSI:

Code	Name	CS	IMS	CSI	Comments
CLIP	Calling Line Identity Presentation	Y	Y	N	If CLI is unavailable, the IMS session is not "combinational"
CLIR	Calling Line Identity Restriction	Y	Y	N	Restriction also must apply to IMS, and as a result it will not be "combinational"
COLP	Connected Line Identification Presentation	Y	Y	Y	If not available, then the calling party's UE can use the provided called party number of the CS call to correlate incoming IMS sessions
COLR	Connected Line Identification Restriction	Y	Y	Y	IMPU also must be restricted. It is possible to use the provided CS called party number to correlate an incoming IMS session
CNAP	Calling Name Presentation	Y		Y	No impact
CFU	Call Forwarding Unconditional	Y		Y	No impact
CFB	Call Forwarding on Busy	Y		Y	No impact
CFNRY	Call Forwarding on No Reply	Y		Y	No impact
CFNRC	Call Forwarding on Not Reachable	Y		Y	No impact
CD	Call Deflection	Y		Y	No impact
ECT	Explicit Call Transfer	Y	Y	N	The transferred call cannot establish CSI because capabilities have not been exchanged. The IMS session cannot be transferred
CH	Call Hold	Y	Y	Y	Combinational IMS session also can be put on Hold, and then resumed
CW	Call Waiting	Y	Y	Y	Accepting CW, capabilities are exchanged and new CSI can be established
CUG	Closed User Group	Y	Y	N	IMS sessions will not be considered "combinational"
AOCI	Advice of Charge — Information	Y	Y	Y	No impact
AOCC	Advice of charge — Charge	Y	Y	Y	No impact
UUS1 UUS2 UUS3	User-to-User Signaling service 1/2/3	Y	Y	Y	No impact
BAOC	Barring of All Outgoing Calls	Y		N	IMS sessions will not be considered "combinational"

Code	Name	CS	IMS	CSI	Comments
BOIC	Barring of Outgoing International Calls	Y		N	IMS sessions will not be considered "combinational"
BAIC	Barring of All Incoming Calls	Y		N	IMS sessions will not be considered "combinational"
EMLPP	Enhanced Multi-Level Precedence Preemption	Y		N	IMS sessions will not be considered "combinational"
CCBS	Call Completion to Busy Subscriber	—	—	—	Not specified for CSI
MC	Multi Call	—	—	—	Not specified for CSI
MPTY	Multi-Party call	—	—	—	Not specified for CSI

Source: From TS 23279.

9.5 IMS-Based Transit

9.5.1 IMS Transit Function

9.5.1.1 Use of IMS for ISUP Transit

Transport of encapsulated ISUP in SIP is implemented already in many softswitch solutions. In the new IMS-based network, if there is no softswitch, this facility can be controlled via IMS.

IMS Transit is utilized in the following business relationships:

■ Within the same operator, providing service to own customers during migration to IMS. IMS Transit can occur between CSN customers across the IP network or a connection between IMS and a CSN destination transits the IP network
■ Providing Transit capability to Enterprise customers, between their IP VPN users or connecting to PSTN or to other IP networks
■ Offering IP Network Transit to other operators and service providers who seek to connect:
  − PSTN users to PSTN users in whatever network, where none of the users belong to the operator
  − IP users to IP users or to other IP networks, where none of the users belong to the operator
  − PSTN to IP and vice versa, where none of the users belong to the operator

The IMS Transit solution can provide management of an IP pipe, optimizing the exit to the CSN and reducing traffic on the traditional internal CSN nodes.

The operator can offer several IMS functions as part of the service, including:

■ Authentication and admission control
■ QoS control for the IP leg
■ CSN interworking with different protocols
■ Charging for wholesale and Enterprise customers

9.5.2 IMS Transit Components

9.5.2.1 IMS Transit Traffic Flow

The IMS Transit is delivered by the standard IMS components, namely the I-CSCF, IBCF, S-CSCF, BGCF, and MGCF. In addition, it needs the ability to encapsulate the ISUP message in SIP, using SIP-ISUP or SIP-Telephony.

Traffic can arrive via the MGCF from PSTN origins, to be converted and encapsulated in SIP so that it can be transported over the IP backbone to remote destinations. SIP based traffic can arrive from the I-CSCF, forwarded from the P-CSCF or from an AGCF towards the operator's own subscribers. Traffic can also arrive from the IBCF, when it originates from other IP networks and it is merely traversing the operator's IP network.

Figure 9.8 shows the IMS Transit Function traffic flow, when originating from either a PSTN or an IP network. In both cases, the IMS Transit can provide "smart" routing over the IP transport network.

9.5.2.2 Stand-Alone IMS Transit

The IMS Transit can be offered as a wholesale service to other operators and service providers that may not have a transport infrastructure or lack presence in a particular region. The IMS Transit, as a separate business, can be a stand-alone system or it can share the operator's IMS system for the Home subscribers.

As a stand-alone system, the IMS Transit will need the IMS components, including the S-CSCF to support any applications. The various elements (e.g., the MGCF and the I-CSCF) will have to produce CDRs for inter-operator accounting.

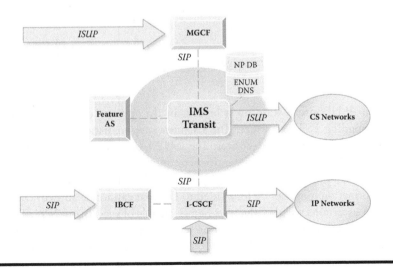

Figure 9.8 The IMS Transit Function.

9.5.2.3 IMS Transit Applications

The Transit Function can be supported by a number of services to enhance the offering. This can be optimal routing service, such as Least-Cost-Routing, which makes transit routing decisions based on time, day, and date, or according to accumulators of volumes toward bonus or discounts per wholesale customer.

Transit applications can also include services based on the destination number, such as non-geographical numbers for call centers that may be routed differently depending on time, day, date, or a load-balancing algorithm between centers, etc.

9.5.3 Conveying ISUP via IMS

9.5.3.1 The Standards for ISUP Transport

The transport of ISUP over SIP networks is performed via encapsulating the message within SIP, and then unpacking it at the other end. The message syntax carried over is defined by the ITU-T Recommendation Q.1980.1: "The Narrowband Signaling Syntax (NSS)."

In TISPAN, the element responsible for sending or receiving ISUP information in NSS messages is the PES (PSTN Emulation System) Interworking Application (IA). This function is performed on the MGCF, or on the AGCF in the case of IMS-based PES.

Services that require manipulation of the ISUP information within NSS message bodies must be implemented on servers acting as a SIP B2BUA rather than a proxy because the PES IA needs to act as the originating UA, terminating UA, or a server performing 3PCC.

9.5.3.2 The ISUP Interworking Application (ISUP IA)

The task of creating SIP messages that encapsulate the ISUP NSS (Narrowband Signaling Syntax) message body is assigned to the PES IA, which resides on a SIP AS, an Outgoing MGCF (O-MGCF), or an Incoming MGCF (I-MGCF). The application operates as:

- An originating UA
- A B2BUA
- A terminating server
- A redirect server relaying the message onward
- A server performing 3PCC

Each piece of ISUP information must be sent in the appropriate SIP message to interwork with the corresponding PSTN services. The IA must support parameter translations and hold the default values specified for these services. The ISUP IA is responsible for the security and the integrity of the ISUP transfer. If network entities communicating to the IMS PES cannot be relied on for information confidentiality and integrity, then the NSS message will not be sent. Similarly, confidential information must not be forwarded to UE that cannot be authenticated.

The ISUP IA ascertains whether the next hop is capable of supporting NSS. This entails seeking an indicator that ensures that the destination server (peer SIP server) supports NSS processing. When constructing an initial INVITE request, an application supporting NSS message includes the "Accept" header in the initial INVITE request with "application/nss" value. Proxy servers in the path retain this indicator for the same process along the path.

Sending the ISUP NSS message must be a reliable transmission. The O-MGCF cannot send the ISUP NSS message until it receives confirmation from the other end. The receiving I-MGCF cannot send an INFO request for the NSS message until a reliable response (provisional or final) has been sent and acknowledged.

9.5.3.3 The Encapsulated ISUP Message

When the ISUP service logic or event matches a basic call control SIP method, such as INVITE, re-INVITE, BYE, UPDATE, the translated ISUP syntax is encapsulated in the corresponding SIP message.

To retain flexibility, some scenarios allow sending an NSS message body marked as "optional." This allows for messages that cannot be translated to be sent forward, and the receiving server ignore them or perform an alternative procedure determined by the ISUP message.

When receiving a SIP message containing an NSS message body, the ISUP IA or I-MGCF will de-encapsulate the ISUP information from the NSS message body, interpret the ISUP message, and trigger the relevant service logic.

The application must be able to insert charging information in the SIP messages sent in the upstream direction and also capable of processing charging information contained in the SIP message received from downstream.

9.5.3.4 Translating ISUP and SIP Basic Call Services

There are many functions that do not have a direct parallel and therefore require further logic to resolve the request message. The following is a listing of ISUP messages that have *no* counterparts in basic SIP call control messages:

- Blocking and blocking acknowledgment
- Call progress
- Charging information (national use)
- Circuit group blocking/unblocking and acknowledgment
- Circuit group query (national use) and response
- Confusion
- Continuity
- Facility, Facility request, Facility accepted, Facility reject
- Forward transfer
- Group reset and acknowledgment
- Identification request and response
- Information (national use) and request
- Loop-back acknowledgment (national use)
- Network resource management
- Reset circuit
- Resume
- Segmentation
- Subsequent address message
- Subsequent directory number (national use)
- Suspend

- Unblocking and acknowledgment
- Unequipped CIC (national use)
- User part test/available
- User-to-user information

Support of supplementary services can be realized by exchanging service information between peer SIP signaling entities via SIP messages or encapsulated ISUP information. The ISUP information necessary to support each individual service is specified by the corresponding ETSI or ITU-T supplementary service specification.

Chapter 10

WLAN Interworking

10.1 This Chapter

This chapter describes how Wireless LANs (WLANs) can be used as an Access Network (AN) to IMS and core services. IMS can be accessed via WLANs by Mobile or NGN terminals. Therefore, descriptions of WLAN network admission according to 3GPP as well TISPAN are given here.

Although IMS authentication procedures, security, QoS management, and charging recording processes are described in other chapters, the equivalent WLAN procedures merit a separate description here.

10.2 WLAN Concepts

10.2.1 WLAN Network Access Overview

10.2.1.1 Parallel Approach from Fixed and Mobile Networks

Wireless networks are already delivering access to computers, in IP packet data networks, using IEEE 802.1x protocol variants. Delivering voice and multimedia is the next step. Both NGN networks and Mobile networks are converging on this technology for a range of UE terminals, including voice handsets, smart phones, PDAs, "pocket PCs," and laptops.

Figure 10.1 illustrates admission to IMS over a selection of operator networks, through a selection of WLAN access networks. The UE may be able to select from more than one WLAN. The Interworking WLAN (I-WLAN) can connect to more than one Visited Network, as shown in the example of I-WLAN-1. Access to the Internet can be achieved via direct IP authentication, as shown in the example of I-WLAN-2.

In a roaming scenario, the UE messages are routed to a Visited IP AN (AAA Server proxy), which links to the Home authentication server. The Visited and Home Networks interact via Proxy and Server arrangements, that is, the Visited AAA Proxy and the Home AAA Server.

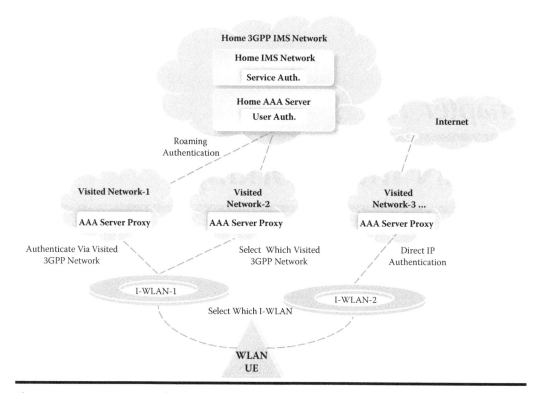

Figure 10.1   WLAN access for both 3GPP and NGN.

10.2.1.2   Methods of Integrating WLANs

When determining how to evolve WLAN access to IMS, it was considered whether to connect I-WLANs via GPRS, reusing GPRS methods of network admission, or via special access servers. Although using GPRS seemed like a neat solution for Mobile operators, it did not comply with the approach to provide access-agnostic services. The IMS standards seek to treat the WLAN as another AN, independent of GPRS or DSL.

Figure 10.2 illustrates the chosen architecture. It supports WLAN access that is a separate structure, reaching IMS without going through GPRS. This enables the WLAN Access technology to develop independently of GPRS and reduce pressure on existing GPRS nodes. It also shows a generic access from DSL, via the AAA Server, allowing convergence of Fixed or Mobile Networks.

In this way, it is easier for another service provider to provide the I-WLAN and increase competition. User mobility should be managed separately but this also extends to other modes of access. This approach encourages faster introduction of WLANs because it does not require extensive upgrades of existing GPRS nodes. It also offers greater flexibility in vendor selection, timing of upgrades, etc.

To enable faster introduction, the standards also support the existing technology of RADIUS AAA Servers, with translation to the Diameter protocol, which is the chosen protocol for IMS management functions. Authentication is performed at access level within the WLAN, as equivalent of the GPRS level or DSL NASS level authentication, and then at the service level, in IMS via the Home Network servers.

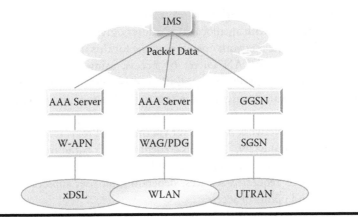

Figure 10.2   Access of WLAN via gateway, not GPRS.

10.2.2   Linking to WLAN

10.2.2.1   WLAN Connection Steps

Setting up a session over the WLAN includes the following processes:

- Network selection by the WLAN UE
- Interworking-WLAN authentication to gain access to the wireless AN
- Service authorization to gain access to the IMS service
- Establishing a secure tunnel to the Home Network with a variety of security measures
- Recording and reporting accounting information for the WLAN sessions

The process of attaching to the network and gaining authorization can be seen as the steps shown in Figure 10.3:

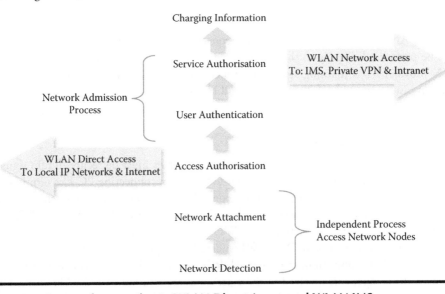

Figure 10.3   Stages of connecting to WLAN Direct Access and WLAN IMS.

■ Discover available networks (SSID) and select one, establishing the WLAN Access Point.

■ Attach to that network with an allocated IP address, giving the subscriber's NAI that enables ascertaining the Home Network operator's name for this user.

■ Authenticate users for WLAN Direct IP Access, using local nodes, according to SLAs between operators. The 3GPP authentication or NASS mechanisms can be used for this, if desired. If successful, the user gains access to the Internet and other locally connected IP networks.

■ Route to the Home Network AAA Server via the AAA Proxy, requesting full service authorization from the remote CN. The user's credentials (user name and password, token, certificates) are authenticated at the Home Network.

■ Authorize access to requested services, depending on user subscriptions as well as interoperator SLAs and points of access.

■ Set up tunneling for secure WLAN connection, via the Visited Network to Home 3GPP network. Gain access to a secure service environment and VPN.

Even when access to the WLAN is required for the purpose of merely connecting to the Internet, the authentication methods can be to utilize the 3GPP or NASS IP Access servers. This is the "WLAN Direct Access" mode. For the purpose of using IMS, the "WLAN 3GPP IP Access" mode is required, where both local access authentication and 3GPP IP Access authentication are needed.

10.2.2.2 WLAN Network Selection

The UE WLAN client selects which I-WLAN to connect to from:

■ Available advertised networks
■ Built-in list
■ List received on request from the APN (Access Point Name) and the W-APN (WLAN APN).

The I-WLAN node can then proceed with requests for authorization of the service. This process can be performed directly to the Home server or via a Visited-Network proxy server if the I-WLAN provider has a roaming agreement. Once the user has service authorization, charging information can be sent to both Intermediate and Home Networks.

10.2.2.3 WLAN Security

The process of obtaining permission to use network resources must be secure. Radio signals on Mobile networks are deemed secure enough, but not so when connecting PC-like terminals with software that is not fully under the operator's control, over airways that are open to everyone. PC WiFi security was previously based on Wired Equivalent Privacy (WEP). Now, WPA (WiFi Protected Access), including WPA802.11 and WPA2, has been created to correct several serious weaknesses that were identified in WEP. WPA (802.11i) implements the majority of the IEEE 802.11i standard, but is only an interim step, while WPA2 implements the full standard.

The network components involved in these processes are a mixture of WLAN-enabled access network nodes and NGN/3GPP AAA Servers that support IMS roaming and charging.

10.3 WLAN Integration with 3GPP IMS

10.3.1 WLAN IMS Functions

10.3.1.1 Scenarios for Usage of WLAN

Several modes for accessing the IMS network via the WLAN have been identified, including:

- Access to Internet only, with local authentication and billing performed by the local AN.
- 3GPP system-based access control and charging, where authentication and billing are performed via 3GPP servers for Internet access in "Direct IP Access" mode as well as other 3GPP data packet services, but no IMS involvement.
- Access to full 3GPP system and PS-based services, such as IMS, SMS, WAP (Wireless Application Protocol), and MMS, from an I-WLAN, on any type of terminal, where service protocols are transported over the WLAN interface to IMS session control and applications.
- Non-transparent Service Continuity where access is available in both WLAN and GPRS/UMTS, but services are separately accessed and separately managed.
- Transparent service provision, where continuous service is maintained across WLAN and GPRS/UMTS access seamlessly, where the same service is offered regardless of the AN, and where the service is continued even while there is a handover between domains.

These scenarios affect the positioning of authentication, according to business borders, and the relationships of the proxy or server that allow for roaming.

10.3.1.2 I-WLAN Main Functions

The following is a brief description of the elements involved in connecting WLAN UE to the 3GPP network, in an Interworking WLAN:

- *WLAN Authentication.* Users can be admitted to a local WLAN without requiring IMS services. In this case, users need only be authenticated locally. To gain access to IMS services, the WLAN UE must authenticate to the IMS Home Network. The user identifiers are relayed to the Home Network server (3GPP AAA Server), which uses them to retrieve the HSS user profile.
- *WLAN Charging.* WLAN communication is charged by the network charging systems, the OCS, and the OFCS, via its CDF and CGF.
- *WLAN QoS.* To facilitate network control for QoS, it is necessary for the Resource Controller to interact with the WLAN nodes (WAG and PDG), which will enforce the QoS policies.

The following network elements deliver these functions:

- *WLAN UE.* A WLAN UE (W-UE) is the user equipment with a UICC utilized by a 3GPP subscriber to access the WLAN network for 3GPP interworking purposes.

- *3GPP AAA Proxy.* The 3GPP AAA Proxy provides forwarding and filtering functionality. It resides in the Visited 3GPP network, and is involved in access and service authentication and authorization procedures for the WLAN UE.

- *3GPP AAA Server.* The 3GPP AAA Server resides in the Home Network and is responsible for access and service authentication and authorization for the WLAN UE. It also maintains state of the user and is involved in producing service-level charging information.

- *WAG.* The Windows Access Gateway connects the WLAN and the 3GPP network. In the roaming case, it resides in the Visited Network. It provides filtering, policing, and charging functionality for the traffic between the WLAN UE and 3G network.

- *PDG.* The Packet Data Gateway connects ito the packet Network. It resides either in the Home (for access to Home services) or in the Visited Network (for access to local services). The PDG is instrumental in monitoring and charging when using Flow Based Charging for WLAN traffic.

Figure 10.4 shows the WLAN components interworking with IMS. It shows the WLAN as a separate infrastructure. Note that the AAA Server, like the PDG, contributes charging information. Also worth noting is that the AAA Server has direct access to the IMS HSS and SLF to obtain User information.

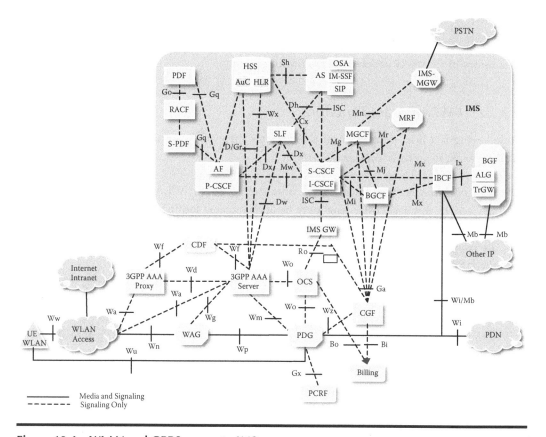

Figure 10.4   WLAN and GPRS access to IMS.

10.31.3   Mobility Management

For 3G Mobile, the HSS contains user profiles and authentication data. Because it also stores dynamic user status and location in the HLR, the HSS/HLR can provide the same information about WLAN login status. The HSS therefore stores the network location of the current 3GPP AAA Server for the WLAN UE, together with the identity and address of the S-CSCF. In addition, the Home AAA Server retains status knowledge of the user when the user is registered via the WLAN access network. It can provide this information to Presence Servers and other applications.

10.3.1.4   Procedures for Home Domain Access

When the user requests access to a WLAN that belongs to the same network that owns the user's account, there is no roaming involved. The user must be registered for 3GPP services, and then it can request access via WLAN.

The user handset or laptop W-UE discovers the local WLAN access point, the WLAN APN (W-APN). This can be a customer-premises router or modem that connects to an xDSL network, or any WLAN gateway.

The W-APN can discover via a DHCP server the addresses of the WAG node or it can be configured with a list of them. In the case of no roaming, the WAG within the Home Network relays signaling messages from the UE to the WLAN AAA Server, without any intervening proxy.

The Home Network AAA Server may already have cached profile data for this user to authenticate the user and make IMS services available. If there is no user data, the AAA Server will first contact the SLF to obtain the address of the appropriate HSS that keeps the profile for this user. Then, the AAA Server will retrieve from the HSS the user's details, profiles, associated identities, and subscribed services.

Figure 10.5 demonstrates the non-roaming scenario, where the PDG, WAG, and AAA Server are all in the same network as the HSS and Home charging systems.

10.3.1.5   Procedures for 3GPP WLAN Roaming

The 3GPP AAA Proxy Server relays messages from the Visited Network where it resides to the target Home Network of the subscriber. It acts as a proxy by sending the messages through, rather than terminating or initiating them.

The 3GPP AAA Proxy functionality can be implemented on a separate physical network node or co-reside on other access nodes. It is also commonplace to have both "Server" and "Proxy" functionality on the same server, acting in different roles depending on the occasion.

In Figure 10.6, the W-APN and the WAG are located in a network that does not have the user's subscription. In this case, the messages are forwarded to a proxy in the local network, which sends them to the Home AAA Server. The local proxy must determine the location of the Home Network server, through DNS inquiry on the realm part of the user ID or through preconfigured lists of affiliated networks.

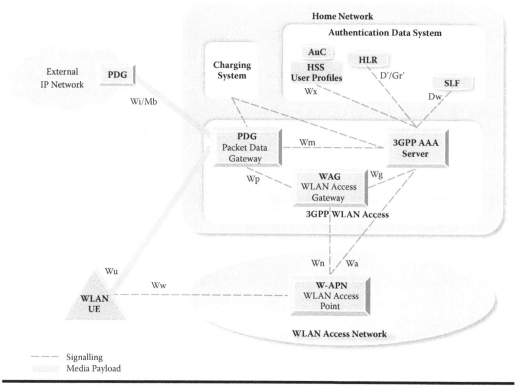

Figure 10.5 No-roaming scenario.

10.3.2   The I-WLAN Elements

10.3.2.1   The WLAN UE

10.3.2.1.1   Network Discovery

WLAN UE is any terminal capable of accessing a wireless network. Some WLAN UE is enabled for simultaneous access to both WLAN and 3GPP radio access (i.e., dual-mode terminals).

Unlike traditional Mobile networks, the WLAN UE may not be under the exclusive control of the network operator, but may be a generic terminal such as a laptop computer or PDA with a WLAN card. This has significant implications on the user's identity management and the security measures that will be implemented. This also has a strong influence on converging 3GPP Mobile and NGN wireless methods.

The WLAN UE searches for beacons of available WLANs and can choose to which it should be attached. This can be done manually, by user decision, or automatically. The WLAN UE can request from the WLAN AP a list of SSIDs (Service Set IDs). Alternatively, the WLAN UE can be configured with a list of WLAN providers that have 3GPP roaming agreements.

The user can indicate a preferred IP network as a Requested W-APN. The Requested W-APN will then indicate a point of interconnection by naming the PDG that connects to that particular external networks.

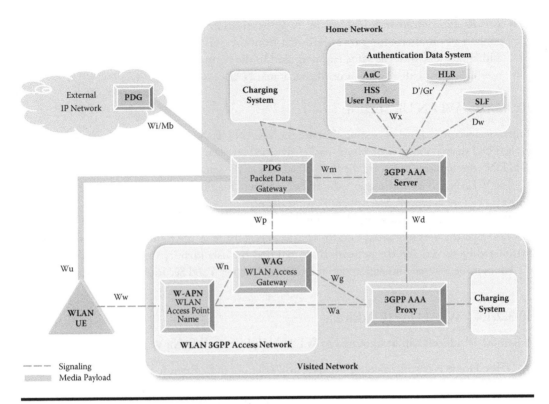

Figure 10.6 Roaming scenario, services provided in Home Network.

10.3.2.1.2 3GPP WLAN User Identifier

WLAN user identification is based on the supplied identifiers and credentials presented by the WLAN UE, in conjunction with prestored credentials on the Home server. The user ID must enable either SIM-based or AKA-based (Authentication and Key Agreement) authentication.

User IDs can take the form of:

■ International telephone numbers (E.164 ITU numbering plan), IMSI or MSISDN
■ Network names, in the format of Internet domain, where the "realm" part is the operator's network identifier, in the format of the NAI

There are several types of NAIs:

■ *Root NAI.* Used when the WLAN UE is trying to authenticate directly to the Home Network.
■ *Decorated NAI.* Used when trying to authenticate via the Visited Network. It allows for both Visited and Home realm names to be sent, in the form of "Homerealm!username@otherrealm."
■ *Alternative NAI.* Used when asking for a list of available networks from the local access node, during a manual selection procedure. It takes the form of "any_username@REALM," where "any_username" is not derived from the IMSI number and the REALM part is "unreachable.3gppnetwork.org."

■ *Fast Re-authentication NAI.* Used for re-authentication in reconnections and handovers. It consists of the username part as returned from the 3GPP AAA Server during the original authentication, with the same realm as used in the permanent user identity. The permanent user identity is either the root or the decorated NAI.

10.3.2.1.3 WLAN UE Security

To ensure a secure channel of communication, the WLAN UE should be capable of creating a tunnel to the PDG that connects to its Home packet-switched services. The WLAN UE needs to use standard DNS mechanisms to resolve the PDG IP address as the remote tunnel endpoint.

The WLAN UE should support the IKEv2 Protocol for IPSec tunnel negotiation. The procedure also can include ESP. See Appendix A for details.

The W-UE can establish trusted relationships after mutual authentication with the PDG takes place, so that the user is assured that the connection is made to the right network server (i.e., not masquerading server) and the network is assured that the user identity is as it is claimed.

For more details, see Authentication Procedures in Chapter 4 and Security Protocols in Chapter 7.

10.3.2.1.4 WLAN UE IP Address

A Remote IP address can be allocated to WLAN UE in four different ways:

1. The Home Network operator permanently assigns a *static* Remote IP address to the WLAN UE.
2. The Home Network operator assigns a *dynamic* Remote IP address to the WLAN UE when the tunnel link is established to the PDG in the Home Network.
3. The Visited Network operator assigns a *dynamic* Remote IP address to the WLAN UE when the tunnel is established to the PDG in the Visited Network.
4. The external IP network operator assigns a *permanent* or *dynamic* Remote IP address to the WLAN UE.

The Home Network operator defines (in the subscription) whether static IP address allocation is used.

10.3.2.2 The 3GPP AAA Server and Proxy

The 3GPP AAA Server facilitates the network attachment process. A local AAA Server detects roaming when analyzing the WLAN UE realm ID while attempting to authenticate the user. This server will then act as a proxy, and will relay the request for authorization to the Home 3GPP AAA Server.

The primary functions of the 3GPP AAA Server include:

■ *Authentication.* It retrieves authentication information and subscriber profile from the HSS of the subscriber's Home 3GPP network and authenticates the 3GPP subscriber based on this data, using authentication protocols. The authentication signaling can pass through AAA Proxies.
■ *Registration and deregistration.* When a user's service subscription is modified, as requested by the HSS, it updates the WLAN access authorization information. When successfully

registered, the name or IP address of this 3GPP AAA Server is stored on the HSS. When the 3GPP AAA Server deletes the information of a subscriber, it initiates the Purge procedure, which effectively deregisters the user.

■ *Maintains 3GPP WLAN status.* The AAA Server keeps registration state and conveys it, on request, to other entities. In particular, this information is sent to Presence servers.

■ *Generates charging information.* For each session per-user, it generates charging and accounting information and sends them to the Offline Charging System in the Home Network.

■ *Supports remote access.* When located in a Visited Network, the 3GPP AAA Server acts as a proxy to the Home AAA Server. It sends the details of the service authorization to the PDG, including the W-APN details, and tunnel establishment parameters, when the PDG is located in the Visited Network. When static remote IP address allocation is used, the AAA Server also provides the PDG with the remote IP address of the WLAN UE as received from the HSS.

■ *Provides policy enforcement.* The AAA Server conveys information to the AAA Proxy and the WAG in the Home Network about the local network capability. It also can provide routing enforcement information to the WLAN AN, following network policy.

The primary functions of the 3GPP AAA Proxy include:

■ Relaying and translating messages between Visited and Home Networks:
  - Enforcing policies according to interpretation of the roaming agreements
  - Communicating to the local WLAN access node any access scope limitations and restrictions on the authorized service that have been received from the Home Network
  - Performing SIP profile adjustments for the purpose of interworking when there are some discrepancies in implementing the Wa and Wd reference points
  - Performing protocol conversion between Diameter and RADIUS where different protocols are used in the Visited and the Home Networks
  - Receiving authorization information for subscriber requests from the Home Network and informing the W-APN
■ Controlling local services:
  - Authorization of access to the Visited Network W-APN according to local policy
  - Receiving policy enforcement information from the Home AAA Server and providing it to the WAG in the Visited Network
  - Transfer service termination instructions to the local gateways, if initiated by the O&M system of the Visited Network operator
  - May provide suitable routing enforcement information to the WLAN AN
■ Reporting charging information:
  - Generating charging records for roaming users
  - Reporting per-user roaming charging information to the Visited Network OFCS
  - Receiving per-tunnel charging information based on the tunnel identifier from the WAG and mapping of user identifiers and tunnel identifiers from the PDG

The AAA Proxy only operates in roaming situations.

10.3.2.3 WLAN Access Gateway (WAG)

The WAG is a gateway between the local WLAN Access Nodes and the 3GPP AN. In the case of roaming, the WAG resides in the Visited 3GPP network. The WAG provides filtering, policing, and charging functionality for the traffic between WLAN UE and the 3GPP network.

The main functions of the WAG include:

- It enforces the routing of packets through the PDG to prevent bypassing billing or misuse of network resources.
- For a roaming scenario only (i.e., when the Wu reference point is between the WLAN UE in a Visited Network and a PDG in the Home Network), it aggregates per-tunnel usage information (e.g., volumes, byte count, and elapsed time). This information is transferred to the 3GPP AAA Proxy in the Visited Network via the Wg reference point, to be used for inter-operator settlements.
- It filters out unwanted packets. These are packets with unencrypted information deemed to be outside the allowed connection. Packets would only be forwarded if they are part of an existing tunnel or are expected dialog messages from the WLAN UE, including service requests and tunnel establishment messages.
- It filters out packets from an unknown IP address, acting as a firewall.
- It performs additional types of message screening that are under the operators' control.
- It allows the Visited Network to generate its own charging information for users accessing via the WLAN AN in the roaming case.

Before tunnel establishment, the WAG is responsible for policy enforcement, to enhance the firewall and prevent unwanted packets from going through the IP network. This may be needed, for example, to forbid the roaming WLAN UE from sending tunnel establishment to other networks in addition to its own Home Network. It also forbids packets from unauthorized WLAN UEs to be allowed through to the network.

After tunnel establishment, the WAG implements policy enforcement. For service provided through a PDG in the Home Network, the WAG ensures that:

- All packets from the WLAN UE are routed to the Home Network.
- Packets from the authorized WLAN UE are only routed to the appropriate Home PDG.
- Packets from sources other than that PDG are not routed to the WLAN UE at all.
- If service is provided through a PDG in the Visited Network, the WAG ensures that:
  - All packets from the WLAN UE are routed to the Visited Network.
  - Packets from the authorized WLAN UE are only routed to the designated Visited Network PDG.
  - Packets from sources other than that PDG are not routed to the WLAN UE at all.

10.3.2.4 Packet Data Gateway (PDG)

The PDG is a media bearing server at the border of the network. The media flow is performed through the PDG, where the Wi reference point of the 3GPP WLAN Interworking architecture can be mapped to the IMS Mb reference point.

The access to 3GPP PS-based services from a WLAN is via a PDG. These services can be accessed via a PDG in the user's Home Network or a PDG in the selected Visited Network. The process of authorization, service node selection (e.g., W-APN), and subscription checking determines whether a service will be provided by the Home Network or by the Visited Network.

The resolution of the IP address for the acting PDG will be performed by the network that is functioning as the serving network (Visited or Home). The choice of using a PDG in the Visited or the Home Network is made by the Home Network. The routing policy applied at WLAN Access can include policy determining whether the user has IP connectivity to the specially assigned PDG that provides a breakout interface to external IP networks.

For general compatibility, the PDG needs to accommodate WLAN UE using both IPv4 and IPv6 local addresses. For this reason, the Breakout PDG must be equipped with a dual IP stack.

When a service is authorized, the serving PDG IP address is determined and sent to the WLAN UE. The WLAN UE uses standard DNS service to resolve the PDG IP address. A remote IP address is allocated to the WLAN UE if one is not already allocated. The local IP address of the WLAN UE is registered with the PDG, and is bound with the WLAN UE allocated remote IP address to enable the PDG to associate traffic with the WLAN UE authorization.

The PDG supports IPSec tunneling and in-place re-establishing of Security Association. The PDG enables the operator to limit the number of IPSec ESP Security Associations (I-WLAN tunnels) per IKE (Internet Key Exchange).

The PDG activities are as follows:-

- It keeps routing information for WLAN/3G connected users. It routes the packet data between the PDN, the WLAN and 3G connected user.
- It performs address translation and mapping.
- It performs encapsulation and decapsulation for tunneling.
- It accepts or rejects the requested W-APN according to the decision made by the 3GPP AAA Server.
- It relays the WLAN UE remote IP address allocated by an external IP network to the WLAN UE, when it is needed.
- It performs registration of the WLAN UE local IP address and binding of this address with the WLAN UE remote IP address.
- On termination of the connection, it provides procedures for unbinding the local address and remote IP address.
- It provides procedures for authentication and prevention of hijacking (i.e., ensuring the validity of the WLAN UE initiating any actions).
- It delivers the mapping of a user identifier and a tunnel identifier to the AAA Proxy.
- It delivers, in the roaming case, information from the 3GPP AAA Server to the 3GPP AAA Proxy.
- It generates charging information related to user data traffic for offline and online charging purposes.

The PDG also has a number of optional functions, including:

- It can apply IP flow-based bearer level charging (e.g., to differentiate or suppress WLAN bearer charging for 3GPP PS-based services).
- It can filter out unauthorized or unsolicited traffic with packet filtering functions. Other types of message screening are left to the operator's control (e.g., through the use of Internet firewalls).

■ It can agree to perform "Multiple Authentication Exchanges" in IKEv2 with an external AAA Server, if this procedure is required by the PDN and supported by the WLAN UE. The "Multiple Authentication and Authorization" takes effect after a successful authorization by the 3GPP AAA Server.

■ It can support the implementation of a VPN server application to assist tunnel establishment toward the WLAN UE. However, the selection of a particular VPN application is dependent on implementation.

10.3.2.5 The WLAN Access Point Name (W-APN)

The Access Point in a WLAN is the first node encountered by the WLAN UE when connecting to the WLAN network. The name of this WLAN-Access Point is used for routing decisions and authorization purposes.

To enable the WLAN UE to use a W-APN in the Visited Network, the 3GPP AAA Server needs to pass to the 3GPP AAA Proxy the details of the authorized W-APN and service related information, which is required by the Visited Network to perform the service. Thus, both the Home and the Visited Networks need to approve the W-APN.

When making service authorization decisions, the 3GPP AAA Server takes into account the identity of the network of the W-APN, and checks against a list of subscribed W-APNs that it retrieves from the HSS. The WLAN UE indicates to the W-APN what service or set of services it wants to access. The W-APN then requests an address for a serving PDG.

The PDG selection is performed under the control of the Home Network, via its DNS server responding to the W-APN request. This response is affected by the identity and location of the W-APN and by the user's subscription information. The PDG also needs to know the identity of the authorized W-APN to select the external network (i.e., the Wi interface).

10.3.3 The I-WLAN 3GPP Reference Points

10.3.3.1 The D'/Gr' Reference Point

These reference points are between the 3GPP AAA Server and a pre-R6 HLR. This allows for interworking between the WLAN AAA Server and GSM. Support for the D'/Gr' reference points should require no modifications to the MAP Protocol at the HLR. The functions provided on the D'/Gr' reference points are a subset of the functions provided on the D/Gr reference points. If a 3GPP AAA Server supports the D' reference point, it will appear to the HLR as a VLR. If a 3GPP AAA Server supports the Gr' reference point, it will appear to the HLR as an SGSN.

The following functions are made available over the D'/Gr' reference point:

■ Retrieval of authentication details (e.g., for USIM authentication) from the HLR.
■ Registration of the 3GPP AAA Server address for an authorized WLAN user on the HLR.
■ Change indication of subscriber profile within the HLR, including termination of WLAN registration.
■ Purge procedure between the 3GPP AAA Server and the HLR.
■ Retrieval of online/offline charging function address from the HLR.
■ Fault recovery procedure between the HLR and the 3GPP AAA Server.

■ Retrieval of service subscriptions and related information (e.g., a list of W-APNs that can be selected by the WLAN UE). This can include restrictions of what services can be delivered in the Visited Network.

10.3.3.2 The Wa Reference Point

The Wa reference point connects the WLAN Access Network (WLAN AN), possibly via intermediate networks, to the 3GPP AAA Server or Proxy. Its prime purpose is to transport AAA information — authentication, authorization, and charging-related information — in a secure manner. To accommodate legacy WLAN ANs, the Wa reference point must interface to either RADIUS or Diameter.

This reference point carries the following information:

■ Authentication data for a connection between the WLAN UE and the 3GPP Network
■ Authorization data for connection between the WLAN AN and the 3GPP Network
■ Operator identifiers for the networks involved in the roaming session connection
■ Security keys for radio interface integrity protection and for encryption
■ Optionally, routing enforcement and policy information to ensure that all packets sent to and from the WLAN UE for PS-based services are routed to the interworking Visited Network (roaming case) or Home Network (no roaming case)
■ Purging-user messages from the WLAN access for immediate service termination when it is instigated by network administrator, applications, or user UE
■ Limitations and restrictions on the access scope to the WLAN, based on the authorized services for each user (e.g., IP address filters)
■ Charging data per WLAN user, to enable offline or online charging

To minimize the load on the WLAN AN, the use of online charging over Wa is optional and depends on the agreement between the operators of the WLAN AN and the 3GPP Network.

10.3.3.3 The Wd Reference Point

The Wd reference point connects the 3GPP AAA Proxy, possibly via intermediate networks, to the 3GPP AAA Server. Its main purpose is to transport AAA data and related information in a secure manner. The Wd is an implementation of EAP authentication. It can be RADIUS or Diameter; however, only a single AAA protocol per WLAN session can be supported.

The functionality of this reference point is to transport AAA messages included for carrying:

■ Data for user authentication between the 3GPP AAA Proxy and 3GPP AAA Server
■ Data for service authorization between the 3GPP AAA Proxy and 3GPP AAA Server
■ Charging messages and data per WLAN user
■ Security key data for the purpose of radio interface integrity protection and encryption
■ Authentication data for the purpose of tunnel establishment, and tunnel data authentication and encryption when the PDG is in the Visited Network
■ Mapping a user identifier and a tunnel identifier sent from the PDG to the AAA Proxy through the AAA Server
■ Data used for purging a user from the WLAN access for immediate service termination

■ Data enabling the identification of the operator networks among which the roaming occurs

10.3.3.4   The Wf Reference Point

The Wf reference point is between the 3GPP AAA Server Proxy and the 3GPP OFCS. This interface is used to forward offline charging information to the CDF and the CGF in the OFCS that is located in the Visited or the Home Network. The Wf reference point is identical to the Rf in the 3GPP OFCS specification.

10.3.3.5   The Wg Reference Point

The Wg reference point applies to the link between the 3GPP AAA Server or Proxy and the WAG. This interface is used to:

■ Send policy information to the WAG to perform policy enforcement functions for authorized users
■ Convey charging information per tunnel from the WAG to the AAA Proxy for a roaming scenario only

10.3.3.6   The Wi Reference Point

This is the reference point between the PDG and a PDN. The PDN can be a Public or Private PDN, or an Intra-operator PDN. The Wi reference point is similar to the Gi reference point provided by 3GPP for general-purpose data connection with the PS domain over IP.

10.3.3.7   The Wm Reference Point

This reference point is between the 3GPP AAA Server and PDG, and between the 3GPP AAA Proxy and the PDG. This interface allows for:

■ The 3GPP AAA Server Proxy and the WLAN UE to exchange tunneling attributes and IP configuration parameters
■ Authentication messages to be carried between the PDG and the AAA Server to authenticate the WLAN UE by the 3GPP AAA Server or Proxy, including requested W-APN identity, etc.
■ The 3GPP AAA Server to provide the PDG with the WLAN UE remote IP address that was received from the HSS, when static remote IP address allocation is used
■ The 3GPP AAA Server to provide the PDG with charging data to enable charging for the WLAN session, using, when necessary, any W-APN charging characteristics
■ Carrying data required for tunnel establishment, and tunnel data authentication and encryption
■ Conveying the mapping of user identifiers and tunnel identifiers sent from the PDG to the AAA Proxy through the AAA Server

10.3.3.8   The Wn Reference Point

This is the reference point between the WLAN AN and the WAG. This interface requires routing of packet media flows from the WLAN UE to the WAG over the initiated tunnel. Implementations of this interface vary and depend on local agreements between the WLAN AN provider and the IP CN provider.

10.3.3.9   The Wo Reference Point

The Wo reference point is used by a 3GPP AAA Server to communicate with 3GPP OCS. This reference point is used for transporting online charging-related information that enables, for example, performing credit control for the WLAN subscriber. The Wo reference point is identical to the Ro interface.

10.3.3.10   The Wp Reference Point

The Wp reference point lies between the WAG and the PDG. It relays messages from the access point on the WAG to the network point, in either the Visited Network or internally. Where the WAG and the PDG are not in the same network, additional security is required. This is usually achieved by tunneling and encryption techniques.

10.3.3.11   The Wu Reference Point

The Wu reference point is between the WLAN UE and the PDG. This is the WLAN UE-initiated tunnel toward the PDG, carrying the media flow. Transport of media through the Wu interface is established by the signaling interfaces of the Ww, Wn, and Wp reference points, which ensure that the data is routed via the WAG, where routing enforcement is applied.

The functionality of the Wu reference point is to enable:

- WLAN UE-initiated tunnel establishment
- User data packet transmission within the WLAN UE-initiated tunnel
- Teardown of the WLAN UE-initiated tunnel

10.3.3.12   The Ww Reference Point

The Ww reference point connects the WLAN UE to the W-APN using IEEE 802.1x protocol specifications. This interface transfers:

- Parameters for authentication between the 3GPP AAA Server and the WLAN UE
- Parameters for identification of the operator networks for roaming purposes

10.3.3.13   The Wy Reference Point

The Wy reference point is used by a PDG to communicate with an OCS when both are in the Home Network. This reference point is used for sending to the OCS real-time charging related information about the WLAN session, so as to perform credit control or pre-pay service. The Wy reference point is identical to the generic Ro reference point.

10.3.3.14   The Wx Reference Point

The Wx reference point is between the 3GPP AAA Server and the HSS, and is used for the transfer of user data stored on the HSS to the WLAN AAA Server. This enables:

- Retrieval of authentication data (e.g., for USIM or ISIM authentication) from the HSS
- Retrieval of WLAN access-related subscriber profile information from the HSS
- Registration of a user with the associated 3GPP AAA Server address
- Indication of change of subscriber profile to update the HSS
- Purge procedure between the 3GPP AAA Server and the HSS, terminating WLAN registration
- Retrieval of online charging or offline charging server addresses from the HSS or DNS
- Fault recovery procedure between the HSS and the 3GPP AAA Server
- Retrieval of service subscriptions and related information (e.g., a list of W-APNs that can be selected by the WLAN UE), which can include restrictions of what services can be delivered in the Visited Network.

10.3.3.15   The Wz Reference Point

The Wz reference point is used by a PDG to communicate with an OFCS. This interface is used to transfer charging-related information about WLAN 3GPP IP Access to the OFCS to create CDRs. This interface transfers the same data as the generic Ga interface.

10.3.4   3G WLAN Authentication

10.3.4.1   Authentication Addresses

WLAN Access authorization between the UE and the 3GPP AAA Server is performed before service authorization and transport IP address allocation. The 3GPP AAA Server reads the User Profile and checks the user's subscription. It can consider the identity of the I-WLAN and its location, and ascertains if there are interworking and roaming agreements.

The authentication procedure can involve challenge–response methods when prestored secret information is used to verify the user's credentials.

The UE requests to be attached to an Access Point in the WLAN radio-based IP-CAN. The UE sends over the User Identity and the Realm of its Home Network, which make up the NAI in the format of "User@realm." The "realm" part is utilized to establish whether the user is roaming, i.e., the realm name is not the same as the local network's name, but can identify the user's Home Network. This can be done via a DNS query (to obtain the public name of the server to contact), or matching against a list of networks with whom the operator has an existing roaming agreement.

10.3.4.2 Authentication Security Measures

Many operators prefer, if not mandate, that their subscribers' identities are not revealed to other operators, even when allowing these subscribers to get attached temporarily to a Visited Network. To that effect, the username part is set to a temporary generic name that is conveyed to external networks and translated back for incoming communications.

To protect identities further, the PEAP (Protected Extensible Authentication Protocol) or TTLS (Tunneled Transport Layer Security) can be used to establish a secure tunnel to the Home Network. When using PEAP or TTLS, the exchanged messages are encrypted and are not revealed to eavesdroppers, but are also not visible to the Visited Network.

When the dialog with the Home Network server has been established, the Home Network provides an alias or a temporary account number so that the charged party's identity is never revealed.

10.3.4.3 Full or Fast Re-Authentication

The 3GPP AAA Server must be able to re-authenticate when the link is severed but the UE wishes to continue with the session. The re-authentication can be "Full" or "Fast," depending on the operator's policies.

- *Full re-authentication.* Full re-authentication is starting the process from the beginning, regenerating all the required keys. Full re-authentication requires the WLAN UE to send pseudonym or permanent identity.
- *Fast re-authentication.* Fast re-authentication repeats only part of the procedures. It reutilizes Master Key and Transient EAP Keys. Fast re-authentication requires the WLAN UE to send re-authentication identity. The operator's policies for fast re-authentication include measures to avoid abuse of their resources. These measures could include limiting the number of fast re-authentications and further restrictions for visiting subscribers.

10.3.4.4 Service Authorization

When the user is authenticated (i.e., the user's identity is verified), the 3GPP AAA Server proceeds to authorize the service. The service authorization procedure is independent of WLAN access authentication. To authorize a specific service, the 3GPP AAA Server takes into account the user's information, credit status, what I-WLAN is involved, what Visited Network, what QoS is required, etc. The 3GPP AAA Server can authorize multiple sessions simultaneously.

Having examined the user details, the requested service parameters, and network policies, the 3GPP AAA Server can set up the session attributes. These attributes include, for example (but are not limited to):

- Access scope or service limitation
- Time limitation
- Bandwidth control values
- User priority
- Blacklist
- Credit limit

Further routing policies also can be applied. Such policies might be, for example, "access only through WAG" or "Do not access through WAG."

10.3.4.5   WLAN Authentication Protocols

10.3.4.5.1   Selecting Methods of Authentication

The 3GPP AAA Server should be able to authenticate in both EAP AKA and EAP SIM-based methods to accommodate both NGN and 3GPP terminals. The use of EAP ensures that user identities are not modified or subverted. Diameter is the preferred authentication protocol but existing RADIUS servers can also be utilized.

The WLAN access node initiates a request, using RADIUS or Diameter, toward the Home AAA Server. When the response is received, a challenge is issued (using EAP) toward the WLAN UE. Responses to the challenge are conveyed back (in RADIUS or Diameter messages). If successful, a success response is then launched and transmitted to the UE.

10.3.4.5.2   ISIM, USIM, or SIM-based Authentication

The WLAN UE, even when installed with USIM, must support USIM EAP AKA-based authentication. AKA (Authentication and Key Agreement) is the method commonly used for non-telephony user identification. The WLAN UE will always first attempt to authenticate using the EAP AKA authentication method.

If the WLAN UE supports the ME-SIM (Mobile Equipment SIM) interface, and if SIM has been inserted, then the WLAN UE will support SIM-based authentication and use the EAP method of negotiation. The EAP SIM-based authentication does not require the ME-SIM interface, and therefore EAP–SIM-based authentication could also be performed using the 2G AKA functions on the USIM application. However, if USIM is present, then by policy, the default method is UMTS-based authentication.

When ISIM is available on WLAN UE, ISIM EAP AKA is the favored method. It operates in the same manner as USIM EAP AKA but utilizes the IMS user identifiers.

Figure 10.7 gives an example of call flow for authentication in a roaming scenario. It shows the first step of admission to the WLAN AN being performed using 802.1x with EAP.

10.3.4.5.3   Using Both RADIUS and Diameter

The protocols used for both authentication and charging information are RADIUS and Diameter. Diameter is a more advanced protocol than RADIUS (see protocol descriptions in Appendix A) and is favored for implementation of new IMS networks. However, there are many RADIUS servers already installed, and therefore the standards allow for reuse of these servers. This entails provision for a translation between RADIUS and Diameter, when necessary.

When receiving a RADIUS Request on the Wa reference point, the 3GPP AAA Proxy Translation Agent translates it into a Diameter Request, to be forwarded on the Wd reference point. When receiving a Diameter Response on the Wd reference point, if the WLAN APN supports only the RADIUS-based Wa reference point, the 3GPP AAA Proxy Translation Agent translates it into a RADIUS Response to be forwarded on the Wa reference point.

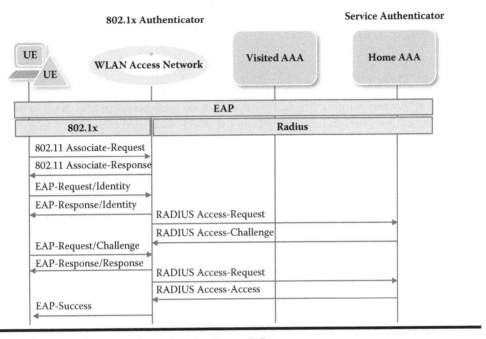

Figure 10.7 EAP and RADIUS in authentication call flow.

Figure 10.8 shows the Translation Agents in the AAA Proxy and the AAA Server. In this example, the Visited Network Proxy in one network can have RADIUS on board, while the Home Network AAA Server can be a Diameter server, or vice versa.

10.3.5 WLAN Charging

10.3.5.1 WLAN Charging Principles

WLAN charging is a two-tier charging system: (1) WLAN Direct IP Access charging and (2) WLAN 3GPP IP Access charging. WLAN Direct IP Access charging is always required for

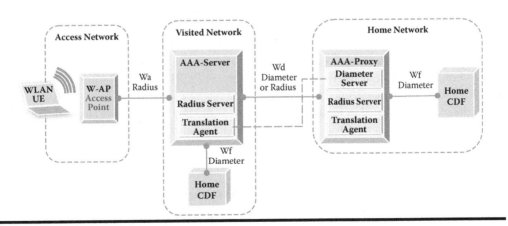

Figure 10.8 RADIUS – Diameter translation.

WLAN connection. On the other hand, WLAN 3GPP IP Access charging is only required when 3G/IMS services are used.

10.3.5.1.1   Direct IP Access Charging

In "Direct IP Access Charging" mode, information from the access network is used to prepare the CDRs. Accounting messages originating from the W-APN are used to report usage and data volumes to charge for events and data flow. Data from the A-WPN is used to construct the WLAN–AN CDR to distinguish it from the 3GPP WLAN PDG CDR.

In generic charging terms, the W-APN acts as a CTF, sending triggers via the AAA Server to start the process of recording charges, on both the OCS and OFCS. If there is an I-WLAN involved, the WLAN–AN CDR can be generated as a result of receiving an Intermediate Accounting Message. Note that the authentication in Direct IP Access Charging mode can still use the 3GPP AAA Servers.

10.3.5.1.2   WLAN 3GPP IP Access

In this mode, charging information is collected from the PDG (not the WLAN APN node) and sent over to the 3GPP AAA Server, via the 3GPP AAA Proxy. The PDG reports charging data, volumes, and durations, per user per connection. The AAA Server also acts as a CTF for particular conditions that are detected only at the AAA Server level.

In this mode, the PDG contains a CTF functionality to trigger the recording of charging details. To capture FBC information or Service Data Flows, the Traffic Plane functionality of the PCEF is built into the PDG. Charging Rules are either pre-defined within the PCEF or downloaded from the PCRF over the Gx interface.

10.3.5.2   Charging for Home Network Services

Where services are provided from the Home Network, both the AAA Server and the PDG are situated in the Home Network, which is in the same network as the OCS and OFCS.

The UE gets attached to the W-APN in the WLAN AN, which links to the WAG, the AAA Server, and the PDG in the Home Network. Both the PDG and AAA Server respond to triggers from the W-APN, which is acting as the CTF.

In parallel, the 3GPP AAA Server also sends charging information, reflecting session-based charging details, to both the OCS and OFCS.

For FBC, the PDG can receive instructions from the PCRF over the Rx interface, or select a charging profile from a predefined built-in list.

In the scenario depicted in Figure 10.9, the PDG is situated in the Home Network. The PDG communicates with both the OCS and OFCS and sends over charging data. This enables FBC to be accomplished in the Home Network.

10.3.5.3   Charging for Visited Network Services

In a roaming situation, the Proxy AAA Server acts as the CTF and the Home AAA Server activates CDR recording according to the received triggers from the proxy AAA.

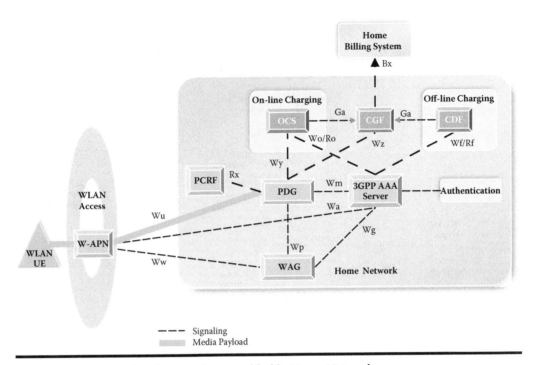

Figure 10.9 WLAN charging, services provided in Home Network.

There is no direct interface between the CDF in the Visited Network and the CDF in the Home Network. In each network, the local CDF forwards the CDRs to the Billing domain where the inter-operator accounting data is exchanged.

The AAA Proxy triggers CDR production in the Visited Network CDF. The AAA Proxy also forwards the Accounting Requests over the Wd reference point to the Home AAA Server, where CDRs are produced on the Home CDF. The transfer of information between the AAA Proxy and the AAA Server over the Wd reference point may necessitate protocol translation between RADIUS and Diameter.

A translation between Diameter and RADIUS may also be required over the Wp reference point between the WAG and the PDG.

Figure 10.10 shows the roaming scenario where the WLAN UE connects via the W-APN to the WAG and to the 3GPP AAA Proxy in the Visited Network. In this scenario, the PDG is situated in the Visited Network. This PDG can create its own CDRs, reflecting Service Data Flow, which then are forwarded to the CGF.

10.3.6 QoS in 3G WLAN Access

10.3.6.1 The Scope for WLAN QoS

Unlike the air interface for Mobile networks, the WLAN airways are not licensed. They are therefore available to anyone and offer additional spectrum and bandwidth, which are particularly useful for expanding Mobile data services. In addition, the WLAN provides mobility for Enterprise solutions for users on the corporate WAN, enabling cordless connections for phones and laptops.

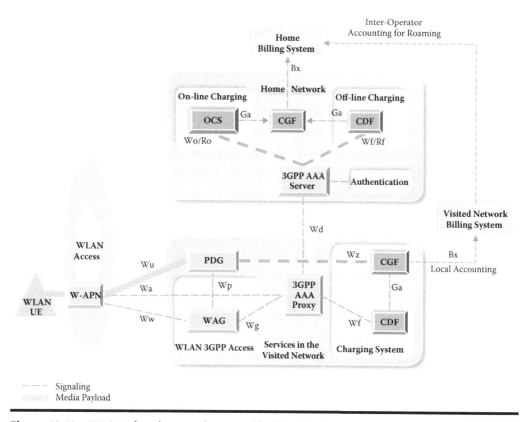

Figure 10.10   WLAN charging, services provided in Visited Network.

The range of multimedia interactive services on WLANs requires QoS management, just as for 3G Mobile handsets and NGN tethered equipment. WLANs can be used without involving QoS management — for example, to connect to the Internet. This is referred to as WLAN IP Direct Access, and uses the "Best-Effort" Class-of-Service. For IMS multimedia sessions, 3GPP WLAN Access can be used to authenticate the user and authorize IMS resources that are needed for a higher level of QoS. The level of QoS also can be reflected in the session charging and is reported within the CDRs.

The QoS requirements for the air interface of WLAN must be compatible with QoS signaling and enforcement on the end-to-end path over the backbone and CNs. This comprises:

■ WLAN IP-CAN between the WLAN UE and the WLAN AN. This is the air interface, with its own constraints of bandwidth capacity.
■ The 3GPP AAA Server and IMS as a signaling network for service based QoS management.
■ The entire 3GPP-WLAN AN between the WLAN UE and PDG, and one or more external bearer networks between the PDG and the Terminating UE. This is the bearer network for the media that needs to enforce the resource management.

In principle, the WLAN QoS must comply with the same rules and policies as other network segments, and therefore should utilize the network's PCC system.

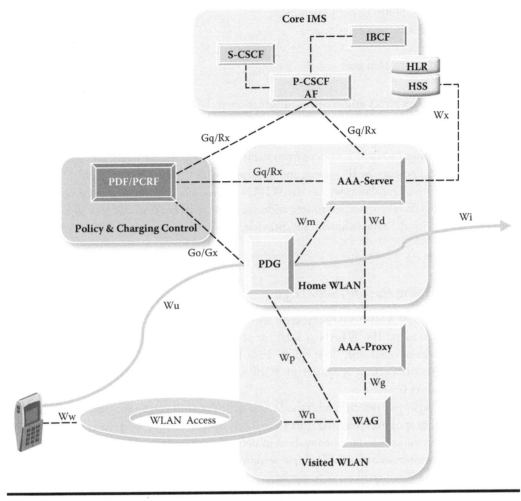

Figure 10.11 WLAN with Multi-Access QoS Management.

Figure 10.11 shows a Mobile WLAN handset attached to a Visited WLAN, through the WAG and the AAA Proxy. The capabilities and restrictions of the WLAN AN are relayed to the AAA Server, to be considered when establishing the session QoS parameters. The PCC system with the PCRF is engaged by the AF in IMS and sends instructions to the PDG to enforce.

10.3.6.2 Enforcing WLAN QoS

Policy enforcement within the Visited Network requires that policy enforcement information is delivered to the WAG. This information includes lists of available PDG servers and WLAN ANs. This information can help in choosing a more secure path.

The Routing-Policy AVP (Attribute Value Pair) is used to transfer policy data to the WAG. The WAG is responsible for enforcing the Routing Policy. The Routing-Policy AVP is used to describe a single IP flow at a time. The AVP defines filters that affect routing according to the following parameters:

- Direction, in (uplink) or out (downlink)
- Source and destination IP address
- Protocol
- Source and destination port

The WAG receives policy enforcement information at the initial authentication, before tunnel establishment. It must be able to associate this information with the authenticated user's traffic. The binding of the policy to a user's traffic flow allows the WAG to recognize authorized packets and drop unauthorized packets sent to or from the user. This can be achieved, for example, by the Visited Network allocating an IP address and binding it with the Home IP address.

Network Address Translators within the WLAN can modify the source address of IP packets from the WLAN UE. The modified source address can be reliably associated to a WLAN UE by the PDG during tunnel establishment and provided to the WAG via the 3GPP AAA Server or Proxy.

10.3.6.3   Utilizing the DiffServ Edge

In WLAN scenarios, it is assumed that there is more than one network to traverse, possibly under different administration. The WLAN is used for access to the Internet in the IP Direct Access mode, which can be satisfied with "Best Effort" basic levels of QoS. In contrast, in 3GPP IP Access mode, the WLAN connection is subject to 3GPP QoS techniques, namely the PCC, which is described in Chapter 6.

The WLAN AN node interacts with the WAG, which in turn connects to the PDG, forming an IP Tunnel. In case of user roaming, this tunnel can be extended beyond the operator's border to the external networks that may be involved in delivering the service. Data flows over WLANs are designed to carry service data for multiple sessions and different services, as long as the general traffic shape and assigned attributes are shared.

Where, for security purposes, packet data within the tunnel is fully encrypted (including packet headers), there is no facility to distinguish between media flows belonging to different services. To distinguish services in the WLAN, DiffServ Edge must be applied at the WLAN UE and at the other end, at the PDG. This mechanism involves an external unencrypted header that contains the DiffServ field with a marker of the QoS classification that is required for the service. Such use of the DiffServ mechanism works well with the GSMA specifications on GRX (IR 34) for exchange of GSM traffic. See details of GRX QoS classification in Chapter 6.

The UE and the PDG are responsible for DiffServ Edge implementation of the correct QoS parameters for the service, that is, setting the DSCP for packets entering the tunnel. Other entities within the WLAN also need to comply.

10.3.6.4   Class-of-Service Tagging for WLAN

The IEEE 802.1d specification is the IEEE standard that addresses, among other things, how to prioritize different classes of user traffic at Layer 2, the Data Link Layer.

The following table provides the traffic priority levels used to distinguish QoS per type of service, according to IEEE 802.1x:

Priority	Class of Service	Example Service
0	Best Effort	Internet
1	Background	File Transfer
2	Standard	Normal Data Service
3	Excellent load	Business Critical
4	Controlled load	Streaming Media
5	Video, Interactive Media	Less than 100-ms Latency and Jitter
6	Voice, Interactive Voice	Less than 10-ms Latency and Jitter
7	Network Control Reserved Traffic	Lowest Latency and Jitter

Source: Based on information in IEEE 802.1d.

Because UMTS identifies four classes and DiffServ identifies four classes, the definitions of seven classes for WLAN should be uniformly translated to achieve interoperability and compatibility. To this end, the WiFi Alliance has defined *four* categories for WiFi Multimedia, based on the IEEE 802.11e standard. These four categories are mapped into the seven classes.

The following table maps these four categories from the WiFi Alliance to the seven IEEE 802.11e classes of service:

WiFi Multimedia Service	IEEE 802.11e Tags
Voice Priority =	7, 6
Video Priority =	5, 4
Best-Effort Priority =	0, 3
Background Priority =	2, 1

See 3GPP TS 23836 for further information. Additional descriptions of QoS management are given in Chapter 6.

10.4   NGN WLAN

10.4.1   *Generic Approach*

As a whole, TISPAN endorses the 3GPP specifications, while making them more generic. The 3GPP Access functions are mapped to the TISPAN generic Access functions.

The requirements for a WLAN station client in an xDSL implementation include:

- Support for either 802.11b or 802.11g
- Support for security mechanisms such as Wireless Protected Access (WPA)/802.1X and WPA2 mode, and WLAN authentication signaling based on EAP, as specified in RFC 3748
- Identifier format compatible with an NAI format as per RFC 2486
- Ability to successfully associate to an open SSID
- Discovery and selection of a Visited or Intermediary network
- User's choice of alternative networks when more than one is available

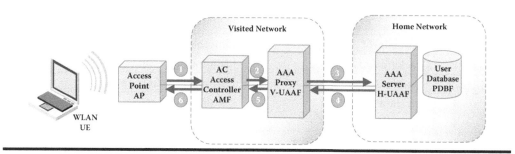

Figure 10.12 DSL WLAN elements.

10.4.2 WLAN Components Mapping in NGN

10.4.2.1 NGN WLAN Network Components

Fixed wireless network clients are implemented on both Mobile handsets and laptop computers. Laptops are considered "fixed" or "tethered" terminals, although they can connect via a wireless interface to an AP (Access Point).

The components of the WLAN connection to an xDSL network can be mapped to the TIS-PAN standard network roles, as shown in Figure 10.12.

The call flow of linking the WLAN terminal to the network follows the steps described below, according to TISPAN-defined call admission procedures:

1. The wireless station (the UE) "discovers" an 802.11x AP and initiates a Connection Request. The AP can be a Home WLAN modem, router, or Edge network receiver. The AP obtains the UE identity from the UE. The AP forwards the UE identity to the AC (Access Controller) that is mapped to the AMF. Either the AP or the AC, or a combination of the two, could implement the AMF functionality.
2. The AMF in the AC formulates an authentication request containing the UE identifier and sends it to the local authentication proxy. This AAA Proxy, which is the same as the 3GPP AAA Proxy, implements the Visited UAAF for nomadic users.
3. If the AAA Proxy is able to authenticate the user credentials, it does so locally, as does the V-UAAF (Visited UAAF). When the AAA Proxy determines that the UE identity does not belong to the local network, it forwards the authentication request on to the AAA Server of the Home Network for the UE, routing to it according to the realm name of the UE.
4. The AAA Server (which also implements the Home UAAF functionality) authenticates the user via an EAP-based challenge–response method that runs end-to-end between the AAA Server and the handset. The AAA Server retrieves the user's special credentials from the database, the PDBF.
5. If the authentication is successful, a session key is returned to the Proxy AAA, the AC, and the AP.
6. The AP blocks any access to the network, or the Internet, until such time that it receives a valid session key, after a successful authentication.

Having made the connection, the AP/AC (the AMF) initiates the charging process via an Accounting "Start" Request, sent through the AAA Proxy to the AAA Server. When the wireless station disconnects and the session terminates, the AP/AC will send an Accounting "Stop" message. Both the AAA-V and AAA-H generate charging records.

For a full description of TISPAN call admission, see Chapter 4.

10.4.2.2 Mapping Generic NGN and WLAN Functions

The WLAN access functions can be mapped to the NGN definition of NASS elements. Both the ARF (the Access Relay Function) that routes the calls in the Transport layer and the AMF (Access Management Function), which negotiates access on behalf of the terminal, perform similar functions to the WAG. The WAG also has the role of filtering unwanted packets and enforcing routing policies. The AMF role also encompasses some of the functions of the PDG, which monitors Service Data Flows and links to external IP networks and the Internet.

The UAAF can operate as Proxy or Server to accommodate roaming. This is equivalent to the relationship between the 3GPP AAA Server and the 3GPP AAA Proxy. This enables the authentication to be performed locally if possible; and if not, the signaling messages are forwarded to the Home Network UAAF for authentication there. The UAAF interrogates the PDBF, which is the NGN Access database containing the authentication details and keys.

10.4.2.3 Mapping NGN and 3GPP WLAN Functions

WLAN access as specified for NGN is very similar to that defined by the 3GPP, and can be mapped to TISPAN NASS. The following is a rough matching of 3GPP and NGN functions:

TISPAN	3GPP WLAN	Functions
AP	W-APN	Access Point, terminating radio signals and connecting to the AN
AC ARF/AMF	WAG/PDG	Provides AAA information for authentication and pre-authorization filtering Routes all media packets, enforces policy, and filters packets
UAAF AAA-V	3GPP AAA Proxy	Proxies user authentication to the Home Network
UAAF AAA-H	3GPP AAA Server	Performs user authentication and service authorization
PDBF/CLF	HSS/HLR	Stores user authentication data and session status

10.4.3 NGN WLAN Authentication

The process of network attachment in NGN includes user identification, gaining admission into the WLAN AN, and then authentication against the user profile and prestored credentials. It is a similar procedure to the process in 3GPP, with its two levels — IP Access and NGN Access.

The WLAN UE is admitted to the network via the W-AP (WLAN Access Point). The particulars of the W-AP through which the UE is connecting can influence the authentication according to whose network it belongs to and the network capabilities. The W-AP sends the information to the WLAN-AC over the e1 interface. The WLAN-AC (mapped to the AMF) contacts the AAA Server (mapped to UAAF), where the authentication is performed.

The process of authentication is performed in NGN, following these steps, as shown in Figure 10.13:

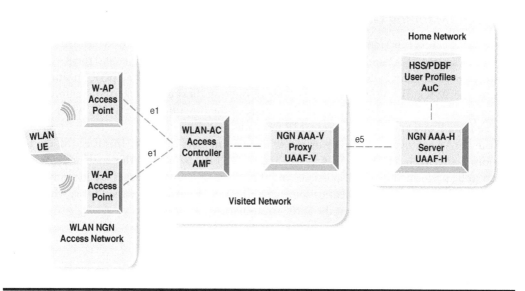

Figure 10.13 NGN WLAN authentication.

■ The UE discovers the AP and requests connection. The W-AP identity can affect the authentication terms.
■ The AP forwards the request via the WLAN AC to the AAA-V.
■ The AAA-V checks credentials. If the user is not found, the request is forwarded to the AAA-H of the user's realm.
■ The AAA-H performs end-to-end authentication using EAP to the UE. Credentials are checked against the data in the PDBF, the NGN authentication database.
■ When the W-AP is notified of a successful authentication, it configures session keys for the Media layer.
■ The W-AP sends, via the AAA-V, accounting messages to the AAA-H.
■ Both the AAA-V and AAA-H generate charging records.
■ The AMF, Access Controller, W-AP and the ARF can be implemented in a single physical node.

10.4.4 NGN WLAN Access Reference Points

10.4.4.1 The e1 Reference Point

The e1 reference point is between the UE and the Visited NGN network and is specific to the type of access. This interface allows the UE to request allocation of an IP address and access Network Configuration servers (DNS, DHCP) to gain entry to the WLAN Acess Network.

This interface also is used to request authentication and authorization from the AMF in the NASS, passing to it user credentials. This interface allows the NASS AMF to provide authentication parameters to the UE for a mutual authentication procedure, if required. The AMF notifies the UE over this interface whether or not the authentication was successful. If successful, this interface provides information on node IP addresses and configuration to the UE.

10.4.4.2 The e5 Reference Point

The e5 reference point is between the Visited Network and Home NGN network, and is network independent. It is intended to connect the UAAF Proxy and UAAF Server when they are in different operators' domains. This is a secure interface that relays Authentication messages to and from the Home Network as well as Accounting data received from the AMF. This interface can be based on either RADIUS or Diameter. Both protocols must be supported to enable easier migration from the existing RADIUS technology toward Diameter.

10.4.4.3 The a3 Reference Point

The a3 reference point is the interface between the AMF and the UAAF Proxy. It transfers Accounting data as well as Authentication messages.

Appendix A

The IMS Protocols

This appendix gives a brief description of the main protocols used in IMS. To understand how the functions interoperate, it is necessary to gain at least a high-level understanding of the protocols.

The IMS protocols have not been created for IMS. Preexisting protocols, defined by the IETF, ITU, and IEEE, are generally used for various areas of IMS. In many cases, special extensions were needed to deal with additional functionality. These extensions have been proposed to and recorded by the originating organizations, and have become part of the standard protocols.

A.1 The Session Control Protocols

A.1.1 SIP

SIP was created with Internet peer-to-peer interactive communication in mind, where much of the intelligence is found at the edge. It was defined as a flexible and extensible protocol that is particularly suitable for multimedia and mixed-media sessions. SIP was established first in 1996 by the IETF as a simple means of exchanging multimedia contents. It was soon adopted for Internet VoIP (peer-to-peer), and the first standard was defined in 1999 via RFC 2543. This was superseded by a more advanced definition in 2002 via RFC 3261. Many extensions followed, including extensions for IMS.

A.1.1.1 The Scope and Structure

SIP is primarily based on HTTP (Hyper Text Transfer Protocol), the Internet protocol. It usually runs over UDP, like most Internet functions, but can also run over TCP, which is more reliable and provides error-correcting facilities.

SIP is a message-based control protocol, leaving media to be handled by other protocols, such as RTP/RTCP (for voice and video session payload) and SCTP (for streaming media).

SIP messages are used to establish consensus regarding session attributes. This is aided by a text-based description message body, utilizing the SDP Protocol.

The SIP protocol is constructed of the following layers:

- *Syntax and encoding.* This is checked first according to syntax defined in RFC 3261.
- *Transport.* This deals with the connection details, sockets, and ports.
- *Transaction.* This ensures the sequence of responses to each request, including errors and time-outs.
- *User transaction.* This is client and server interaction dialog, creating and handling transactions to manage the session.

A.1.1.2   SIP Server Functions

SIP message sequences define roles of network entities, as follows:

- *User Agent Client (UAC).* This element is also called the User Agent Client — the application that initiates requests, usually from a handset client or a soft client in the UE.
- *User Agent Server (UAS).* This is a network server that receives SIP messages from SIP clients, and accepts, rejects, or redirects messages. In essence, this server terminates messages and acts upon their instructions. This server is responsible for the session and maintains its state.
- *Proxy Server.* This is the network's "well-known node," which is first to receive communication messages into the network and then forwards them to the SIP Server. All incoming messages to the endpoint also go through the Proxy. Proxies can interpret and check information but generally forward messages to other servers and do not keep session status. A Forking Proxy is one that relays the messages to several servers at once.
- *B2BUA.* This special server acts as an initiating and terminating server at the same time (i.e., acts as UAC and UAS together). It is often used to reinitiate messages and replace the original requests as appropriate or as a result of an intermediary service logic or data processed. This type of server usually maintains session state and remains in the call flow even when forwarding it further.
- *Redirect Server.* This is a specialized network server that maps requested addresses to network addresses and defines the route to take.
- *Location Server.* This server contains the IP addresses of registered SIP users.
- *Registrar Server.* This server accepts registration requests from UACs, and gets the Location Server to allocate an IP address. The Registrar also maintains knowledge of user status.
- *Presence Server.* This server allows external entities to inquire on the status of users and entities.

A.1.1.3   Using Domain Names

The domain can be referred to by a commonly known name, such as "internet.com" or "network. org." This domain name can be resolved by a DNS server that is able to provide the routes to the appropriate domain, and using DNS "SRV" to locate the particular server.

SIP allows entering a specific domain IP address or a server identity such as "sip:user. name@123.456.78.90" or "sip:user.name@server123," where "123.456.78.90." and "server123" identify the domains uniquely. This type of addressing means that a DNS server enquiry is not required. However, in doing so, the flexibility to use alternative servers (for redundancy and load balancing) is denied.

A.1.1.4   SIP URI and AOR

SIP addressing uses the Internet URI — user.name@domain. This term has become synonymous with URL (Uniform Resource Locator). For SIP messages, either "sip" or "sips" (secure SIP, using TLS) must precede the URI, e.g., "sips:user.name@domain." URIs can have additional parameters that help recognize the user and set preferences.

The username can be a user-selected identifier as agreed to with the Service Provider, or a network given name, for example, allocated to a server. The domain identifies an "area" within the Service Provider's network. Once the domain is identified, the user part of the URI can be determined within the destination network; therefore, the domain reference must be a globally unique address.

Where the URI is a long-term address that is not tied to a single endpoint device and is published as a known address, it is referred to as an Address of Record (AOR). This can be the contact name by which the user is known and therefore used by callers to launch calls toward that user. Other URIs may be only known internally by the network. An AOR is the contact name that can be enquired upon in DNS, to find any addresses that are associated with this user.

A.1.1.5   Telephone Numbers in IP Networks

Telephone numbers can be utilized in IP networks using either SIP URI or Tel URI.

In the SIP URI, the telephone number appears as a user part identifier, where the domain must be specified separately and the fact that this is a number must be conveyed in a parameter. For example, "sip:+44111133445566@Homenetwork.com;user=phone." The user part "+44111133445566" can be used to retrieve an address via the ENUM algorithm that generates a unique IP address. However, this user part can be any digit-based identifier, while the domain name must always exist.

Alternatively, telephone numbers can be used via the format of "tel:" as described in RFC 3999. This format is independent of SIP and can be used by other protocols, such as H.323. This format allows the use of:

- *Global numbers* using the ITU-T defined E.164 scheme preceded by a "+" in the format of, for example, "tel:+44-1111-123450" (where the "-" is a separator).
- *Local numbers*, usually short codes, when they are understood within that network, accompanied by a "phone-context" parameter that provides routing information toward that network, for example, "tel:888;phone-context=mydomain.com." Local numbers are, for example, service codes (fault line, directory inquiries) or emergency numbers (112, 911, 999) that cannot be described in the globally routable E.164 format.

The SIP UA (in the handset or a CPE gateway) interprets the dialed numbers according to a digit map and numbering plans to create the Tel URI.

A.1.1.6   SIP Session State

SIP servers maintain state at several levels:

- *Dialog state.* This is a process that retains awareness of the entire dialog with multiple messages between the parties, from initiation (INVITE) to final teardown (BYE).
- *Transaction state.* This process maintains "state," that is knowledge of a sequence of related messages that may constitute a logical part of a session, usually ending in "OK" or a failure message.
- *Stateless.* This process retains no knowledge of sequences of messages.

A.1.1.7   The SIP Session

A session consists of dialogs and media flows to all participants. A session is initiated by an INVITE request. The type of connection between the parties must be negotiated according to respective equipment capabilities as well as applications and preferences. This is performed by SIP carrying SDP between the parties.

When the INVITE finally receives an ACK, the media can start flowing. The originator's UAC can CANCEL at any time, and either party can terminate by sending BYE. An INVITE request can result in multiple dialogs being created, with each dialog defining its own media flow through a separate SDP negotiation.

A.1.1.8   The SIP Dialog

The SIP dialog manages the relationship between session parties. The dialog is tracked through the Call-ID unique identifier, together with the local tag (in the originator's FROM header) and the remote tag (from the responder's TO header).

Record-Route headers can be added as the initial request is processed, in order to add servers that are responsible for intervening applications to be aware of the session. The serving server identity can be changed when the CONTACT header contains a different SIP URI. These headers, in their specific order, constitute the "route set" for the dialog.

Destinations can be altered as a result of additional logic that causes a server to issue an UPDATE request with another SIP URI target or an alternative INVITE.

The level of security is determined according to TLS being specified in the headers and Secure SIP in the Request start line (SIPS URI), and the security flag is set accordingly.

New requests within the dialog are populated with this "dialog state" data to achieve continuity.

A.1.1.9   The SIP Message

The SIP Message consists of the Start Line, several Headers, and the Message body.

- *The Start Line.* The Start Line contains:
  - Method name
  - Request URI
  - Protocol version, separated by a space
  - Methods to define the type of Request: INVITE (to start a session), BYE (to terminate a session), CANCEL (the request), ACK (acknowledge), REGISTER (log in user), and

OPTIONS (to query servers and their capabilities). The main methods and the defining RFC are shown below:

Method	References
ACK	RFC 3261
BYE	RFC 3261
CANCEL	RFC 3261
INFO	RFC 2976
INVITE	RFC 3261
MESSAGE	RFC 3428
NOTIFY	RFC 3265
OPTIONS	RFC 3261
PRACK (Provisional ACK)	RFC 3262
PUBLISH	RFC 3903
REFER	RFC 3515
REGISTER	RFC 3261
SUBSCRIBE	RFC 3265
UPDATE	RFC 3311

- Responses have a different Start Line from those in SIP Requests — they are Status Lines. They contain 3 digits Status Code (e.g., 180 or 200) and Reason that describes the status response.
- Status codes are defined within classes of codes, as follows:

Code	Class	Function	Examples
1xx	Provisional information	Request has been received and is being processed	100 Trying 180 Ringing 183 Session in progress
2xx	OK	Successful, understood, accepted	200 OK 202 Accepted
3xx	Redirect	Redirect or revise request	300 Moved 305 Use Proxy
4xx	Client error	Error detected by the Client, syntax error	401 Unauthorized 404 Not found 415 Unsupported media type
5xx	Server error	Error detected by the Server, cannot fulfill a valid request	501 Not implemented 504 Server timeout
6xx	Global network error	Request cannot be fulfilled by any server	600 Busy everywhere 603 Decline

Source: Based on IETF RFC 3261.

- *The Headers.* Each message has a header with navigational information. Headers can be:
  - General Headers: used generally in all messages (e.g., TO: and FROM:)
  - Request Headers: used to add information about the request (e.g., PRIORITY for emergency, etc.)
  - Response Headers: used to pass information that cannot be included in the status line (e.g., UNSUPPORTED).
  - There can be several Headers in each message. Some headers are mandatory for a specific dialog stage and others are optional. The following table shows some of the main headers:

Header	Syntax/Examples	Comments
To	**To: SIP-URI; (parameters):** To: sip:+4412345678@mydomain.com To: sips:secure.server@mydomain.com>	Sent to the AOR
From	**From: SIP-URI; tag=123456; (parameters):** From: sip:myname@mydomain.com; tag=1234	Sender's address (AOR) with the server's address plus a tag to identify the dialog
Call-ID	**Call-ID: xxxxx; (parameters):** Session123456@999.999.999.99	Unique ID that identifies the session in subsequent dialogs
Cseq	**Cseq: digit Method; (parameters):** Cseq: 6 INVITE	The digit identifies the transaction within the Request for the same method, and therefore marks the point in the dialog
Via	**Via: SIP/2.0/[bearer] sent-by;** **branch=z9hG4bK.....;received=999.9.99:** Via: SIP/2.0/TCP example1server.com; branch=z9hG4bOfficename;received:192.0.2.1	Protocol version and bearer protocol (UDP, TCP, TLS), plus address of servers that are required to be in the path, in the order of the given "via" headers
Max-Forward	**Max-Forward: (digit):** Max-Forward: 70	Limits the number of hops per Request to avoid endless looping
Contact	**Contact: SIP-URI; (parameters e.g.** **expires=99):** Contact: sip:myname@999.999.99.9.99	Originator's address, or multiple AORs to register several devices
Request-URI	**Request-URI: SIP-URI ; (parameters):** Request-URI: sip: to-name@ server-address	Requesting a URI for a function (e.g., for registration, the Registrar's address is returned)
Allow	**Allow: Method, Method, Method...:** Allow: INVITE, ACK, BYE, CANCEL, REFER, NOTIFY, SUBSCRIBE	Restricting what methods can be allowed; for example, may omit the INFO message that is deemed too open and risky
Record-Route	**Record-Route:<sip:SIP-URI; (parameter)>:** Record-Route:sip:server1.name@ domain1;lr	Determining which servers remain in the signaling path, for security filtering or for monitoring events that may trigger an action

Source: Based on IETF RFC 3261.

■ *The Message body.* The payload of the message can carry any textual information. The Start Line, Headers, and Status Code influence how the message body is used. The SIP message body for establishing a session is governed by the SDP.

A.1.1.10 Session Description Protocol (SDP)

SDP is used to negotiate multimedia session attributes between the parties. These attributes can be the type of media to be used, what codecs, etc. This is a textual format rather than a full protocol. Its format is a parameter identifier, which is always a single alpha character = <value><parameters>.

SDP works in offer/counteroffer style (RFC 3264) until an agreement is reached. The Offer contains media stream type, available codecs, and IP addresses and ports to be used. The Answer indicates which media types are acceptable, which codecs are preferred, and the IP address and ports to be used for receiving media. Answers must be given, even if only to reject, and must be received before either party can generate an alternative offer.

SDP manages two types of information:

1. Session level description, including session identifiers and parameters, subject, contact information, and timing description, such as start and stop times and repeat times
2. Media type and format, including transport protocol and port number, encryption key, bandwidth information, and other media attributes

The parameters s-lines (session name), t-lines (time the service is active), and m-lines (media description) must appear in this sequence in SDP lines.

The m-lines format is:

$$m=<media> <port><transport><format-list>$$

Each media type (audio, video) is specified with its own port number and transport type, e.g., UDP or RTP (Audio/Video Profile). The Format-list contains further media information.

The a-line enables defining further attributes. Through its open nature, it helps extend the current parameter range. The format is simply:

$$a=<attribute field>: <value>.$$

A.1.1.11 SIP Extensions for IMS

3GPP developers have contributed many new RFCs and extended RFCs to the IETF as part of the collaboration between the two organizations. These extensions cover, among others, the following topics:

■ Aspects of mobility
■ Location plus Presence
■ Subscription and admission control
■ Interfaces to circuit-switched networks and CAMEL services
■ End-to-end QoS and Dynamic Policy Decision
■ Specific services such as Messaging
■ Conferencing and POC
■ Management and charging interfaces
■ Security and filtering

A good mechanism for carrying additional information that is needed for IMS functions (such as security, session routing, and charging) is the use of "private headers" — the P-Header (as specified in RFC 3455). Examples of P-Headers are:

- P-Associated-URI
- P-Called-Party-ID
- P-Visited-Network-ID
- P-Access-Network-Info
- P-Charging-Function-Addresses
- P-Charging-Vector

Using these fields enables the passing of essential information between processing nodes. For example, P-Assert-Identity is used for ensuring that the right identity is forwarded, and P-Charging-Vector is used to propagate the charging association.

Another useful header is the Service-Route header, which is used to establish the path of the session through specific proxies, to enable provision of a specific service. This Service Route discovery is performed during registration.

The following SIP extensions of P-header require support on both IMS terminals and IMS servers:

- P-Access-Network-Info, which carries information of the AN from UE to IMS and from IMS to AS, but only to "trusted" AS
- P-Asserted-Identity, which delivers the ID to be asserted where IDs must not be changed or where a different ID should be presented
- P-Called-Party-ID, which carries the target public user identity from IMS to UE, is revealed to the destination AS (of the called party) but not to proxies
- P-Charging-Vector, which helps to correlate associated records for charging
- P-Charging-Function-Addresses, which provide charging servers addresses

Support is required for the following P-headers in the IMS UE, but is not needed in the AS:

- P-Associated-URI, which carries associated URIs to the registered public user identity from IMS to UE
- P-Media-Authorization, which carries media authorization token from IMS to the UE
- P-Preferred-Identity, which carries identity preferred by the user from the UE to IMS

Support for the following P-headers is needed in the AS and Server, but not in the UE:

- P-Asserted-Identity, which carries valid and authenticated public user identity from IMS to AS
- P-Charging-Vector, which carries charging correlation information from IMS to the AS
- P-Charging-Function-Addresses, which carries information of offline and online charging function addresses from IMS to the AS.

The IMS facilities for authentication and security also are based on IETF RFC extensions. This includes Digest AKA authentication. The extension in RFC 3310 provides mapping of AKA parameters onto HTTP Digest authentication method, which enables a one-time password generation mechanism for Digest Authentication.

The following table provides a summary of Modified SIP messages for IMS in the IETF RFCs:

Message	RFC
UPDATE	RFC 3311
Preconditions	RFC 3312
PRACK	RFC 3262
QoS/Media Auth	RFC 3313
Events	RFC 3265
Tel URI	RFC 2806
P-Header (private)	RFC 3455
Service route	RFC 3608
Asserted ID	RFC 3325
DNS support	RFC 3262
SigComp	RFC 3320-2/3485/3486
ENUM	RFC 2915/2916
REFER	RFC 3515
Digest AKA	RFC 3310
Path-header	RFC 3327
Sec-agree	RFC 3329
IPv6 support	RFC 3314/3315/3316
Diameter 3GPP AAA	RFC 3589
RTP payload 3GPP format	RFC 4396
3G transition Scenarios	RFC 3574
MIME for 3GPP	RFC 3839/3842

Note that this list is a snapshot in time, as work is continuing in this area.

A.1.2   MGCP and Megaco/H.248

Media Control protocols (MGCP and H.248) provide media-specific commands that can manage media resources more precisely than general session control protocols such as SIP or H.323. Instead of treating the media processor (media servers or media gateways) artificially as another party in the session, the media resources are seen as subservient to the session logic.

A.1.2.1   Media Gateway Control Protocol (MGCP)

MGCP has been favored by many softswitches as a means of controlling POTS phones behind CPE gateways. MGCP defines the interaction of terminals (with built-in media processors) in a call model that expects the service logic to be external to it. The terminal media processor is, therefore, driven by the external logic of the media session controllers in a master/slave mode.

MGCP defines "endpoints" as objects that are logical (virtual entities such as soft phones) or physical entities (such as phones or channels). Endpoints can "source" (generate) or "sink" (receive) media streams. The endpoint detects events and signals (i.e., phone "off-hook" or phone ringing).

MGCP also defines "connections" for the media flows. These connections may be point-to-point between two endpoints, or multi-point with several endpoints. Connections are established over different types of bearers (IP, ATM), using RTP for media and UDP/TCP over IP as signaling protocols.

A.1.2.2 Megaco/H.248

Megaco/H.248 is the media control protocol adopted by IMS. This protocol evolved from MGCP. It is the product of cooperation between the IETF and ITU. Megaco is defined by IETF RFC 3525, while H.248.1 is an ITU recommendation.

H.248 has formalized the concepts of MGCP and extended the protocol to multimedia and conferencing. The Transport layer can use TCP as well as UDP, and command syntax can be provided in binary as well as in text.

H.248 is the protocol used to control media functions for session interactions with CSNs. In IMS, H.248 is used by the MGCF to control the media gateways in the Mc reference point, and also in the Mn reference point between MRFC and MRFP.

The terms in Megaco/H.248 are similar to MGCP. These terms include "terminations" that represent objects involved in the media flow, "context" that represents the media stream, and "descriptors" that define parameters.

A termination (phone or an MRF "object") can source (originate) or sink (terminate) a media stream. The H.248 commands manipulate logical entities (terminations) within a "context." Descriptors contain variable parameters that describe the terminations and are included in the protocol command messages.

A "context" is a logical association of two or more terminations involved with a media stream on a media processor (media gateway or media server). The context describes the topology and media mixing parameters for the associated terminations. The context is uniquely identified by a ContextID. A context can have more than one media stream flowing between terminations, corresponding to the descriptors.

A "topology descriptor" specifies various aspects of the media flow, such as the flow direction, i.e., one-way, both ways, or isolate ("no way"). By default, all terminations associated in a context receive the same media flow.

A.1.2.3 H.248 Commands and Descriptors

The following commands are available in H.248:

Command	Usage
Add	Add a termination to a context, or implicitly create a context for the first termination
Modify	Modify the state of the termination or the characteristics of the media stream
Subtract	Subtract (remove) a termination from the context, and delete the context with the last termination

Command	Usage
Move	Move a termination from one context to another
AuditValue	AuditValue to obtain current termination properties, events (e.g., off-hook, on-hook), signals (generated media, e.g., tones, announcements)
AuditCapabilities	AuditCapabilities to return all allowable values for the termination's parameters, signals, and events
Notify	Notify to report events to the media controller.
ServiceChange	ServiceChange to inform status changes of the media resource to the media controller (e.g., when the MGW is taken offline or put online again).

Source: Based on information from RFC 3525.

For each command, there may be a number of descriptors (parameters). They include:

Descriptor	Usage
DigitMap	A pattern for sequence of events that, when detected, is notified
Audit	To define what information is requested by the AuditValue
EventBuffer	For using a buffer to capture events
Local	The properties of a media flow received from a remote end
LocalControl	Properties that will be exchanged between MGW and MGCF
Modem	Type and information of the modem capability
Mux	Information about terminations forming the input mux
ObservedEvents	To report on events
Packages	To enable grouping together information of events, signals, and statistics for a termination
Remote	Media flow properties for media sent toward the remote end
ServiceChange	Information of the changed service status
Signals	Signals applied to a termination (e.g., tones, ringing)
Statistics	Historical data kept for a termination
Stream	Descriptors for a single media stream, including Local Remote LocalControl
TerminationState	The termination properties
Topology	Information of the context (e.g., direction)

Source: Based on information from RFC 3525.

A.2 Data Management Protocols

A.2.1 Diameter

Diameter is the protocol used for IMS/NGN for AAA (Authentication, Authorization, and Accounting). It is intended to work both in local networks and in roaming situations.

Diameter is specified primarily by the IETF in RFC 3588. The name is a pun on the name of the predecessor protocol, RADIUS — a diameter is twice the radius. Diameter is not directly backward compatible but does provide an upgrade path for RADIUS.

A.2.1.1 Diameter Usage Areas

Diameter usage is defined for the following types of applications:

- IPv4 Mobile: access for mobile handsets in IPv4 packet networks (i.e., IMS).
- NASREQ (Network Access Server Request): access to terminals via a server, generally used for peer-to-peer IP data.
- Basic Accounting: recording accounting information and exchanging such information between networks and service providers.

A.2.1.2 The Requirements Beyond RADIUS

While RADIUS is widely implemented and operating successfully, its inherent shortcomings for multi-access advanced networks and for roaming scenarios have necessitated the evolution toward Diameter, a more powerful protocol.

RADIUS falls short of providing full AAA for the NGN IMS network, which is access-agnostic and globally interconnected. RADIUS is vulnerable to attack from external parties and is susceptible to fraud perpetrated by users, service providers, and roaming partners alike. On the other hand, Diameter enables the new networks to facilitate roaming with an appropriate level of security and with the essential means of cross-accounting.

Diameter introduces the concept of proxy chaining via intermediate servers. By providing explicit support for inter-domain roaming, message routing, audit ability, and transmission-layer security features, Diameter addresses the major limitations of its predecessor, RADIUS.

The following are requirements satisfied in Diameter that were not satisfied by RADIUS:

- *Standard failover:* to achieve common behavior at failure time through the same implementation. Diameter supports application-layer acknowledgments, and defines failover algorithms and the associated state machine.
- *Transmission-level security:* to provide universal support for transmission-level security and enable both intra-network (roaming) as well as inter-network AAA deployments. IPSec support must be mandatory in Diameter, and TLS support is optional.
- *Reliable transport protocol:* to achieve consistent reliability and predictable behavior, Diameter uses the more reliable transport mechanisms of TCP or SCTP rather than UDP (used by RADIUS), which tolerates errors without correction. A dependable error-correcting transport protocol is essential for accounting data where packet loss can translate directly into revenue loss.

- *Agent support:* to streamline AAA server behavior, it should be standardized. Diameter supports defined types of agents, including Proxy servers, Redirect servers, and Relay servers.
- *Server-initiated messages:* this feature is essential for unsolicited messages, disconnections, re-authentication, and re-authorization on demand, across a heterogeneous deployment.
- *Audit ability:* data-object security mechanisms should be supported to prevent untrusted proxies modifying attributes or even packet headers without being detected. Audit ability is also important to track what transactions occurred in the event of a dispute.
- *Capability negotiation:* enabling clients and servers to negotiate capabilities to determine a mutually acceptable service, and be aware of what service has been implemented.
- Dynamic peer discovery and configuration: This automates configuration and reduces the administrative burden. Manual configuration is laborious and creates the temptation to reuse the RADIUS shared secret, which can result in major security vulnerabilities, if the Request Authenticator is not globally unique. Through DNS SRV and NAPTR, Diameter enables dynamic discovery of peers. Derivation of dynamic session keys is enabled via transmission-level security.

A.2.1.3 New and Enhanced Diameter Features

Diameter also includes:

- Larger address space for AVP and identifiers (32-bit instead of 8-bit)
- A client/server protocol, also supporting some server-initiated messages
- Use of both stateful and stateless models
- Support for application layer acknowledgments
- Definition of failover methods and state machine
- Error notification
- Extensive roaming support
- Basic support for user sessions and accounting
- Backward compatibility with RADIUS, to smooth over migration to Diameter

A.2.1.4 Diameter Accounting

Both the Accounting and Authentication facilities are native parts of Diameter. The accounting information handled by Diameter is conveyed in real-time. Batch accounting is not a requirement for Diameter. CMS security may be optionally applied to the accounting messages to secure this sensitive data.

The basic protocol comes with a set of AVPs (Attribute Value Pairs), although every application can specify additional ones. The basic AVPs provide means of associating accounting messages via the Session-ID AVP and perhaps the Accounting-Sub-Session-ID AVP. Accounting-Multi-Session-Id AVPs can also be used.

A.2.1.5 The NASREQ Diameter Application

The NASREQ application requirements include addressing transport, scalability, server failover, security, authentication, authorization, policy management, resource management accounting, and supporting a roaming environment.

Diameter is coupled with EAP to provide flexible and secure authentication, even when roaming. The NASREQ application defines the Diameter-EAP-Request and Diameter-EAP-Answer messages that allow encapsulation of the EAP payload within the Diameter protocol. The NASREQ application's AA-Request message corresponds to the RADIUS Access-Request. The AA-Answer message corresponds to the RADIUS Access-Accept and Access-Reject messages.

The NASREQ application also supports interaction between RADIUS and Diameter services via a RADIUS-Diameter protocol gateway, that is, a server that receives a RADIUS message that is to be translated and transmitted as a Diameter message, and vice versa.

A.2.1.6 Diameter Mobile IPv4

The Diameter Mobile IPv4 application adjusts Diameter for a mobile network use. The Diameter Mobile IPv4 application cannot be used with the Mobile IPv6 protocol.

The Diameter Mobile IPv4 is compatible with CDMA2000 Wireless Data Requirements for AAA. This includes Mobile IP AAA Requirements. In particular, the Diameter Mobile IPv4 application, with its extensions, can support roaming facilities across administrative domain boundaries with dynamic assignment of Visited Network and Home Network agents and better scaling of security associations.

The Mobile IPv4 application also enables the Diameter AAA Server to act as a Key Distribution Center, which creates and distributes session keys to mobile network entities during a secure Mobile IP Registration procedure. The procedure involves sharing a secret between the AAA Server and the Network Entity, which forms a security association and is used to derive the security key.

A.2.1.7 Diameter CMS

For ensuring security between two Diameter peers or hop-by-hop, the Diameter protocol can run over IPSec or TLS. However, these security functions cannot operate when the Relay or Proxy Diameter agents are utilized.

The Diameter CMS (Cryptographic Message Syntax) provides end-to-end protection at the AVP level, with authentication, integrity, confidentiality, and non-repudiation. The Diameter CMS security application defines the Diameter messages and AVP used to establish a security association between two Diameter nodes, and the AVP used to subsequently carry secured data within Diameter messages.

CMS is delivered via two primary techniques:

1. Digital signatures, along with digital certificates, provide authentication, integrity, and non-repudiation.
2. Encryption provides confidentiality.

Each AVP can be digitally signed or encrypted. The techniques can be used simultaneously. Diameter proxies can add, delete, or modify unsecured AVPs in a message.

A.2.1.8   Diameter Agent Servers

Diameter defines a set of agents (servers) that are involved with passing Diameter messages. These agents can distribute Diameter processing and can direct messages according to specific network topology and special security associations. The various server types can help in load balancing, managing multiple authentication sources, and improving the overall performance of Diameter.

Diameter specifies that agents keep transaction state but not all of them need to keep session state. Maintaining transaction state is important for errors and failover processes. Transaction state is updated at every step and the hop-by-hop identifier is saved. The request's state is released upon receipt of the answer. A stateless agent is one that only maintains transaction state at any point.

Four types of servers have been introduced: Relay, Proxy, Redirect, and Translation agents:

1. *Relay agents* route messages to other Diameter agent servers. Rerouting is performed based on received information (e.g., destination realm processed against stored lists in the Diameter Routing Table).
2. *Proxy agents* also route Diameter messages using the Diameter Routing Table but they modify the Diameter messages to implement policy enforcement. Proxy agents, therefore, must maintain knowledge of the downstream peer nodes (e.g., access devices) to enforce resource usage and provide provisioning and admission control.
3. *Redirect agents* can be used for centralizing some processes, hosting AAA services, etc. A Redirect agent does not relay the messages but merely returns information that enables direct linking with the target server. Such servers do not need to maintain state and are said to be stateless.
4. *Translation agents* are introduced to enable interworking between RADIUS and Diameter, and coexistence of both in one network. Translation Agents can be used as aggregation servers to communicate with the new Diameter infrastructure, linking existing RADIUS servers that are yet to be updated. Because Diameter caters for long-duration multimedia sessions, the Translation agents must be session stateful.

A.2.1.9   Diameter Message Format

A Diameter message consists of a Message Header and a Message Payload. The header contains the command code and message parameters, length, version number, and flags. The message payload contains the AVP codes, parameters, and flags as well as any data to be transferred.

There are two types of messages: Requests and Answers. Every answer message carries a Result-Code AVP. The data value of the Result-Code AVP is an AVP identifier that conveys whether the request was completed successfully or whether an error occurred.

The header must contain a valid Diameter Command Code. The Application-ID can be Mobile IP, NASREQ, or Basic Accounting. Figure A.1 maps out the structure of the Diameter message.

The Command Flags field is 8 bits. The following bits are assigned:

0	1	2	3	4	5	6	7
R	P	E	T	r	R	R	R

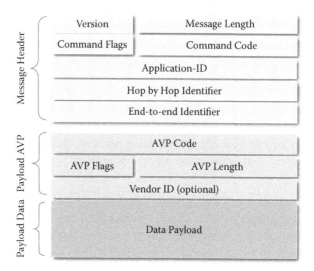

Figure A.1 Diameter message structure.

- *R(equest)*. If set, the message is a request. If cleared, the message is an answer.
- *P(roxiable)*. If set, the message can be proxied (forwarded), relayed, or redirected. If the message is not forwarded in any manner, it must be processed locally.
- *E(rror)*. If set, the message contains a protocol error. This means that the message is not conformant for this command (i.e., as commonly referred to, it is an error message).
- *T(ransmit)*. Potentially retransmitted message. This flag is set after a link failover procedure, to help recognize a retransmission as a duplicate and not a new request. It is set when resending requests that are not yet acknowledged, after a link failure.
- *r(eserved)*. These flag bits are reserved for future use, and must be set to zero and ignored by the receiver.

The AVP part of the message in the payload describes a variable-length string. The AVP Code, combined with the Vendor-Id field, identifies a specific attribute. AVP numbers 1 through 255 are reserved for backward compatibility with RADIUS, without setting the Vendor-Id field. AVP numbers 256 and above are used for Diameter and are allocated by IANA (Internet Assigned Numbering Authority). The AVP Flags field informs the receiver how each attribute must be handled. The "r" (reserved) bits are unused and set to 0.

0	1	2	3	4	5	6	7
V	M	P	r	r	R	r	R

The "P" (proxiable) bit indicates the need for encryption for end-to-end security. The "M" bit, known as the Mandatory bit, indicates whether support of the AVP is mandatory. It is set according to the rules defined for the AVP containing it. The "V" bit, known as the "Vendor-Specific" bit, indicates whether the optional Vendor-ID field is present in the AVP header. When set, the AVP Code belongs to the specific vendor code address space. This allows extending the AVP with the vendor's specific features.

A.2.1.10 Diameter Commands

Diameter commands come in pairs, where one ends in "R" = "Request" and the other ends in "A" = "Answer."

Each command is assigned a command code, which is used for both requests and answers. The following table lists Diameter commands:

Abbr.	Command Name	Code
ASR	Abort-Session-Request	274
ASA	Abort-Session-Answer	274
ACR	Accounting-Request	271
ACA	Accounting-Answer	271
CER	Capabilities-Exchange-Request	257
CEA	Capabilities-Exchange-Answer	257
DWR	Device-Watchdog-Request	280
DWA	Device-Watchdog-Answer	280
DPR	Disconnect-Peer-Request	282
DPA	Disconnect-Peer-Answer	282
RAR	Re-Auth-Request	258
RAA	Re-Auth-Answer	258
STR	Session-Termination-Request	275
STA	Session-Termination-Answer	275

Source: Data from IETF RFC 3588.

A.2.1.11 The AVP (Attribute Value Pair)

The Diameter protocol is designed to be extensible through the following functions:

- Defining new AVP values
- Creating new AVP
- Creating new authentication or authorization applications
- Creating new accounting applications
- Creating application authentication procedures

Although Diameter is extensible, it is highly recommended to reuse existing AVP values and Diameter applications as much as is feasible. This simplifies standardization and implementation, and avoids potential interoperability issues.

The AVP is the vehicle of data delivery by Diameter. These AVP values are used by the Diameter protocol itself for its dialogs, but also to deliver data associated with particular applications that employ Diameter. There is no restriction on adding AVP to Diameter messages, as long as the necessary AVPs are correctly included and those that must be excluded for a particular function are not present.

AVPs are used by the basic Diameter protocol to support the following required features:

- Transporting of user authentication information to the server for authentication

■ Transporting of service-specific authorization information, allowing peers to decide whether a user's access request should be granted
■ Exchanging resource usage information, possibly for accounting purposes or planning
■ Relaying, proxying, and redirecting of Diameter messages through a server chain

A.2.1.12   Diameter for IMS

The IMS uses Diameter as the protocol for AAA as well as user subscription management.

Diameter is extended to support further functions identified by IMS. Diameter AVPs for IMS include:

■ User ID management
■ Authentication
■ Accounting records
■ Charging information
■ Credit control/failure handling (real-time)
■ Event management
■ Destination, origin, and domain information
■ Time stamps

A.2.1.13   Diameter for Credit Control

The need for credit control for real-time charging has spawned a separate set of Diameter commands with their matching AVPs. The Credit control application is defined in RFC 4006. It contains the CCR/CCA pair, with a Command Code of 272.

There are several scenarios for using the Accounting commands in Diameter:

1. *Event based.* The Diameter dialog is in a context of a stand-alone event with a single CCR/CCA exchange. This type is used when the transaction completion is certain, and therefore the service will be charged.
2. *Session based.* The Diameter dialog is performed in the context of a session where multiple CCR/CCA exchanges can occur. This type is used when there is a need to reserve credits before providing the service.
3. *State based.* The Diameter dialog is in a context of a session where the server must maintain state. This is used when the server first reserves the credits, then debits against them accordingly after receiving the subsequent CCR.

The Credit Control commands are summarized below:

CCR	Credit Control Request	Sent from client to server	Request authorization for a given service
CCA	Credit Control Answer	Sent from server to client	Carries the result of the corresponding authorization request
RAR	Reauthorization Request	Sent by server to trigger a new CCR	For example, after successful credit replenishment during a service
RAA	Reauthorization Answer	Sent by client as an answer to RAR	

Source: Based on RFC 4006.

The Diameter Credit Control commands are accompanied by AVPs, as in the list below:

AVP	Use	Possible Values
C-Request-Type AVP	Indicates type of the request for a CCR	For session-based scenarios: INITIAL_REQUEST, UPDATE_ REQUEST, and TERMINATION_REQUEST For event-based scenarios: EVENT_REQUEST
CC-Request-Number AVP	Identifies a request within a session	
Requested-Action AVP	Used to indicate type of requested action for event-based scenarios	DIRECT_DEBITING, REFUND_ ACCOUNT, CHECK_BALANCE, and PRICE_ENQUIRY

Source: Based on RFC 4006.

A.2.2 RADIUS

RADIUS is an authentication protocol for IP network access. It is capable of handling tethered users or roaming (nomadic) users but with limited security. It was originally developed for access to the Internet where the service provider requires user identification and authentication via entry of username and password.

Before submission of the user's credentials can occur, the messages are transmitted over PPP to find the RADIUS server. The RADIUS server applies authentication schemes by which it validates the user credentials. Such schemes are defined further in protocols such as PAP, CHAP, and EAP (discussed below).

If the authentication is successful, the user is able to obtain an IP address and L2TP configuration, and bind it with the user ID. This binding is then used in subsequent communication.

RADIUS is currently widely installed and in use for data services and Internet access. For this reason, the Early IMS standards allow for its use for authentication instead of Diameter, which is the required protocol for IMS. However, RADIUS cannot provide all the functionality that Diameter is designed to deliver in User ID management and accounting.

A.2.3 Password Authentication Protocol (PAP)

PAP, a straightforward protocol to evaluate passwords, is widely used in IP network authentication. It allows sending unencrypted ASCII passwords over PPP.

As it is entirely lacking in any security measures, it is only used for low-risk authentication or as a last resort, where the remote server does not support anything else. It can be followed by further authentication to enhance the level of security.

A.2.4 Challenge-Handshake Authentication Protocol (CHAP)

CHAP is an authentication scheme used over PPP. It provides measures to challenge the remote user by requiring a successful "handshake" to take place. The handshake involves a dialog where the server "challenges" the UE, and the UE responds by sending a value calculated according to a formula, including checksum, etc. The same calculation is performed by the challenging serve and, if matched, the user is authenticated. This procedure is performed at the initial user registration and at random time intervals during the life of the connection.

A.2.5 Extended Authentication Protocol (EAP)

EAP is a flexible authentication framework that enables the deployment of several different authentication methods. It is defined in RFC 3748.

EAP is used to select a suitable authentication mechanism, having obtained enough information to ascertain which method should be used (i.e., after an initial dialog has taken place). Rather than updating the EAP server to support each new authentication method, EAP permits the use of a back-end authentication server, with the EAP server acting as a pass-through. EAP authenticates the "back-end" server to the user, providing mutual authentication both ways.

Among EAP compliant protocols are the AAA protocols needed for IMS and WLAN: EAP-RADIUS, as described in RFC 3579, and EAP-Diameter, described as DIAM-EAP. An "AAA Server" is then a "back-end authentication server" for EAP.

EAP typically runs directly over data-link layers such as PPP or IEEE 802, without requiring IP. EAP is a lock step protocol that only supports a single packet in flight. As a result, EAP cannot efficiently transport bulk data.

EAP is particularly favored by WLAN providers. The requirements for EAP methods used in wireless LAN authentication are described in RFC 4017.

As a framework, not a protocol, it provides a range of EAP methods that can suit different circumstances. Such methods include the following, to name but few:

- *EAP-AKA.* This method deals with authentication and session key distribution in UMTS for the AKA Protocol, and is defined in RFC 4187
- *EAP-SIM.* This method is used for authentication and session key distribution, primarily in GSM, and is defined in RFC 4186.
- *EAP-IKEv2.* This method supports the IKE Protocol. It provides mutual authentication for both the user client and the network server. It supports asymmetrical key pairs, where the private key is only known to one of the parties. The public/private key pair is inserted into the issued digital "certificate," which is subsequently used in authentication of session requests. EAP enables the use of different techniques and different credentials in each direction. For example, the EAP server authenticates itself using the public/private key pair, but the EAP client can use symmetric keys.
- *EAP-TLS.* This method, as defined in RFC 2716, supports TLS. It is a secure connection utilizing PKI to link to a RADIUS server. This method provides a high security level but is considered rather complicated to implement. Its strength is in the client-side certificates but this is also an operational overhead.

A.2.6 Protocol for Authentication for Network Access (PANA)

PANA (Protocol for Authentication for Network Access), as defined within the IETF, can be used to carry DHCP-based authentication signaling at the IP layer. PANA can transport EAP between a PANA client on the user equipment and a PANA authentication agent on a network node in the transport plane.

The rights of a PANA client to transport PANA are determined by the PANA authentication agent, which verifies the credentials. If the authentication server and the PANA agent share the same platform, this interface can be merely an API. If they reside on separate nodes, RADIUS or Diameter is used for this purpose.

When a user is authorized to access the network, the PANA authentication agent sends configuration information to the enforcement point so that policies and data filters can be applied to all subsequent data packets on this connection.

A.3   QoS Protocols

A.3.1   DiffServ

A.3.1.1   DiffServ Principles and Uses

DiffServ, or **Diff**erentiated **Serv**ices, is a method widely used for large private IP networks, for data and VoIP, and is defined in RFC 2475 plus associated RFCs. DiffServ has been recommended by the 3GPP for implementing QoS management for Telecom IP networks.

DiffServ can provide support, within a single IP network, for applications requiring high-level QoS, such as videoconferencing or VoIP, alongside tolerant applications that require no more than mere "Best Effort." Unlike IntServ, which is flow based, DiffServ is class based. That is, DiffServ aggregates the flow into given classes of service, each with a class-based traffic profile.

Using DiffServ, each packet is marked according to the class of service QoS it requires. All network elements in the path of the media flow must recognize the classification and apply methods of queuing to comply with the specifications of QoS. Routers can vary in their queuing behavior, allowing configuration of, for example:

- Number of queues
- Relative priorities
- Bandwidth per queue

A.3.1.2   Classification

Elaborate and flexible behavior can be assigned to the traffic flow per hop on the path to the destination, but classification is usually performed for each DiffServ domain at the ingress point only and it remains the same throughout, until reaching the egress to another network.

The classifier makes its decisions according to network policies. These policies affect how traffic is marked and conditioned upon entry to the DiffServ Domain, and how that traffic is forwarded within that network.

The classifications of the different services occupy a special 4 bits for "Type Of Service" and 3 bits for "Precedence" in each message. These bits describe the DSCP (Differentiated Services Code Point).

A.3.1.3 The Class-of-Service Method

DiffServ organizes traffic into streams that have the same traffic requirements, that is, within separate classes of service. This means that the media flow profile is determined for an entire class rather than an individual media flow. As such, it is not as refined as the IntServ method but is better suited for large volumes of traffic. Because there are only a relatively small number of QoS profile variants, it is argued that individual characterization is not needed.

In DiffServ, every packet is marked with class of service parameters. All participating routers must recognize these classes of service and manage the traffic flow for each class independently, ensuring preferential treatment for higher-priority traffic on the network. The bearer network makes no decisions as to how to treat the packets — that is decided by the process of allocating the class of service. DiffServ also provides standard definitions of classes of service, which allow for easy interworking across router vendors and different carriers.

A.3.1.4 DiffServ Implementation

The implementation of DiffServ is built on behavior at each hop between routers. The PHB (Per Hop Behavior) specifies the packet forwarding properties associated with a class of traffic. DiffServ defines the PHB for each Class of Service.

All the traffic flowing through a router that belongs to the same class is referred to as a Behavior Aggregate. The Behavior Aggregate combines media flows with the same QoS profile. Routers that share the same configuration for Classes of Service and Network Policies make up a DiffServ Domain. An operator network can contain several such domains.

A.3.1.5 Classification

Classification of packets into classes of service is based on:

- Sender address or ID
- Destination
- Service type
- Traffic type
- Received marking for a Class of Service

The classifier can accept the requested Class of Service or choose to modify it. Operators often choose to reclassify packets at the ingress into their DiffServ Domain, to maintain efficient use of their network and prevent unnecessary high QoS rating. Within each class, the traffic profile can be different than that of the adjacent networks.

A.3.1.6 Per-Hop Behavior (PHB)

The popular PHBs most commonly in use include:

- *Default PHB*. This is the lowest QoS level (i.e., Best Effort). It is used when no other standard PHB fits. It exists even if no PHBs have been defined.
- *Expedited Forwarding (EF) PHB*. This profile is characterized by low delay, low loss, and low jitter that are needed for real-time communications such as Voice and Video services. EF PHB queuing packets are given higher priority over other classes. To avoid packet loss, delay, and jitter, links carrying EF traffic are strictly controlled to maintain adequate free capacity.

■ *Assured Forwarding (AF) PHB Behavior Group.* This traffic behavior enables the assured delivery of the media flow under the subscribed traffic rate. Above this rate, there is higher probability of congestion and therefore, dropping packets and deterioration of service quality. AF traffic can be assigned to one of four classes combined with one of three Packet Drop levels. Therefore AF contains twelve DSCPs (DiffServ Code Points).

■ *Class Selector PHB.* The Class Selector PHB aims to provide compatibility with the previous way of selecting Precedence, which never gained much popularity. The same TOS (Type Of Service) byte is now utilized for the DSCPs. The Class Selector PHB allows sharing the same byte with the older system by retaining the leading three bits as IP precedence bits, in the format of "xxx000." Each IP precedence value can be mapped into a DiffServ class. If a packet is received from a non-DiffServ-aware router that used IP precedence markings, the DiffServ router can still understand the encoding as a Class Selector code point.

The table below gives the mapping of PHBs to DSCP codes. It also shows the mapping of the old precedence levels to the PHB CS.

PHB	Detail PHB	DSCP
AF	AF11	001010
	AF12	001100
	AF13	001110
	AF21	010010
	AF22	010100
	AF23	010110
	AF31	011010
	AF32	011100
	AF33	011110
	AF41	100010
	AF42	100100
	AF43	100110
CS	CS1 (precedence1)	001000
	CS2 (precedence2)	010000
	CS3 (precedence3)	011000
	CS4 (precedence4)	100000
	CS5 (precedence5)	101000
	CS6 (precedence6)	110000
	CS7 (precedence7)	111000
Default	Default	000000
EF	EF	101110

Source: Based on RFC 2475 and other associated RFCs.

The message header can accommodate up to 64 different classes of service, each characterized by a PHB. This allows operators to tailor the PHB to their network-particular capabilities. However, easier cross-operator interworking is achieved when the standard PHBs, as recommended (but not mandated) by the IETF standards, are followed.

A.3.1.7 Marking

The IP message header contains a TOS byte, where each bit can be set to flag a service attribute. Originally, the first three bits where intended to convey service precedence, however, this has not been utilized widely. Therefore, the IETF agreed to reuse the ToS byte as the DS Field for DiffServ networks. In case there are some networks that still use the Precedence TOS bits, DiffServ enables the Class Selector PHB to define precedence in the first three bits. Each IP precedence value can be mapped into a DiffServ class. If a packet is received from a non-DiffServ-aware router that used IP precedence markings, the DiffServ router can still understand the encoding as a Class Selector Code Point.

A.3.1.8 Advantages and Weaknesses

Although perfectly suitable for the Internet, service providers of Internet peer-to-peer services have not rushed to deploy DiffServ, perhaps to avoid complex peering agreements of policy. However, the 3GPP recommends DiffServ for carriers of IP networks.

DiffServ policing of traffic (e.g., dropping excess packets) and classifying the traffic (defining PHB) is performed at the borders of the network, saving processing time throughout the internal hops. However, because PHBs are defined per hop, end-to-end behavior is not easy to predict, especially when crossing more than one DiffServ domain.

The fact that each network can override the received PHB settings means that operators cannot guarantee network performance in a commercial SLA. In practice, packets are often classified at the highest level even when the application does not justify it, therefore rendering the classification useless when crossing borders.

DiffServ is effectively a mechanism of selectively delay or drop packets of some service classes and not others. Because this will only occur at near-congestion level, the actual benefit is limited. Dropping packets causes retransmission that increases the pressure on the network. Hence, DiffServ can contribute to the congestion.

For these reasons, it is argued that it is more cost effective to increase network capacity in an effort to avoid reaching the level at which DiffServ operates.

A.3.2 RSVP

A.3.2.1 RSVP Principles

RSVP (Resource Reservation Protocol) is a resource reservation setup protocol designed for an integrated services network. It is used by session controllers to request resources from the IP Bearer node, specifying the QoS profile suitable for the type of session.

RSVP is used by bearer nodes and routers to convey the QoS requirements along the path of the media packets, requesting to establish appropriate resources to be maintained throughout the session. RSVP requests will generally result in resources being reserved in each node along the data path.

The following features characterize RSVP:

- RSVP is able to reserve resources for both unicast and many-to-many multicast.
- RSVP is simplex. It makes reservations for unidirectional data flows.
- RSVP is recipient-oriented. The data flow recipient initiates the resource reservation.
- RSVP maintains "soft" state in routers and hosts, for dynamic membership changes and automatic adaptation to routing changes.
- RSVP transports traffic control and policy control parameters that are opaque to RSVP.
- RSVP provides transparent operation through routers that do not support it.
- RSVP supports both IPv4 and IPv6.

A.3.2.2 The RSVP Method of Operation

RSVP is a one-way protocol. Therefore, it is necessary to initiate a recipient-to-sender RSVP at the same time. RSVP operates over IP and is capable of working over IPv4 as well as IPv6. Although it is a Transport level protocol, it is not regarded as a routing protocol because it deals only with resource reservation. It runs in background mode, like other management protocols.

For greater flexibility in scenarios involving multi-party and heterogeneous media types, RSVP places the responsibility of requesting QoS parameters on the recipient, not the sender. The resourcing requirements are carried back to the nodes (routers and hosts) along the reverse path, but only up to the point of joining the multicast distribution. As a result, RSVP's reservation overhead is lower than in a case of linear connections to all recipients.

RSVP reserves resources for a flow using two main concepts — Flowspec and Filterspec. An RSVP reservation request contains both a Flowspec and a Filterspec. They both constitute the Flowdescriptor.

Flowspec. For each flow, RSVP identifies the particular quality of service. This QoS-specific information is called a Flowspec. The Flowspec sets the parameters of the packet scheduler at a node, defining the flow characteristics.

RSVP carries the Flowspec from the application to the host nodes and routers along the path to the destination. These network elements extract details of the QoS requirements from the Flowspec to ascertain their ability to support this media flow and to reserve the appropriate resources for it.

A Flowspec contains the following:

- Service class
- Reservation spec (defines the QoS)
- Traffic spec (describes the data flow)

Filterspec. The Filterspec determines what packets will be subjected to the Flowspec QoS settings. That is, the Filterspec sets the parameters at the packet classifier.

A Filterspec sets up the filters to distinguish the particular media flow packets from the overall total of packets processed by a node. This filtering is based on a number of attributes, including the sender's IP address and port. These filters form reservation "styles":

- Fixed filter: reserves resources for a specific flow.
- Shared explicit filter: reserves resources for several flows that share the same resources.
- Wildcard filter: reserves shared resources for a general type of flow, partially specified.

A.3.2.3   RSVP Messages

There are two primary types of messages:

1. *The PATH message.* The PATH message is sent from the sender host along the communication path, and stores the path state in each node. The path state includes the IP address of the previous node, plus the format of the sender data, the sender's TSPEC for traffic characteristics and the ADSPEC that carries additional data about the QoS properties in the path, advertised externally.

    The PATH message starts the process of resource reservation by identifying the path. If it reaches a non-RSVP aware router, the message will be sent on without interpretation or reservation.

2. *The RESV message.* The RESV message is sent from the recipient back to the sender host along the reverse path, as a response to the PATH message. On the way, the IP address for the destination is updated to reflect the hop to the previous node on the reverse path.

    The RESV message contains the Flowspec to identify the resources that the media flow needs. At each node, the RESV message Flowspec can be modified. For example, in the case of multicast transmission, flow reservation can be merged. Each node in the path can either accept or reject the request.

A.3.3   IntServ

A.3.3.1   Flow Specs

IntServ (Integrated Services) is a mechanism to specify parameters for service QoS, for example, it can specify QoS definition for video or voice. IntServ is said to provide fine-tuning of QoS requirements, in contrast to DiffServ, which is a coarse-grained control system.

In IntServ, "Flow Specs" describe what the reservation is for, while "RSVP" is the underlying mechanism to signal it across the network. IntServ needs all traffic-bearing elements to implement and comply with it, and every application, unless it is highly tolerant, to make an individual reservation.

Flow Specs define the IntServ parameters. This entails two types of specifications:

1. *TSPEC (Traffic SPECification):* defining the traffic profile.
2. *RSPEC (Reservation SPECification):* defining the type of reservation of resources.

A.3.3.2   TSPEC

TSPEC uses the Token Bucket technique to define the traffic shape and ensure that packets arrive at the right rate. Using TSPEC, it is possible to specify the token rate and the bucket depth per service.

For example, a video with a refresh rate of 75 frames per second, with each frame taking 10 packets, might specify a token rate of 750 Hz, and a bucket depth of only 10. The bucket depth is therefore designed to contain an entire frame. Using a Token Bucket (rather than a Leaky Bucket) allows releasing a "burst" that consists of the whole frame.

Another example is a voice conversation. This requires a lower token rate but a much higher bucket depth for buffering unpredictable bursts of sound and silent pauses between words and sentences. Fewer tokens are needed where silent gaps are not conveyed in the payload. However, the depth of the "bucket" buffering must be greater to accommodate traffic peaks and avoid discarding excessive packets.

A.3.3.3   RSPEC

RSPEC allows defining the levels of reservation for the service:

- *Best Effort.* An Internet service is regarded as "Best Effort," that is, no reservation at all. "Best Effort" status is also given to File Transfer that does not require real-time interaction.
- *Controlled Load.* This reservation setting is utilized for a lightly loaded network and for applications that have a certain level of tolerance but still require a fairly constant packet rate.
- *Guaranteed Setting.* This highest level of reservation is required for services that do not tolerate delay and packet loss, that is, interactive real-time sessions, such as conversational video.

A.3.3.4   RSVP for IntServ

Enforcing IntServ is achieved using RSVP, according to RFC 2210. RSVP is carried over by all participating network nodes capable of sending QoS data. The requesting Network Element sends a "PATH" message every 30 seconds toward the destination. Available nodes on the path respond by sending back an "RESV" message to denote a request for a reservation with particular settings of Flow Specs to be supported by the reservation.

Routers between the sender and target recipient must decide if they can support the requested reservation in their current loading status. If they cannot, they send a reject message. If they have enough spare capacity, they accept the reservation and commit to carry the media flow for this session.

The committed routers record the traffic flow details and process the traffic for the session according to these parameters. The individual routers can, at their option, police the traffic to check that it conforms to the Flow Specs.

If no traffic arrives in either direction for a predefined length of time, the reservation will be cancelled automatically. This provides a self-clearing network function that recovers from failures of any endpoint and terminates the session gracefully.

A.3.3.5   IntServ Scalability Issues

IntServ requires each router to maintain status and Flow Specs for every media flow. This can ramp up quickly in large networks and take too much space. IntServ is therefore ideal for small-scale networks that need individual media flow reservations such as small Enterprise VPN.

Alternatively, IntServ can still be used in larger networks where media flows are aggregated according to common Flow Specs for transmission in the CN and are only handled individually at the edge. The additional requirements for new media flow are added to the aggregated Flow Spec of the core routers by the "ADSPEC" message.

A.3.4 COPS (Common Open Policy Service)

A.3.4.1 COPS Principles

COPS is a dialog protocol providing a query and response to exchange policy and resource status information as defined in RFC 2748. This protocol can be used between the PDP (Policy Decision Point) and the PEP, although the more recently defined PCRF interface is based on Diameter.

COPS is deployed between a policy decision-maker element and a policy enforcer element. The function making a policy decision is the PDF, and the point from where it is served is the PDP. The point where the policy is implemented, in terms of filtering packets and prioritizing them, is the PEF that operates in the bearer network elements, referred to as the PEP.

The COPS Protocol also defines a local PDP that operates in the local network, whether it is a visited network or a separate access network. The LPDP (Local Policy Decision Point) can contribute more policies to the PEP to be addressed as well as the CN policies from the remote PDP. A local PDP is usually a static configuration, whereas the remote PDP is driven by session events from applications and session controllers. The PDP remains the final decision-making authority. Information from the local PDP is relayed to the PDP, where conflicts are resolved. This means that the PEP must obey the PDP's final decision.

A.3.4.2 COPS Design

The COPS Protocol is based on a client/server model, where the server is deemed to be at the PDP, communicating with multiple clients in the PEP. COPS is designed to be extensible and generic, able to accommodate future needs for exchange of data of administration and configuration, and in particular for conveying policies. To achieve that, COPS can distinguish between different kinds of clients, depending on the type of message used. These clients can transport varied data related to the QoS request, and expect a different type of decision making.

The PEP can also initiate certain COPS actions. PEP can initiate a request for decision making from the PDP. It can also notify the PDP when the bearer has closed down the session, without CN elements involved.

The PEP maintains a "named unit of configuration" that configures a particular profile of media stream. That named unit can be used by more than a single session.

COPS is said to be stateful. The server and the involved client keep track of the running session and can relate new events to it. The PDP is able to receive and support mid-session events, which may cause renegotiation of QoS.

The remote PDP can send unsolicited decisions to the PEP to force changes in previously approved request states. Because COPS is stateful, the PDP must instruct the PEP to terminate the kept state of the session when the call ends. Alternatively, when the PEP is made aware of a call termination, it will notify the PDP to delete the request state.

A.3.4.3 COPS Resilience

The process of allocating resources is essential to the operation of the network, and therefore must be secure and resilient. To ensure a reliable exchange of information, COPS runs over the self-correcting TCP, and not UDP.

The exchanged contents may be protected by means of IPSec and/or TLS measures. The integrity of the PEP-PDP link is constantly tested by a "heart beat" method, where the PEP and PDP confirm that the link is still live, obeying the command of "Keep Alive."

Continuous availability is provided by redundancy. If PDP fails, the PEP will seek an alternative remote PDP to get attached to, or operate temporarily through a local PDP.

A.3.4.4 COPS Messages

The content of a COPS request or decision message depends on the context. The message header identifies the type of request. The COPS Context is defined in the COPS message body (i.e., the COPS "object").

COPS identifies the following types of outsourcing events:

- Incoming message arrival
- Allocation of local resources
- Forwarding of an outgoing message
- Request for downloading of configuration from the PDP

COPS message headers contain Version, Flags, Op Code, Client Type, and Message Length, as follows:

Field	Field Length	Description
Version	4 bits	COPS protocol version
Flags	4 bits	For example, for Solicited Message
Op Code	8 bits	Operation type (e.g., REQ)
Client Type	16 bits	Defines which client PEP
Message Length	32 bits	Defines variable length for header plus object

Source: Based on information in RFC 2748.

The following table of Operation Codes shows the type of actions taken via COPS:

	Type	From PEP	From PDP	Description
1	REQ	√		Request
2	DEC		√	Decision
3	RPT	√		Report State
4	DRQ	√		Delete Request State
5	SSQ		√	Synchronize State Request
6	OPN	√		Client-Op
7	CAT		√	Client-Accept
8	CC	√	√	Client-Close
9	KA	√	√	Keep-Alive
10	SSC	√		Synchronize Complete

Source: Based on information in RFC 2748.

The COPS object (i.e., the message body) is preceded by descriptors:

■ Message length
■ C-Num, which identifies the class of information contained in the object
■ C-Type, which identifies the subtype or version of the information contained in the object

The C-Num contains the following identifiers:

C-Num Values	COPS Objects
1 = Handle	An identifier of installed state
2 = Context	Type of triggering event, used with REQ and DEC
3 = In Interface	REQ Identifier of incoming interface on PEP
4 = Out Interface	REQ Identifier of outing interface on PEP
5 = Reason code	DRQ reason for deletion of request state
6 = Decision	PDP Decision description, sent with DEC
7 = LPDP Decision	LPDP Decision description, sent with DEC
8 = Error	CC particular error code
9 = Client Specific Info	Client info in REQ/DEQ/RPT/OPN
10 = Keep-Alive Timer	Value for the timer in CAT
11 = PEP Identification	Identifies the PEP in OPN
12 = Report Type	RPT value for report type for the specific Handle
13 = PDP Redirect Address	PDP alternative address notified to PEP in CC
14 = Last PDP Address	The last PDP address to be linked with the PEP, notified in OPN
15 = Accounting Timer	Time between reporting of accounting type in CAT
16 = Message Integrity	Sequence number and checksum for the request integrity protection

Source: Based on information in RFC 2748.

A.3.4.5   COPS-PR

COPS-PR (Provisioning) is an extension of COPS defined by RFC 3084. While the original COPS defined a model of "Outsourcing" operation between the PEP and the PDP, the extensions in COPS-PR identify the "Configuration" model or "Provisioning" mode. The Outsourcing mode provides instantaneous policy decision for session requests where resources are allocated dynamically at a granted QoS level. Alternatively, due to lack of resources, sessions can be rejected. The PEP is said to outsource the decision making to an external PDP, requesting it to make the decision and obey it when it is received.

In the Provisioning model, PEP parameters are configured rather than individual sessions. The policies are configured independently of incoming sessions or events. Provisioning can be performed in bulk, to alleviate scalability and management issues, for network sections or for the entire router population. The downloaded "decisions" are, therefore, not mapped directly to requests of policy decisions per session. They are driven by PDP responding to events such as

SLA updates. As such, these updates are infrequent and are not performed under time-critical conditions.

The Provisioning model allows for different types of configuration profiles to be handled — for example, for Security and Charging policies as well as QoS. This means that multiple policy functions can be deployed to manage disparate areas of policies in different Client Types.

This policy data block is kept in a named data structure, a PIB (Policy Information Base). The PIB concept is similar to the MIB (Management Information Base) used by SNMP. Equally, a PIB is assigned to each area of policies that can be supported by COPS-PR. The PIB contains a tree-like structure with branches of PRCs (Provisioning Classes) and sub-branches of multiple PRIs (Provisioning Instances).

Typically, a Client Type specified by the PEP to the PDP is unique for the area of policy being managed. A single Client Type for a given area of policy (e.g., QoS) will be used for all PIBs that exist in that area. COPS-PR uses the PIB to identify type and purpose of unsolicited policy information that is "pushed" from the PDP to the PEP. The PIB is also used for notifications sent from the PEP to the PDP.

A.4 Media Transport Protocols

A.4.1 The Role of Transport Protocols

The Transport layer facilitates data flowing between two hosts, which can be endpoints, handsets, or network servers. This requires end-to-end connection, often over several access and backbone networks. The purpose of the separate Transport layer is to provide transparent data transfer between end users, so that the higher layers will not have to be concerned with the transmission, but provide service logic control. Therefore, the Transport protocols governing the media flow are responsible for the level of flow control and data transfer resilience applied to the media transfer.

In the IP protocol suite, the popular protocols deployed are UDP and TCP, and the more recent addition of the SCTP. These protocols differ on a fundamental characteristic — the type of connection they support:

- Connection-oriented transmission describes the means of transmitting data in which the endpoints use a preliminary protocol to establish an end-to-end connection before sending any data. The protocol is aware of the state of the transmission and acts upon this knowledge to provide connection resilience. This type of service guarantees that data will arrive in the proper sequence. Circuit-switched networks are said to be connection oriented. TCP and SCTP are connection-oriented protocols for packet networks.

- Connectionless transmission describes the means by which data is sent from one endpoint to another without prior arrangement (i.e., no handshake). This type of transmission is described as "stateless" because there is no record of the ongoing dialog or data stream, and hence no adjustments based on this knowledge. IP, UDP, and ICMP (for "ping" and sending error messages) are connectionless protocols.

A.4.2   UDP

UDP (User Datagram Protocol) provides fast and simple data transfer but lacks error recovery and flow control, relying on the transmission to be understood despite some lost or delayed "datagrams." This means that UDP avoids the overhead of reliable transmission and is better at avoiding congestion.

Unlike TCP, UDP supports broadcasting (sending to all subscribers where each copy of the media flow is sent from the source all the way to the destination) and multicasting (send to multiple specific subscribers, where media flows are duplicated only when the paths to the destinations splits). It is considered more versatile and flexible for high-volume, non-real-time or near-real-time communications, but lacks congestion avoidance facilities.

For these reasons, UDP has been favored for Internet data transfer such as Web browsing but cannot ensure by itself (without additional mechanisms) a high-quality Voice or Video service.

A.4.3   TCP

TCP (Transmission Control Protocol) supports applications requiring guaranteed data integrity for the data transfer. This is often the case for enterprise services such as e-mail and FTP file transfer. The protocol guarantees reliable and in-order delivery of data from sender to receiver.

TCP is a more resilient protocol that is aware of the connection and performs error detection and error correction. This entails acknowledgments of received data and continuous retransmission of failed deliveries, until the acknowledgments have been received.

Before starting to send data, TCP performs a three-way handshake, with the server first binding a port for this connection, then acknowledging the requesting client.

TCP performs the following functions to ensure reliable data transfer:

■ Ordered data transfer using unit sequencing
■ Retransmission of lost packets
■ Discarding duplicate packets
■ Error-free data transfer with error correction
■ Congestion and flow control

TCP has been updated to handle both IPv4 and IPv6. TCP has been optimized for wired networks but there are some concerns about its efficiency for wireless connection.

A.4.4   SCTP

Neither TCP nor UDP satisfy entirely the requirements for a reliable multimedia transport protocol for multimedia services. The SCTP is a more recently defined IP transport protocol which is specified in RFC 4960. SCTP enables a connection that is message-based and supports reliable transmission of multiple media streams.

Similar to TCP, SCTP provides the following features:

■ SCTP provides a connection-oriented transport service between two endpoints.
■ SCTP provides reliable transmission, ensuring that data is delivered in the correct sequence, without loss or duplication.

- SCTP is full duplex (bi-directional).
- SCTP employs a windowing mechanism to provide flow control.

SCTP, however, delivers some capabilities not provided by TCP:

- SCTP provides multiple data streams between the two endpoints. Within each data stream, messages are delivered reliably, in sequence, and without loss or duplication.
- Independent data exchanges can be delivered over different streams. Message loss in any one stream does not affect data delivery within other streams. This is different from TCP, which can handle only a single stream of data, where any data loss results in delaying delivery of all subsequent packets. Minimizing this "head-of-line blocking" problem is of great benefit for Diameter-based signaling, for example.
- SCTP is message oriented. This means that SCTP maintains message boundaries and delivers complete messages (called "chunks") between the upper-layer protocols employing SCTP. In contrast, TCP is byte oriented, so it does not preserve data units within a transmitted byte stream, requiring the upper-layer protocol to count and accumulate the bytes of each message. This makes SCTP more suitable for multimedia transmission.
- SCTP supports the higher-resilience techniques of multi-homed hosts. A multi-homed host has more than one IP interface, where a hot-standby server can take over from the prime server and retain ongoing sessions. To do that, at initialization time, the SCTP communicating parties exchange lists of their IP interface addresses. An SCTP message requiring retransmission can then be sent to an alternate IP address at the same time.
- SCTP enables bundling of "chunks" (data units), into a single message. This is a more efficient transmission that is automatically deployed at the time of nearing congestion, but is normally optional. SCTP reconstitutes the messages in the right order at the receiving end.

Where there is no support on both sides of the transmission, SCTP can still be used by tunneling over TCP. This can also be more secure, using IPSec and TLS.

A.4.5 Comparing Media Transport Protocols

The following table summarizes the differences between the three main transport protocols:

Feature	Description	TCP	UDP	SCTP
Packet header size	Length of header	20 Bytes	8 Bytes	—
Packet entity	Definition of the service data unit	Segment	Datagram	Chunk
Multi-streaming	Allowing parallel transmission of several streams that can deliver associated data or independent data	No	No	Yes
Ordered delivery	Enforcing packet arrival in the right order	Yes	No	Yes
Unordered delivery	Allowing arrival of packets out of sequence but restore sequence where possible	No	Yes	Yes

Feature	Description	TCP	UDP	SCTP
Data checksum	Detecting errors by sending a calculated sum of bytes and comparing it at the receiving end	Yes	Yes	Yes
Automatic repeat request (ARQ)	Reliable transmission feature that sends acknowledgments of a successful delivery, otherwise will timeout and flag it for re-transmission	Yes	No	Yes
Port assignment	Mapping data packet flow to a particular process or protocol (e.g., 389/TCP, UDP for LDAP, 220/TCP, UDP for IMAP-interactive e-mail, 990/TCP, UDP for FTP over TLS)	Yes	Yes	Yes
Connection oriented	Connecting end-to-end by prior agreement, unlike connectionless communication	Yes	No	Yes
Multi-homing redundancy for resilience	Enabling servers to send data to a hot standby that can take the place of the primary server at failure time and achieve fast failover recovery	No	No	Yes
Preserve message boundary	The connection is aware of the messages, not merely bytes or packets	No	Yes	Yes
Flow control	Managing the rate of data transmission	Yes	No	Yes
Congestion avoidance	Algorithm to shape traffic, use Active Queue Management (AQM) or RED	Yes	No	Yes

Source: Multiple articles, including Wikipedia.

A.5 Security Protocols

A.5.1 *TLS (Transport Layer Security)*

A.5.1.1 *About TLS*

Secure communication is now high priority for the Internet as well as private IP networks. To achieve this, the process of identifying users and the following data transmission must be protected. This means that the transmission of cleartext passwords over unencrypted channels should be eliminated.

The TLS Protocol, previously known as SSL, was originally designed to prevent eavesdropping, tampering, or message forgery for the Internet. It has gained popularity for any IP-based communication. TLS 1.1 is the approved version of the TLS Protocol, as described in RFC 4346.

The great advantage of TLS is that it is independent of the application protocol. It can protect higher-level protocols transparently. For this reason, TLS is used to protect protocols such as HTTP, SIP, IMAP, POP, SCTP, EAP, COPS, and others. The TLS standard, however, does not specify how to initiate TLS handshaking and how to interpret the authentication certificates that have been exchanged. This is left to the particular implementation of the protocols that run on top of TLS.

TLS involves three main elements:

1. Peer negotiation between the communication parties for a cryptographic algorithm that they can both support
2. Exchanging keys, based on public key encryption and certificate-based authentication of both ends
3. Symmetric cipher-based traffic encryption

A.5.1.2   The TLS Record Protocol

TLS operates with applications in client/server mode. It consists of a TLS handshake protocol and a TLS record protocol, where the handshake protocol enables setting up and commencing secure communication, and the TLS Record Protocol, which protects the subsequent communication messages.

The TLS Record is in a format for encapsulation of various higher-level protocols. The encapsulation and the encryption ensure that the data cannot be read or tampered with.

The TLS Record Protocol Format contains a protocol type identifier. Protocol types can be:

20 = ChangeCipherSpec
21 = Alert
22 = Handshake
23 = Application (The application protocol is the high-level protocol that TLS protects, such as SIP.

The TLS Record also contains the TLS version, to enable negotiation of compatible versions. This can be SSLv3, TLS 1.0, TLS 1.1, or TLS 1.2. This record can contain more than one protocol message. Optionally, the MAC (Message Authentication Code) also is appended. The MAC is computed from the Protocol message, with additional key material.

A.5.1.3   The TLS Handshake

Before the communication of data begins, the client and the server exchange handshake messages that perform mutual authentication. The client or server's identity can be authenticated using the asymmetric method of public key with cryptography. Once authenticated, the client and server agree on the encryption algorithm and the cryptographic keys.

The negotiation of a shared secret should be in itself private, to prevent an eavesdropping intruder obtaining the negotiated secret. This negotiation is also protected from tampering, and any attempt to modify the communicated data cannot go undetected.

A typical handshake proceeds as follows:

■ A client sends a ClientHello message suggesting cipher suites and compression methods and other parameters to negotiate compatible TLS version, etc.
■ The server responds with a ServerHello message containing the chosen cipher, compression method, and TLS protocol version that is agreeable to the client.

- The client and server also exchange random numbers that are used in the algorithm later.
- Depending on the particular method, the server sends its certificate, currently based on X.509.
- If mutual authentication is required, the server requests a certificate from the client, using a CertificateRequest and the client responds.
- The server sends a ServerHelloDone message to complete the handshake negotiation.
- The client responds with a ClientKeyExchange, which may contain a PreMasterSecret and public key.
- The client and server then use the exchanged random numbers and PreMasterSecret to compute a common secret — the "master secret" — from which key data is derived.
- The client now sends a ChangeCipherSpec message to denote the start of encrypted transmission.
- The client sends an encrypted Finished message, containing a hash and MAC over the previous handshake messages. The server attempts to decrypt it and verify the hash and MAC. If this fails, the handshake is abandoned.
- Finally, the server sends a ChangeCipherSpec message and its encrypted Finished message, which the client needs to decrypt and verify the contents.
- If this is successful, encrypted transmission of application data can commence, with content type 23.

A.5.1.4   TLS Features

There are several security measures in TLS/SSL, including:

- The client verifying the CA (Certification Authority) using its public key. The client also verifies that the issuing CA is on its list of trusted CAs.
- The client checking that the server's certificate validity period has not expired.
- Protecting against Man-in-the-Middle (undetected intruder) attacks, the client compares the actual DNS name of the server to the DNS name on the certificate.
- Protecting against several known attacks, including a downgrade of version that is less secure or a weaker cipher suite.
- Numbering all the Application records with a sequence number, and using this sequence number in the MAC.
- Using a message digest enhanced with a key, so only a key-holder can check the MAC.
- Additional checks when the "Finished" message sends a hash of all the exchanged handshake messages seen by both parties.
- Splitting the input data in half and process each one with a different hashing algorithm.

RFC 4346 specifies additional capabilities that allow the TLS client to:

- Declare the name of the server to be linked to the TLS server. This facilitates secure connections to servers that host multiple "virtual" servers at a single underlying network address.
- Negotiate the maximum fragment length to send. This helps in managing constraints of client memory and availability of bandwidth in some access networks.
- Negotiate the use of client certificate URLs, to save memory on clients.

- Indicate root keys to TLS servers to prevent multiple handshake failures that may overload the limited memory in some clients.
- Negotiate the use of truncated MACs to conserve bandwidth in constrained access networks.

A.5.2 IPSec

A.5.2.1 The IPSec Concepts

IPSec is a suite of protocols that allows for secure authentication, encryption of the IP packets in the application data stream, and establishing cryptographic keys.

IPSec protocol operates at Layer 3 of the OSI model while TLS/SSL operates above the Transport layer. IPSec operates in transport mode and in tunnel mode. It is used primarily to create IP-VPN connections, with varying degrees of security.

IPSec with tunneling mode is the 3GPP standard for communication across security zones. IPSec is optional for IPv4 connections but is mandated for IPv6.

The IP security architecture uses the concept of an SA between the two communicating parties. SA is an established set of security algorithms, keys, and other parameters that are associated with a particular one-way channel connection. The SA is identified by the SAID in the Security Parameter Index (SPI), which is provided along with the destination address. In the case of multicast, the SA is propagated to all the receiving parties. More than one SA can be used within the group.

The IP Authentication Header provides integrity, authentication, and non-repudiation, with the appropriate choice of cryptographic algorithms.

A.5.2.2 Transport Mode

The Transport mode operates as a connectionless transmission for host-to-host connection. In Transport mode, the encryption is performed on the message body (payload) only, not the message headers. This allows the network nodes to see all the information they need for routing but is more vulnerable to attacks. For added security, the Authentication Header is used to protect any header information that should not be altered on the way. This is achieved by a hash sum that is kept within the encrypted data.

A.5.2.3 Tunnel Mode

In Tunnel mode, the entire IP packet is encrypted. It is encapsulated into a new IP packet in the encrypted data stream. In this case, no packets can traverse normal NAT because the IP addresses are not known to intermediate servers. Tunnel mode requires connection-oriented transmission. It is used primarily for network-to-network communications (i.e., secure tunnels between routers), and also host-to-network and host-to-host communications over the Internet.

A.5.2.4   Authentication Header (AH)

The AH enhances the ability to secure data integrity and data origin authentication in a connectionless IP transmission. The AH helps protect the connection from Replay Attack (repeating transmitted data units to masquerade as the other party), and can be used to provide acknowledgment delay (sending additional data to confuse intruders).

The AH protects the IP payload and all header fields of an IP datagram except for those that are modified during transit. In IPv4, such fields include TOS (to define QoS classification, now used for Diffserv), other flags including TTL (Time-To-Live expiration time) and Header Checksum (to confirm correct transmission).

The AH carries the SPI, which includes the security parameters that, together with the IP address, identify the Security Association. The AH also carries the Sequence Number that can be used to prevent replay attacks, and Authentication Data that contains the Integrity Check Value.

A.5.3   L2TP

A.5.3.1   L2TP History

L2TP is used for creating tunnels to transport communications involving other protocols. L2TP amalgamates tunneling protocols of Cisco (L2F) and Microsoft (PPTP). The more recent definition (version 3) in RFC 3931 provides additional security features, improved encapsulation, and the ability to carry not only PPP, but also Frame Relay, Ethernet, ATM, etc.

A.5.3.2   Tunneling Features

The tunnel is initiated by an L2TP Access Concentrator (LAC) towards a passive L2TP Network Server. Once a tunnel is established, the network traffic between the peers is bi-directional, where either the originator or the recipient can initiate sessions. Each session can carry a different protocol, such as PPP. L2TP isolates and separates traffic for each session to support multiple virtual networks across a single tunnel.

L2TP supports reliability features for control Packets (i.e., signaling only), but not for the Data Packets (i.e., the media flow). Additional resilience may be provided by other nested protocols.

A.5.3.3   L2TP with IPSec

L2TP does not provide confidentiality or strong authentication by itself. Therefore, IPSec can be used to secure L2TP packets by providing confidentiality, authentication, and integrity. The combination of these two protocols is generally known as L2TP/IPSec.

To set up a secure L2TP tunnel, first negotiate an IPSec tunnel, which creates a SA and establishes security keys, typically using IKE. This establishes a secure connection using ESP (Encapsulating Security Payload). Finally, negotiate the L2TP tunnel parameters between the SA endpoints. This means that the tunnel negotiation itself is fully protected. The subsequent data transmission is also encapsulated by IPSec, protecting both Control and Data packets. The IPSec packets need not go through packet filtering firewall because they are decrypted and stripped of the encapsulation only at the endpoint. In summary, L2TP provides the tunnel for the communication and IPSec provides the secure association for the tunnel.

Appendix B

Abbreviations

2G: Second-Generation Mobile.

3G: Third-Generation Mobile.

3GPP: 3rd Generation Partnership Project (http://www.3gpp.org).

3GPP2: 3rd Generation Partnership Project 2.

3PCC: Third-Party Call (Connection) Control.

AAA: Authentication, Authorization, and Accounting.

ABCF: Access Bearer Control Function.

A-BGF: Access Border Gateway Function.

ACD: Automatic Call Distribution.

ACL: APN Control List; Access Control List.

ADMF: Administration Function (for Lawful Interception).

ADSL: Asynchronous Digital Subscriber Loop; Asynchronous Digital Subscriber Line.

AE: Applications Environment.

AF: Application Function.

AF: Assured Forwarding (a subclass of the DiffServ).

AGCF: Access Gateway Control Function.

AHE: Application Hosting Environment.

AIN: Advanced Intelligent Network.

AKA: Authentication and Key Agreement.

ALG: Application Layer Gateway.

AMF: Authentication Management Function.

AMF: Access Management Function.

AMR: Adaptive Multi-Rate, in Mobile networks.

AMS: Agent Management System.

AN: Access Network.

ANSI: American National Standards Institute.

AoC: Advice of Charge.

AOR: Address Of Record (SIP).

API: Application Programming Interface.

APN: Access Point Name.

A-RACF: Access – Resource and Admission Control Function.
ARF: Access Relay Function.
ARPU: Average Revenue Per User.
AS: Application Server.
ASDH: Access Synchronous Digital Hierarchy.
AS-ILCM: Application Server Incoming Leg Control Model.
ASN: Access Service Network (WiMAX).
ASN.1: Abstract Syntax Notation One.
AS-OLCM: Application Server Outgoing Leg Control Model.
ASP: Application Service Provider.
ASR: Automatic Speech Recognition.
ASS: Ancillary Service Suite.
ATCA: Advanced Telecom Computing Architecture.
ATI: Alliance for Telecommunications Industry Solutions.
ATM: Asynchronous Transfer Mode.
AuC: Authentication Center.
AUTH: Authentication.
AuthML: Authentication Markup Language.
AUTN: Authentication Token.
AUTS: Authentication Token Sync (Resynchronization).
AV: Authentication Vector.
AVP: Attribute Value Pair.
B2BUA: Back-to-Back User Agent.
BAC: Bandwidth Allocation Controller.
BB: Broadband.
BCF: Bearer Control Function.
BD: Billing Domain.
BGCF: Border Gateway Control Function.
BGF: Border Gateway Function.
BGP: Border Gateway Protocol (IETF RFC 1771).
BICC: Bearer Independent Call Control.
B-ISDN: Broadband ISDN.
bit/s (or bps): bits per second (digital transmission speeds corresponding to bandwidth).
BLES: Broadband Loop Emulation Service.
BRAS: Broadband Remote Access Server/System.
BRI: Basic Rate Interface (ISDN).
BS: Base Station (GSM).
BS: Billing System (IMS).
BSC: Base Station Controller.
BSF: Bootstrap Server Function (security).
BSS: Base Station Subsystem.
BSS: Business Support System.
BWM: Bandwidth Management/Manager.
BwoD: Bandwidth on Demand.
C2D: Click-to-dial.
C7: CCITT No. 7 signaling.
CA: Certification Authority.

CAC: Connection Admission Control.

CAMEL: Customized Applications for Mobile network Enhanced Logic.

CAP: CAMEL Application Part.

C-BGF: Core-Border Gateway Function.

CC: Call Control.

CCAF: Call Control Agent Function.

CCCF: Call Continuity Control Function (VCC).

CCF: Call Control Function (ITU-T CS-1).

CCF: Charging Collection Function.

CCITT: Comite Consultatif Internationale de Telegraphie et Telephonie (replaced by the ITU-T).

CCR: Credit Control Request.

CCXML: Call Control XML.

CCXML: Call Control XML (Voice Browser Call Control, W3C).

CDF: Charging Data Function.

CDMA: Code-Division Multiple Access.

CDMC: Cross-Domain Maintenance Center.

CDR: Charging Data Record.

CE: Customer Edge (in VPN context); Customer Environment (Architecture Node).

CEF: Capability Exposure Function (OMA).

Centrex: CENTRalized EXtension service ("virtual" PBX).

CFM: Charging Function Manager.

CGF: Charging Gateway Function.

CGI: Cell Global Identification.

CGI: Common Gateway Interface (e.g., Perl Script).

CGW: Charging Gateway.

CGW: Customer Gateway (router, NAT, firewall, etc.).

CHAP: Challenge Handshake Authentication Protocol.

CK: Cipher Key.

CLF: Connectivity session Location and repository Function.

CLI: Calling Line Identification.

CM: Connection Management.

CMIP: Common Management Information Protocol.

CN: Core Network.

CNG: Customer Network Gateway (in NASS).

CNGCF: Customer Network Gateway Configuration Function (in NASS).

CoD: Class of Device (in UMA).

Codec: Coder/decoder, or compressor/decompressor.

COPS: Common Open Policy Service (RFC 3084 and 3159).

COPS-PR: Common Open Policy Provisioning.

CORBA: Common Object Request Broker Architecture.

COTS: Commercial Off-The-Shelf.

CPE: Customer Premises Equipment.

CPL: Call Processing Language.

CPM: Customer Profile Management.

CPS: Call Processing Syntax.

CPS: Common Profile Storage.

CPU: Central Processing Unit.
CRA: Connectivity Resources Architecture.
C-RACF: Core – Resource and Admission Control Function.
C-RACS: Core – Resource and Admission Control Subsystem.
CRF: Charging Rules Function.
CRM: Customer Relationship Management.
CS: Circuit switched.
CSCF: Call Session Control Function.
CSE: CAMEL Support Environment.
CSI: Combination of CS and IMS Services.
CSICS: Circuit-Switched IMS Combinational Services.
CSN: Circuit-Switched Network.
CSS: Customer Service System.
CSTA: Computer Supported Telecommunication Application.
CSV: Comma-Separated Variables.
CTF: Charging Trigger Function.
CTI: Computer Telephony Integration.
CUF: Charging Usage Function.
CUG: Closed User Group.
CW: Call Waiting.
DAB: Digital Audio Broadcasting (digital radio standard).
DCE: Distributed Computing Environment.
DCN: Data Communications Network.
DHCP: Dynamic Host Configuration Protocol (IETF RFC 2131).
Diameter: Management Protocol (IETF RFC 3588).
DiffServ: Differentiated Services.
Digest: Digital Geographic Exchange Standard.
DLE: Digital Local Exchange.
DMZ: De-Militarized Zone (between the Internet and the secure corporate network).
DNS: Domain Name System/Server (IETF RFC 1035).
DOCSIS: Data Over Cable Service Interface Specification.
DoS: Denial-of-Service (attacks).
DP: Detection Point.
DPI: Deep Packet Inspection.
DPNSS: Digital Private Network Signaling System.
DR: Digital Right.
DRM: Digital Rights Management.
DSA: Denial-of-Service Attack.
DSCP: Differentiated Services Code Point (in MPLS QoS).
DSL: Digital Subscriber Loop/Line (variants are ADSL, HDSL, SDSL, VDSL).
DSLAM: Digital Subscriber Loop Access Multiplexer (DSL to ATM).
DSML: Directory Services Markup Language.
DSP: Digital Signal Processor.
DTM: Dual Transfer Mode.
DTMF: Dual Tone Multiple Frequency.
DVB-H: Digital Video Broadcasting to a Handset (IPTV).
DVD: Digital Versatile Disk.

DWDM: Dense Wavelength Division Multiplexing.

E.164: ITU-T E.164 recommendation for the international public telecommunication numbering plan.

E1: ITU-T 2-Mbps line system specification in ISDN/PSTN.

EAI: Enterprise Application Integration (unrestricted data sharing).

EAP: Extensible Authentication Protocol.

ECF: Event Charging Function.

ECT: Explicit Call Transfer.

ECUR: Event Charging with Unit Reservation.

EDGE: Enhanced Data in GSM Evolution.

EDP: Event Detection Point.

EF: Expected Forwarding (a subclass of the DiffServ).

EFM: Ethernet in the First Mile.

EIR: Equipment Identity Register.

EGPRS: Enhanced GPRS (EDGE).

EJB: Enterprise Java Beans.

ENUM: Electronic NUMbering (mapping E.164 numbers to SIP URIs).

E-OTD: Enhanced–Observed Time Difference.

EPA: Event Publication Agent.

ERAN: EDGE Radio Access Network.

ESI: Enabler Server Implementation (OMA)

ESP: Encapsulating Security Payload.

ETI: Enabler Terminal Implementation (OMA).

ETSI: European Telecommunications Standards Institute.

FAN: Fixed Access Network.

FBC: Flow Based Charging.

FC: Filter Criteria.

FDD: Frequency Division Duplex.

FDDI: Fiber Distributed Data Interface.

FMC: Fixed Mobile Convergence.

FMCA: Fixed Mobile Convergence Alliance.

FQDN: Fully Qualified Domain Name.

FQPC: Fully Qualified Partial CDR.

FR: Frame Relay.

FTAM: File Transfer, Access and Management.

FTP: File Transfer Protocol.

FTTP: Fiber to the Premises.

FW: Firewall.

G.711, G.729: ITU-T voice codec standards.

GAA: Generic Authentication Architecture.

GBA: Generic Bootstrapping Architecture.

GCC: Generic Call Control (Parlay).

GCID: GPRS Charging ID.

GCP: Gateway Control Protocol.

GERAN: GSM EDGE Radio Access Network.

GGSN: Gateway GPRS Support/Service Node — 3GPP.

GigE: Gigabit Ethernet.

GK: Gatekeeper.

GLMS: Group List Management Server.

GMLC: Gateway Mobile Location Center.

GMPLS: Generalized Multi-Protocol Label Switching.

GMSC: Gateway Mobile Switching Center.

GOCAP: Generic Overload Control Activation Protocol.

GoS: Grade-of-Service.

GPC: Geographic/Global Position Control Function.

GPRS: General Packet Radio Service.

GPRS-CID: GPRS Charging Identifier.

GPRS-CSI: GPRS CAMEL Subscription Information.

GPS: Global Positioning System.

GRUU: Globally Routable User Agent URI.

GRX: GPRS Roaming eXchange.

GSM: Global System for Mobile communications.

GSMA: GSM Association.

GSMA IREG: GSMA International Roaming Experts Group.

GsmSCF: GSM Service Control Function (INCAMEL).

GSM: Global System for Mobile (2G radio-based communications).

GSN: GPRS Support Node (SGSN, GGSN).

GSSF: Gateway SSF.

GUI: Graphical User Interface.

GUP: Generic User Profile.

GW: Gateway.

H.248: ITU-T H.248 Gateway Control Protocol.

H.323: ITU-T Recommendation H.323.

HDSL: High-Speed Digital Subscriber Line.

HLR: Home Location Register.

HO: Handover (UMTS-WLAN).

H-PLMN: Home PLMN.

HSCSD: High-Speed Circuit-Switched Data.

HSDPA: High-Speed Downlink Packet Access.

HSS: Home Subscriber Server.

HTML: HyperText Markup Language.

HTPP: High Touch Packet Processing.

HTTP: HyperText Transfer Protocol (IETF RFC 2068).

HTTPS: Secure HTTP (over TLS).

IAD: Integrated Access Device (home or office hubs).

IANA: Internet Assigned Numbering Authority.

IBCF: Interconnection Border Control Function.

I-BGF: Interconnection-Border Gateway Function

IBGP: Inter-network Border Gateway Protocol.

IC: Integrated circuit (ICC: IC Card).

ICE: Interactive Connectivity Establishment (Media NAT methodology).

ICID: IMS Charging ID/IDentifier).

ICM: Intelligent Contact Management.

ICS: IMS Centralized Services.

I-CSCF: Interrogating Call Session Control Function.

ICT: Information Communication Technologies.

ICW: Internet Call Waiting.

ID: Identity (or Identifier), for a user, device, or service endpoint.

IDL: Interface Definition Language.

IEEE: Institute of Electronic and Electrical Engineers.

IETF: Internet Engineering Task Force.

iFC: Initial Filter Criteria.

IGMP: Internet Group Multicast Protocol.

IK: Integrity Key.

IKE: Internet Key Exchange.

ILCM: Incoming Leg Control Model.

IM: IP Multimedia (usually in reference to IMS).

IM: Instant Messaging.

IM CN: IP Multimedia Core Network Subsystem (IMS).

IMAP: Internet Message Access Protocol (IETF 3501).

IM-BCSM: IP Multimedia – Basic Call State Model.

IMCN: IP Multimedia Core Network.

IM-CSI: IP Multimedia – CAMEL Subscription Information.

IMEI: International Mobile-station Equipment Identity.

IMGI: International Mobile Group Identity.

IMPI: IP Multimedia Private-user Identity.

IMPU: IP Multimedia Public User identity.

IMR: Interactive Media Response.

IMRN: IP Multimedia Routing Number.

IMS: IP Multimedia Subsystem.

IMSI: International Mobile Subscriber Identity.

IM-SSF: IP Multimedia Service Switching Function.

IN: Intelligent Network.

INAP: Intelligent Network Application Part.

INE: Intercepting Network Elements (Lawful Interception).

IOI: Inter-Operator Identification.

IP: Internet Protocol.

IP: Intelligent Peripheral, in intelligent networks.

IP-CAN: IP-Connectivity Access Network.

IPDR: Internet Protocol Detail Record.

IPSec: Internet Protocol Security (IETF RFC 2401).

IPTV: Internet Protocol TV (television).

IPv4: Internet Protocol Version 4.

IPv6: Internet Protocol Version 6.

IP-VPN: Internet Protocol Virtual Private Network.

ISC: IP Multimedia Service Control.

ISDN: Integrated Services Digital Network.

ISIM: IMS Subscriber Identity Module.

ISO: International Standards Organization.

ISP: Internet Service Provider.

ISUP: ISDN User Part protocol (SS7/C7) for TDM services.

ITSP: Internet Telephony Service Provider.

ITU-T: International Telecommunication Union – Telecommunication.

IVR: Interactive Voice Response.

IWF: Interworking Function.

I-WLAN: Interworking-Wireless Local Area Network.

J2EE: Java 2 Platform, Enterprise Edition.

J2ME: Java 2 Platform, Micro Edition.

JAIN: Java API for Integrated Networks.

KC: Key Ciphering.

L1: Transmission Layer 1.

L2: Transmission Layer 2.

L2TP: Layer 2 Tunneling Protocol.

L2VPN: Layer 2 Virtual Private Network.

L3: Layer 3.

L3VPN: Layer 3 Virtual Private Network.

LAC: L2TP Access Concentrator.

LAN: Local Area Network.

LBS: Location-Based Services.

LDAP: Lightweight Directory Access Protocol (IETF RFC 3494).

LE: Local Exchange.

LEMF: Law Enforcement Monitoring Facility.

LI: Lawful Interception.

LLU: Local Loop Unbundling.

LPDP: Local Policy Decision Point.

LRF: Location Retrieval Function.

LTE: Long-Term Evolution (after 3G).

M3UA: IP Protocol for transport of any SS7 MTP3-User signaling (e.g., ISUP, SCCP, and TUP messages).

MAC: Message Authentication Code.

MACinMAC: Encapsulation of a MAC by another MAC.

MAP: Mobile Application Part.

MBMS: Multimedia Broadcast/Multicast Services.

MBS: Multicast Broadcast Service.

MCC: Mobile Country Code.

M-CSI: Mobility Management event CAMEL Subscription Information.

MCU: Multi-Conferencing Unit; Multi-point Conference-control Unit.

ME: Mobile Equipment (part of UE).

Megaco: Media Gateway Control protocol.

MG: Media Gateway.

MGC: Media Gateway Controller.

MGCF: Media Gateway Control Function.

MGCP: Media Gateway Control Protocol.

MGF: Media Gateway Function.

MGPF: Mobile Geographic Control Function.

MGW: Media Gateway.

MIB: Management Information Base (IETF).

MIDCOM: Middleware Communications (for traversing firewalls).

MIME: Multipurpose Internet Mail Extensions (IETF RFC 2045 — e-mail attachments format).

MIP: Mobile IP (IETF RFC 2002).

MIPv6: Mobile IPv6.

MIS: Management Information System.

MISP: Mobile Internet Service Provider.

MLC: Mobile Location Center.

MM: Multimedia.

MME: Mobile Management Entity.

MMI: Man-Machine Interface.

MMS: Multimedia Messaging Service.

MNC: Mobile Network Code/Control point.

MO: Managed Object (X.700).

MOML: Media Objects Markup Language.

MoU: Memorandum of Understanding.

MP3: MPEG-1 Audio Layer 3, (the *de facto* Internet music standard).

MPCC: Multi-Party Call Control (Parlay).

MPEG: Moving Picture Experts Group.

MPLS: Multi-Protocol Label Switching.

MRCP: Media Resource Control Protocol.

MRF: Media Resource Function.

MRFC: Media Resource Function Controller.

MRFP: Media Resource Function Processor.

MS: Mobile Station (ME + USIM).

MSAD: Multi-Service Access Device.

MSAN: Multi-Service Access Node.

MSC: Mobile Switching Center.

MSCML: Media Server Control Markup Language.

MSF: Multiservice Switching Forum.

MSIN: Mobile Subscriber Identification Number (PLMN).

MSISDN: Mobile Subscriber (Integrated Services Data Network (international number format).

MSML: Media Session Markup Language.

MSRN: Mobile Station Roaming Number.

MSRP: Message Session Relay Protocol.

MT: Mobile Terminating.

MTNM: Multi-Technology Network Management (including SONET, SDH, DWDM, ATM, Ethernet).

MTP: Message Transfer Part, Layer 1, 2, or 3 (PSTN/ISDN).

MTU: Maximum Transmission Unit.

MVNO: Mobile Virtual Network Operator.

NACF: Network Access Configuration Function.

NAF: Network Application Function.

NAI: Network Access Identifier as defined (RFC 2486).

NAIF: Network Authorization and Interconnect Facility.

NAP: Network Access Point.

NAP: Network Access Provider (WiMAX).

NAPT: Network Address and Port Translation.

NAPTR: Naming Authority Pointer (DNS Resource Record RFC 2915).
NAS: Network Access Server (IETF RFC 2881).
NAS: Network Attached Storage.
NASS: Network Attachment Sub-System (TISPAN).
NAT: Network Address Translation.
NAT-PT: Network Address Translation–Protocol Translation.
NDS: Network Domain Security.
NETANN: Network Announcement (SIP APIs).
NGN: Next Generation Network.
NM: Network Management.
NMS: Network Management System.
NNI: Network-Network Interface.
NP: Number Portability.
NSS: Narrowband Signaling Syntax (e.g., SIP-I).
NT: Network Termination.
NTE: Network Termination Equipment.
NTP: Network Time Protocol (IETF RFC 1305).
NTU: Network Termination Unit.
NUP: National User Part (SS7 Signaling).
NW: Network.
O&M: Operations and Maintenance.
OASIS: Organization for the Advancement of Structured Information Standards.
OCF: Online Charging Function.
OCS: Online Charging Server, Online Charging System.
OFCS: Offline Charging Server, Offline Charging System.
OLCM: Outgoing Leg Control Model.
OMA: Open Mobile Alliance.
OS: Operating System.
OSA: Open Service Architecture (Parlay).
OSE: OMA Service Environment (OMA).
OSF: Operations System Function.
OSI: Open Systems Interconnection.
OSPF: Open Shortest Path First (IETF RFC 2328).
OSS: Operational Support System.
P2P: Peer-to-Peer.
PA: Presence Agent.
PABX: Private Automatic Branch eXchange.
PAM: Presence and Availability Management.
PANA: Protocol of Authentication for Network Access (RFC 4058).
PAP: Password Authentication Protocol.
Parlay: International API specification equivalent to OSA.
PAT: Port Address Translation.
PBB-TE: Provider Backbone Bridge–Traffic Engineering (Ethernet).
PBT: Provider Backbone Transport (Ethernet).
PBX: Private Branch eXchange.
PC: Personal Computer.
PCC: Policy and Charging Control.

PCEF: Policy Control Enforcement Function.
PCF: Policy Control Function.
PCRF: Policy and Charging Rules Function.
P-CSCF: Proxy Call Session Control Function.
PCU: Packet Control Unit.
PDA: Personal Digital Assistant.
PDB: Presence Database.
PDBF: Profile Database Function (TISPAN NASS).
PDC: Packet Data Context.
PDF: Policy Decision Function.
PDG: Packet Data Gateway.
PDN: Packet Data Network.
PDP: Policy Decision Point.
PDP Context: Packet Data Protocol Context (GPRS/UMTS).
PDU: Packet Data Unit.
PEAP: Protected EAP, using TLS to protect EAP.
PEEM: Policy Evaluation, Enforcement, and Management (OMA).
PEP: Policy Enforcement Point.
PES: PSTN/ISDN Emulation Subsystem (TISPAN).
PHB: Per-Hop Behavior (DiffServ, QoS parameters).
PIB: Policy Information Base.
PIDF: Presence Information Data Format.
PIM: Protocol-Independent Multicast.
PIN: Personal Identification Number.
PINT: PSTN and INTernet interworking.
PKCS: Public Key Cryptography Standard.
PKI: Public Key Infrastructure.
PKM: Privacy Key Management.
PLMN: Public Land Mobile Network.
PN: Presentation Number.
PNA: Presence Network Agent.
PNO: Public Network Operator.
PoC: Push-to-Talk over Cellular.
PoP: Point of Presence.
POP3: Post Office Protocol version 3 (IETF RFC 0937 e-mail).
POTS: Public-Owned Telephone Service.
PPP: Point-to-Point Protocol (IETF RFC 1661).
PPTP: Point-to-Point Tunneling Protocol.
PRI: Primary Rate Interface (ISDN).
PS: Packet Switched.
PSAP: Public Safety Answering Point (emergency agency).
PSCF: Packet Service Control Function.
PSI: Public Service Identifier, Public Service Identity.
PSK: Preshared Key.
PSN: Public-Switched Network (was PSTN).
PSS: PSTN Simulation Subsystem.
PSTN: Public-Switched Telephone Network.

PTT: Push to Talk.
PUA: Presence User Agent.
QoS: Quality-of-Service.
QSIG: PBX signaling for Private Networks carried over PSTN.
RAB: Radio Access Bearer.
RACF: Resource and Admission Control Function (ITU).
RACS: Resource and Admission Control Subsystem (TISPAN).
RADIUS: Remote Authentication Dial-In User Service (IETF RFC 2865).
RAN: Radio Access Network.
RAND: Random challenge (security).
RAS: Route Analysis System.
RCAF: Radio Control Access Function.
RCEF: Resource Control Emulation Function.
RES: RESponse to challenge (expected from the user in authentication procedure).
RFC: Request For Comments.
RFID: Radio Frequency Identity.
RGW: Residential Gateway.
RIP: Routing Information Protocol.
RLS: Resource List Server.
RMI: Remote Method Invocation (Java).
RNC: Radio Network Controller.
RSA: Rivest, Shamir, and Adleman (asymmetric encryption system).
RSVP: Resource ReSerVation Protocol (RFC 2205).
RSVP-TE: Resource ReSerVation Protocol with Traffic Engineering.
RTCP: Real-time Transport Control Protocol.
RTF: Real-Time Framework.
RTP: Real-time Transport Protocol (IETF RFC 3550).
RTSP: Real-Time Streaming Protocol (IETF RFC 2326).
S2ML: Security Services Markup Language.
SA: Security Association (in IPSec).
SAE: System Architecture Evolution.
SALT: Speech Application Language Tags.
SAML: Security Assertion Markup Language.
SAN: Storage Area Network.
SAT: SIM Card Application Toolkit.
SBLP: Service Based Local Policy.
SCCF: Subscriber Content Charging Function.
SCCP: Signaling Connection Control Part.
SCD: Session Control Domain.
SCE: Service Creation Environment.
SCF: Service Control Function (IN).
SC-GF: Service Control–Gateway Function.
SCIM: Service Capability Interaction Manager.
SCN: Switched Circuit Network.
SCORM: Sharable Content Object Reference Model.
SCP: Service Control Point (IN).
SCR-SP: Session Control and Routing–Service Provider.

SCR-SS: Session Control and Routing–Special Services.
SCS: Service Capability Server.
S-CSCF: Serving Call Session Control Function.
SDA: Synchronization Distribution Amplifier.
SDF: Service Data Function.
SDH: Synchronous Digital Hierarchy.
SDK: Software Development Kit.
SDL: Service Description Language.
SDP: Session Description Protocol (SIP).
SDSL: Synchronous Digital Subscriber Loop/Line.
SDU: Service Data Unit (i.e., data packet).
SEF: Service Execution Function.
SEG: Security Gateway.
SFC: Subsequent Filter Criteria.
SG: Signaling Gateway.
SGF: Signaling Gateway Function.
SGSN: Serving GPRS Support Node.
SGW: Signaling Gateway.
SigTrans: Signaling Translation.
SIM: Subscriber Identity Module (GSM).
SIMPLE: SIP and IM Presence Leveraging Extensions.
SIP: Session Initiation Protocol (IETF RFC 3261).
SIP UA: SIP User Agent.
SIPS: SIP Secure (SIP over TLS/Transport Layer Security).
SIP-T: Session Initiation Protocol for Telecommunications.
SLA: Service Level Agreement.
SLF: Subscription Locator Function (on the HSS database).
SLP: Service Logic Program.
SME: Small to Medium Enterprise.
SMPP: Short Message Peer-to-Peer Protocol.
SMS: Session Management Server.
SMSC: Short Message Service Center.
SMTP: Simple Mail Transfer Protocol (IETF RFC 2821e-mail).
SNMP: Simple Network Management Protocol (IETF RFC 1157).
SNMP MIB: SNMP Management Information Base.
SNMS: Signaling Network Management System.
SOA: Service Oriented Architecture.
SOAP: Simple Object Access Protocol (exchanging XML-based information).
SOHO: Small Office/Home Office.
SONET: Synchronous Optical NETwork.
SP: Service Provider.
SPDF: Service Policy Decision Function.
SPF: Service Provider Framework.
SPML: Service Provisioning Markup Language.
SPR: Subscription Profile Repository (for IP-CAN internal data).
SPT: Service Point Trigger (in S-CSCF).
S-PVC: Soft Permanent Virtual Circuit.

SQL: Structured Query Language.

SQN: Sequence Number.

SS7: Signaling System No. 7 (ITU-T).

SSF: Service Subscription Function.

SSF: Service Switching Function.

SSID: Service Set ID (for WLAN).

SSL: Secure Sockets Layer (superseded by TLS).

SSML: Speech Synthesis Markup Language.

SSO: Single Sign-On.

SSP: Service Switching Point (Intelligent Networks).

SSU: Synchronization Supply Unit.

STB: Set-top box.

STM1: Synchronous Transport Module level 1 (SDH), as is STM16/64.

STP: Service (platform) Trigger Point (in AS).

STT: Speech-to-Text.

T.38: ITU-T T.38: (Procedures for real-time Group 3 facsimile communication over IP networks).

TCAP: Transaction Capability Application Part.

TCP: Transmission Control Protocol.

TCP/IP: Transmission Control Protocol/Internetworking Protocol.

TDM: Time Division Multiplex (common name for traditional circuit-based networks).

TE: Terminal Equipment.

TFT: Traffic Flow Template.

TGW: Trunking Gateway.

THIG: Topology Hiding IP Gateway.

TINA: Telecommunications Information Networking Architecture.

TISPAN: Telecommunications and Internet Services and Protocols for Advanced Networks (ETSI).

TLD: Top-Level Domain.

TLS: Transparent LAN Service.

T-MGW: Trunking Media Gateway.

TMN: Telecommunication Management Network.

TMSI: Temporary Mobile Subscriber Identity.

TOS: Type Of Service (DiffServ).

TPF: Traffic Plane Function.

TrGW: Transition/Translation Gateway (IPv4 – IPv6).

T-SGW: Transport Signaling Gateway.

TTLS: Tunneled TLS.

TTS: Text-To-Speech.

UA: User Agent.

UAAF: User Access Authorization Function.

UAC: User Agent Client.

UAS: User Agent Server.

UDDI: Universal Description, Discovery, and Integration (Web services).

UDP: User Datagram Protocol (IETF RFC 768).

UE: User Equipment.

UI: User Interface.

UICC: UMTS Integrated Circuit Card.
UMA: Unlicensed Mobile Access (also MSP).
UMTS: Universal Mobile Telecommunications System.
UNI: User-Network Interface.
UPS: User Profile Server.
UPSF: User Profile Server Function (TISPAN).
URI: Uniform Resource Identifier (IETF RFC 2396).
URL: Uniform Resource Locator (IETF RFC 1738).
USIM: UMTS Subscriber Identity Module (User SIM), or User Service Identity Module.
USSD: Unstructured Supplementary Service Data.
UTRAN: UMTS Terrestrial Radio Access Network.
VAS: Value-Added Service.
VCC: Voice Call Continuity.
VDSL: Very high speed Digital Subscriber Loop/Line.
VHE: Virtual Home Environment.
VLAN: Virtual LAN.
VLR: Visitor Location Register (3GPP).
VMNO: Virtual MNO.
VMSC: Visited Mobile Switching Center.
VNO: Virtual Network Operator.
VoATM: Voice over Asynchronous Transfer Mode.
VoBB: Voice over Broadband.
VoD: Video on Demand.
VoDSL: Voice-over-Digital Subscriber Loop.
VoiceXML: Voice eXtensible Markup Language.
VoIP: Voice-over-Internet Protocol.
VoP: Voice-over-Packet.
V-PLMN: Visited Public Land Mobile Network.
VPLS: Virtual Private LAN Service.
VPN: Virtual Private Network.
VR: Voice Response.
VXML: Voice extensible Markup Language.
WAG: WLAN Access Gateway.
WAN: Wireless Access Network.
WAP: Wireless Application Protocol.
W-APN: WLAN APN.
WAV: WAV form format.
WCDMA: Wideband Code-Division Multiple Access.
WDM: Wavelength Division Multiplexing.
WGP: Wireless Gateway Proxy.
WiFi: Wireless Fidelity (802.11 standards, including 802.11b, 802.11a, and 802.11g).
WiMAX: Worldwide Interoperability for Microwave Access (IEEE 802.16).
WLAN: Wireless Local Area Network (Wi-Fi 802.11 series).
WLAN AN: Wireless Local Area Network Access Network.
WLAN UE: Wireless Local Area Network User Equipment.
WLL: Wireless Local Loop.
WML: Wireless Markup Language.

WPA: WiFi Protected Access.
WPP: Wireless Push Proxy.
WS: Web Services.
WSDL: Web Service Description Language.
WSID: WLAN Specific Identifier.
X.500: ITU-T X.500. Directory Service.
XACML: eXtensible Access Control Markup Language.
XCAP: XML Configuration Access Protocol.
XDMS: XML Document Management Server.
xDSL: Digital Subscriber Line variants include Asymmetric, High Speed, Symmetric, Very High Speed.
XMAC: eXpected Message Authentication Code.
XML: eXtensible Markup Language.
XMPP: eXtensible Messaging and Presence Protocol.
XRES: Expected Response.

Appendix C

Sources and References

C.1 IETF RFCs

Spec	Description
RFC 768	UDP-User Datagram Protocol
RFC 1321	The MD5 Message-Digest Algorithm
RFC 2205	RSVP
RFC 2210	RSVP with IETF Integrated Services
RFC 2246	TLS/SSL
RFC 2396	Uniform Resource Identifiers (URI): Generic Syntax
RFC 2401	IPSec, ESP AH
RFC 2462	IPv6 Address Autoconfiguration
RFC 2486	The Network Access Identifier
RFC 2616	Securing apps
RFC 2617	Using AKA as Digest Auth for SIP Registration
RFC 2617	HTTP Authentication: Basic and Digest Access Authentication
RFC 2663	IP Network Address Translator (NAT) Terminology and Considerations
RFC 2748	The COPS (Common Open Policy Service) Protocol
RFC 2766	Network Address Translation-Protocol Translation (NAT-PT)
RFC 2833	RTP Payload for DTMF Digits, Telephony Tones and Telephony Signals
RFC 2915	The Naming Authority Pointer (NAPTR) DNS Resource Record
RFC 2976	The SIP INFO method
RFC 3041	Privacy Extensions for Stateless Address Autoconfiguration in IPv6
RFC 3084	COPS and COPS-PR

Spec	Description
RFC 3174	US Secure Hash Algorithm 1 (SHA1)
RFC 3261	SIP: Session Initiation Protocol
RFC 3262	Reliability of Provisional Responses in Session Initiation Protocol (SIP)
RFC 3263	Session Initiation Protocol (SIP): Locating SIP Servers
RFC 3264	An Offer/Answer Model with Session Description Protocol (SDP)
RFC 3265	Session Initiation Protocol (SIP) Specific Event Notification
RFC 3310	Hypertext Transfer Protocol (HTTP) Digest Authentication Using Authentication and Key Agreement (AKA)
RFC 3311	The Session Initiation Protocol (SIP) The Session Initiation Protocol (SIP) UPDATE Method
RFC 3312	Integration of resource management and Session Initiation Protocol (SIP)
RFC 3313	Private Session Initiation Protocol (SIP) Extensions for Media Authorization
RFC 3315	Dynamic Host Configuration Protocol for IPv6 (DHCPv6)
RFC 3320	Signaling Compression (SigComp)
RFC 3323	A Privacy Mechanism for the Session Initiation Protocol (SIP)
RFC 3325	Private Extensions to the Session Initiation Protocol (SIP) for Network Asserted Identity within Trusted Networks
RFC 3326	The Reason Header Field for the Session Initiation Protocol (SIP)
RFC 3327	Session Initiation Protocol Extension Header Field for Registering Non-Adjacent Contacts
RFC 3329	Security Mechanism Agreement for the Session Initiation Protocol (SIP)
RFC 3428	Session Initiation Protocol (SIP) Extension for Instant Messaging
RFC 3455	Private Header (P-Header) Extensions to the Session Initiation Protocol (SIP) for the 3rd-Generation Partnership Project (3GPP)
RFC 3485	The Session Initiation Protocol (SIP) and Session Description Protocol (SDP) Static Dictionary for Signaling Compression (SigComp)
RFC 3524	Single Reservation Flow in SDP
RFC 3524	Mapping of Media Streams to Resource Reservation Flows
RFC 3556	Session Description Protocol (SDP) Bandwidth Modifiers for RTP Control Protocol (RTCP) Bandwidth
RFC 3588	Diameter
RFC 3608	Session Initiation Protocol (SIP) Extension Header Field for Service Route Discovery during Registration
RFC 3646	DNS Configuration options for Dynamic Host Configuration Protocol for IPv6 (DHCPv6)
RFC 3680	A Session Initiation Protocol (SIP) Event Package for Registrations
RFC 3761	The E.164 to Uniform Resource Identifiers (URI) Dynamic Delegation Discovery System (DDDS) Application (ENUM)
RFC 3840	Indicating User Agent Capabilities in the Session Initiation Protocol (SIP)
RFC 3841	Caller Preferences for the Session Initiation Protocol (SIP)
RFC 3948	UDP Encapsulation of IPSec ESP Packets

Spec	Description
RFC 3966	The tel URI for Telephone Numbers
RFC 4960	SCTP – Stream Control Transmission Protocol
RFC 4457	The SIP P-User-Database Private-Header (P-Header)
RFC 4566	SDP: Session Description Protocol
RFC 4662	A Session Initiation Protocol (SIP) Event Notification Extension for Resource Lists

C.2 3GPP

Type	Spec	Description
TS	21.905	Vocabulary for 3GPP Specifications
TS	22.086	Advice of Charge (AoC) supplementary services; Stage 1
TS	22.090	Unstructured Supplementary Service Data (USSD); Stage 1
TS	22.091	Explicit Call Transfer (ECT) supplementary service; Stage 1
TS	22.093	Completion of Calls to Busy Subscriber (CCBS); Service description, Stage 1
TS	22.094	Follow Me service description - Stage 1
TS	22.096	Name identification supplementary services; Stage 1
TS	22.097	Multiple Subscriber Profile (MSP) Phase 2; Service description; Stage 1
TS	22.115	Service aspects; Charging and billing
TS	22.121	Service aspects; The Virtual Home Environment (VHE); Stage 1
TS	22.129	Handover requirements between UTRAN and GERAN or other radio systems
TS	22.141	Presence service; Stage 1
TS	22.173	IMS Multimedia Telephony Communication Enabler and supplementary services
TS	22.228	Service requirements for the Internet Protocol (IP) multimedia core network subsystem (IMS); Stage 1
TS	22.234	Requirements on 3GPP system to Wireless Local Area Network (WLAN) interworking
TS	22.240	Service requirements for 3GPP Generic User Profile (GUP); Stage 1
TS	22.242	Digital Rights Management (DRM); Stage 1
TS	22.250	IP Multimedia Subsystem (IMS) Group Management; Stage 1
TS	22.258	Service requirements for an All-IP Network (AIPN); Stage 1
TS	22.279	Combined Circuit Switched (CS) and IP Multimedia Subsystem (IMS) sessions; Stage 1
TS	22.340	IP Multimedia Subsystem (IMS) messaging; Stage 1
TS	22.903	Study on Video-telephony tele-service
TS	22.934	Feasibility study on 3GPP system to Wireless Local Area Network (WLAN) interworking

Type	Spec	Description
TS	22.935	Feasibility study on Location Services (LCS) for Wireless Local Area Network (WLAN) interworking
TS	22.936	Multi-system terminals
TS	22.940	IP Multimedia Subsystem (IMS) messaging; Stage 1
TS	22.953	Multimedia priority service feasibility study
TS	22.967	Transferring of emergency call data
TS	22.973	IMS Multimedia Telephony service; and supplementary services
TS	22.978	All-IP network (AIPN) feasibility study
TS	22.979	Feasibility study on combined Circuit-Switched (CS) calls and IP Multimedia Subsystem (IMS) sessions
TS	23.002	Network architecture
TS	23.003	Numbering, addressing and identification
TS	23.008	Organization of subscriber data
TS	23.009	Handover procedures
TS	23.011	Technical realization of Supplementary Services
TS	23.012	Location management procedures
TS	23.014	Support of Dual Tone Multi Frequency (DTMF) signaling
TS	23.016	Subscriber data management; Stage 2
TS	23.018	Basic Call Handling; Technical realization
TS	23.031	Fraud Information Gathering System (FIGS); Service description; Stage 2
TS	23.082	Call Forwarding (CF) Supplementary Services; Stage 2
TS	23.083	Call Waiting (CW) and Call Hold (HOLD) Supplementary Service; Stage 2
TS	23.084	Multi-Party (MPTY) Supplementary Service; Stage 2
TS	23.085	Closed User Group (CUG) Supplementary Service; Stage 2
TS	23.086	Advice of Charge (AoC) Supplementary Service; Stage 2
TS	23.087	User-to-User Signaling (UUS) supplementary service; Stage 2
TS	23.088	Call Barring (CB) Supplementary Service; Stage 2
TS	23.09	Unstructured Supplementary Service Data (USSD); Stage 2
TS	23.091	Explicit Call Transfer (ECT) Supplementary Service; Stage 2
TS	23.093	Technical realization of Completion of Calls to Busy Subscriber (CCBS); Stage 2
TS	23.094	Follow Me Stage 2
TS	23.096	Name identification supplementary service; Stage 2
TS	23.097	Multiple Subscriber Profile (MSP) Phase 1; Stage 2
TS	23.107	Quality of Service (QoS) concept and architecture
TS	23.125	Overall high level functionality and architecture impacts of flow based charging; Stage 2
TS	23.135	Multi-call supplementary service; Stage 2
TS	23.141	Presence service; Architecture and functional description; Stage 2

Type	Spec	Description
TS	23.167	IP Multimedia Subsystem (IMS) emergency sessions
TS	23.171	Location Services (LCS); Functional description; Stage 2 (UMTS)
TS	23.203	Policy and charging control architecture
TS	23.206	Voice call continuity between Circuit Switched (CS) and IP Multimedia Subsystem (IMS); Stage 2
TS	23.207	End-to-end Quality of Service (QoS) concept and architecture
TS	23.218	IP Multimedia (IM) session handling; IM call model; Stage 2
TS	23.221	Architectural requirements
TS	23.227	Application and user interaction in the UE; Principles and specific requirements
TS	23.228	IP Multimedia Subsystem (IMS); Stage 2
TS	23.234	3GPP system to Wireless Local Area Network (WLAN) interworking; System description
TS	23.240	3GPP Generic User Profile (GUP) requirements; Architecture (Stage 2)
TS	23.271	Functional stage 2 description of Location Services (LCS)
TS	23.278	Customized Applications for Mobile network Enhanced Logic (CAMEL) Phase 4; Stage 2; IM CN Interworking
TS	23.279	Combining Circuit Switched (CS) and IP Multimedia Subsystem (IMS) services; Stage 2
TS	23.802	Architectural enhancements for end-to-end Quality of Service (QoS)
TS	23.803	Evolution of policy control and charging
TS	23.805	Selective disabling of User Equipment (UE) capabilities; Report on technical options and conclusions
TS	23.806	Voice call continuity between Circuit Switched (CS) and IP Multimedia Subsystem (IMS) Study
TS	23.808	Supporting Globally Routable User Agent URI in IMS; Report and conclusions
TS	23.815	Charging implications of IMS architecture
TS	23.816	Identification of communication services in IMS
TS	23.836	Quality of Service (QoS) and policy aspects of 3GPP - Wireless Local Area Network (WLAN) interworking
TS	23.837	Location Services (LCS) architecture for 3GPP system - Wireless Local Area Network (WLAN) interworking
TS	23.841	Presence service architecture
TS	23.867	Internet Protocol (IP) based IP Multimedia Subsystem (IMS) emergency sessions
TS	23.871	Enhanced support for user privacy in Location Services (LCS)
TS	23.882	3GPP system architecture evolution (SAE): Report on technical options and conclusions
TS	23.934	3GPP system to Wireless Local Area Network (WLAN) interworking; Functional and architectural definition

Type	Spec	Description
TS	23.941	3GPP Generic User Profile (GUP); Stage 2; Data Description Method (DDM)
TS	23.979	3GPP enablers for Open Mobile Alliance (OMA) Push-to-talk over Cellular (PoC) services; Stage 2
TS	23.981	Interworking aspects and migration scenarios for IPv4-based IP Multimedia Subsystem (IMS) implementations
TS	24.141	Presence service using the IP Multimedia (IM) Core Network (CN) subsystem; Stage 3
TS	24.206	Voice call continuity between Circuit Switched (CS) and IP Multimedia Subsystem (IMS); Stage 3
TS	24.228	Signaling flows for the IP multimedia call control based on Session Initiation Protocol (SIP) and Session Description Protocol (SDP); Stage 3
TS	24.229	Internet Protocol (IP) multimedia call control protocol based on Session Initiation Protocol (SIP) and Session Description Protocol (SDP); Stage 3
TS	24.234	3GPP system to Wireless Local Area Network (WLAN) interworking; WLAN User Equipment (WLAN UE) to network protocols; Stage 3
TS	24.241	3GPP Generic User Profile (GUP) Common objects; Stage 3
TS	24.247	Messaging using the IP Multimedia (IM) Core Network (CN) subsystem; Stage 3
TS	24.279	Combining Circuit Switched (CS) and IP Multimedia Subsystem (IMS) services; Stage 3
TS	24.819	Protocol impact from providing IMS services via fixed broadband; Stage-3
TS	24.841	Presence service based on Session Initiation Protocol (SIP); Functional models, information flows and protocol details
TS	24.879	Combining Circuit-Switched (CS) calls and IP Multimedia Subsystem (IMS) sessions
TS	24.930	Signaling flows for the session setup in the IM CN subsystem (IMS) based on Session Initiation Protocol (SIP) and Session Description Protocol (SDP); Stage 3
TS	26.141	IP Multimedia System (IMS) Messaging and Presence; Media formats and codecs
TS	29.207	Policy control over Go interface
TS	29.208	End-to-end Quality-of-Service (QoS) signaling flows
TS	29.209	Policy control over Gq interface
TS	29.21	Charging rule provisioning over Gx interface
TS	29.211	Rx Interface and Rx/Gx signaling flows
TS	29.212	Policy and charging control over Gx reference point
TS	29.213	Policy and charging control signaling flows and Quality of Service (QoS) parameter mapping
TS	29.214	Policy and charging control over Rx reference point
TS	29.228	IP Multimedia (IM) Subsystem Cx and Dx Interfaces; signaling flows and message contents
TS	29.229	Cx and Dx interfaces based on the Diameter Protocol; Protocol details

Type	Spec	Description
TS	29.230	Diameter applications; 3GPP specific codes and identifiers
TS	29.232	Media Gateway Controller (MGC) - Media Gateway (MGW) interface; Stage 3
TS	29.234	3GPP system to Wireless Local Area Network (WLAN) interworking; Stage 3
TS	29.240	3GPP Generic User Profile (GUP); Stage 3; Network
TS	29.328	IP Multimedia Subsystem (IMS) Sh interface; Signaling flows and message contents
TS	29.329	Sh interface based on the Diameter protocol; Protocol details
TS	29.332	Media Gateway Control Function (MGCF) - IM Media Gateway (IM-MGW) Mn interface
TS	29.333	Multimedia Resource Function Controller (MRFC) - Multimedia Resource Function Processor (MRFP) Mp interface; Stage 3
TS	29.414	Core network Nb data transport and transport signaling
TS	29.415	Core network Nb interface user plane protocols
TS	29.846	Multimedia Broadcast/Multicast Service (MBMS); CN1 procedure description
TS	29.847	Conferencing based on SIP, SDP, and other protocols; Functional models, information flows and protocol details
TS	29.962	Signaling interworking between the 3GPP profile of the Session Initiation Protocol (SIP) and non-3GPP SIP usage
TS	31.102	Characteristics of the Universal Subscriber Identity Module (USIM) application
TS	31.103	Characteristics of the IP Multimedia Services Identity Module (ISIM) application
TS	31.133	IP Multimedia Services Identity Module (ISIM) Application Programming Interface (API) for Java Card
TS	32.225	Telecommunication management; Charging management; Charging data description for the IP Multimedia Subsystem (IMS)
TS	32.251	Telecommunication management; Charging management; Packet Switched (PS) domain charging
TS	32.252	Telecommunication management; Charging management; Wireless Local Area Network (WLAN) charging
TS	32.296	Telecommunication management; Charging management; Online Charging System (OCS): Applications and interfaces
TS	32.815	Telecommunication management; Charging management; Online Charging System (OCS) architecture study
TS	33.102	3G security; Security architecture
TS	33.103	3G security; Integration guidelines
TS	33.105	Cryptographic algorithm requirements
TS	33.106	Lawful interception requirements
TS	33.107	3G security; Lawful interception architecture and functions
TS	33.108	3G security; Handover interface for Lawful Interception (LI)

Type	Spec	Description
TS	33.12	Security Objectives and Principles
TS	33.141	Presence service; Security
TS	33.203	3G security; Access security for IP-based services
TS	33.204	3G Security; Network Domain Security (NDS); Transaction Capabilities Application Part (TCAP) user security
TS	33.210	Network Domain Security (NDS)
TS	33.220	Generic Authentication Architecture (GAA); Generic bootstrapping architecture
TS	33.220	Generic Bootstrapping
TS	33.221	Generic Authentication Architecture (GAA); Support for subscriber certificates
TS	33.222	Generic Authentication Architecture (GAA); Access to network application functions using Hypertext Transfer Protocol over Transport Layer Security (HTTPS)
TS	33.222	TLS
TS	33.234	3G security; Wireless Local Area Network (WLAN) interworking security
TS	33.901	Criteria for cryptographic Algorithm design process
TS	33.902	Formal Analysis of the 3G Authentication Protocol
TS	33.908	3G Security; General report on the design, specification and evaluation of 3GPP standard confidentiality and integrity algorithms
TS	33.918	Generic Authentication Architecture (GAA); Early implementation of Hypertext Transfer Protocol over Transport Layer Security (HTTPS) connection between a Universal Integrated Circuit Card (UICC) and a Network Application Function (NAF)
TS	33.919	3G Security; Generic Authentication Architecture (GAA); System description
TS	33.941	Presence service; Security
TS	33.978	Security aspects of early IP Multimedia Subsystem (IMS)
TS	33.980	Liberty Alliance and 3GPP security interworking; Interworking of Liberty Alliance Identity Federation Framework (ID-FF), Identity Web Services Framework (ID-WSF) and Generic Authentication Architecture (GAA)

C.3 TISPAN

Spec	Description
ES 282 001	NGN Functional Architecture Release 1. Overall architecture
ES 282 002	PSTN/ISDN Emulation Sub-system (PES); Functional architecture and PES architecture
ES 282 003	Resource and Admission Control Sub-system (RACS); Functional Architecture
ES 282 004	NGN Functional Architecture; Network Attachment Sub-System (NASS)

Spec	Description
ES 282 007	IP Multimedia Subsystem (IMS); Functional architecture
ES 282010	Charging
ES 283 003	IP Multimedia Call Control Protocol based on Session Initiation Protocol (SIP) and Session Description Protocol (SDP) Stage 3, [3GPP TS 24.229 (Release 7), modified]
ES 283 018	H.248 Profile for controlling Border Gateway Functions (BGF) in the Resource and Admission Control Subsystem (RACS) Protocol specification Ia Interface H.248
ES 283 024	H.248 Profile for controlling Trunking Media Gateways in the PSTN/ISDN Emulation Subsystem (PES); Protocol specification
ES 283 026	Protocol for QoS reservation information exchange between the Service Policy Decision Function (SPDF) and the Access-Resource and Admission Control Function (A-RACF) - Protocol specification Rq Interface
ES 283 031	H.248 Profile for controlling Multimedia Resource Function Processors (MRFP) in the IP Multimedia System (IMS); Protocol specification H.248 and MRFP
ES 283 034	Network Attachment Sub-System (NASS); The e4 interface based on the DIAMETER protocol
ES 283 035	Network Attachment Sub-System (NASS); The e2 interface, based on the Diameter protocol
TR 180 000	NGN terminology
TR 181 003	Services Capabilities, Requirements and strategic direction for NGN services. NGN strategy
TR 181 004	NGN Generic capabilities and their use to develop services.
TR 182 005	Organization of user data
TR 183 013	Analysis of relevant 3GPP IMS specifications for use in TISPAN_NGN Release 1 specifications
TR 183 014	PSTN/ISDN Emulation; Development and Verification of PSTN/ISDN Emulation
TR 183 032	Feasibility study into mechanisms for the support of encapsulated ISUP information in IMS
TR 187 002	TISPAN NGN Security (NGN_SEC); Threat and Risk Analysis
TS 181 002	Multimedia Telephony with PSTN/ISDN simulation services. NGN simulation services
TS 181 010	Service requirements for End-to-End session control in multimedia networks
TS 182 006	IP Multimedia Subsystem (IMS); Stage 2 description (3GPP TS 23.2 v7.2.0, modified)
TS 182 012	IMS-based PSTN/ISDN Emulation Subsystem; Functional architecture of IMS-based Emulation
TS 183 017	Resource and Admission Control: Diameter protocol for session based policy set-up information exchange between the Application Function (AF) and the Service Policy Decision Function (SPDF); Protocol specification Gq' interface stage 3

Spec	Description
TS 183 019	Network Attachment; Network Access xDSL and WLAN Access Networks; Interface Protocol Definitions, WLAN and xDSL
TS 183 020	Network Attachment: Roaming in TISPAN NGN Network Accesses; NGN Roaming interface
TS 183 022	MGC Information Package
TS 183 028	Common Basic Communication procedures; Protocol specification
TS 183 033	IP Multimedia Diameter based interfaces between the Call Session Control Function and the User Profile Server Function/Subscription Locator Function; [3GPP TS 29.228 V6.8.0 and 3GPP TS 29.229 V6.6.0, modified]
TS 183 043	IMS-based PSTN/ISDN Emulation, Stage 3 specification PES Stage 3
TS 184 002	Identifiers (IDs) for NGN
TS 185 001	Quality of Service (QoS) Framework and Requirements
TS 187 001	NGN SECurity (SEC); Requirements
TS 187 003	NGN Security; Security Architecture
TS 188 002-2	NGN Subscription Management; Part 2: Information Model (Draft)

C.4 ITU

Spec	Description
ITU Recommendation H.248	Gateway control protocol
CCITT Recommendation E.164	Numbering plan for the ISDN era
ITU Recommendation Y.1541	QoS classes
ITU Recommendation Q.735	MLPP (Multi-Level Precedence and Pre-emption) and GVNS (Global Virtual Network Service) descriptions
ITU Recommendation Q.736	REV (Reverse charging) service description
ITU Recommendation P.800	MOS – Mean Opinion Score, measuring perception of voice quality

C.5 OMA

Spec	Description
OMA-Arch-2003-0254R01	Functional Application Layer Architecture
OMA-Policy V1-0-20060312-D	Evaluation Enforcement Management
OMA-Service-Environment_V2_0_0_20051114-D	Service Environment
OMA-AD-Presence_SIMPLE-V2_0-20060412-D	Presence

Appendix D

Standard Reference Points

The following table lists reference points (Ref) defined by both 3GPP and TISPAN. As many such reference points are shared, the table provides a combined list. Note that reference points do overlap sometimes, and may evolve between standards releases.

Ref	Between Elements	Main Functions
a1	AMF (Access Management Function) – NACF	Configuration and DHCP-based IP Address allocation in NGN
a2	NACF – CLF (Communications Location Function)	Transfer of physical and logical locations for storage in NGN location server
a3	AMF – UAAF (User Access Authorization Function)	Requests and dialogs of authentication with NGN access point
a4	UAAF – CLF	NGN location information for the authentication process
Bc	xMSC/HLR – BS (Billing System)	Charging data collection
Bc	CDR file transfer from the circuit-switched CGF to the BD	CDR type
Bg	CDR file transfer from the GPRS CGF to the BD	CDR type
Bi	IMS CGF (or CCF) – BD (Billing Domain)	Charging data collection
Bl	GMLC CGF – BD	CDR file transfer
Bm	CDR file transfer from the MMS CGF to the BD	CDR type
Bmb	CDR file transfer from the MBMS CGF to the BD	CDR type
Bmp	From the MBMS CGF to Billing	CDR file transfer

Ref	Between Elements	Main Functions
Bo	CDR file transfer from the OCF CGF to the BD	CDR type
Bp	CDR file transfer from the PoC CGF to the BD	CDR type
Bs	CDR file transfer for CAMEL services to the BD	CDR type
Bw	CDR file transfer from the I-WLAN CGF to the BD	CDR type
Bx	CDR file transfer between any (generic) 3G domain, subsystem, or service CGF and the BD	Generic CDR file
C	HLR – MSC server	MAP Protocol
Cx	HSS – I-CSCF or S-CSCF	CSCF inquiries of HSS data, for user profiles, applications filters, and server locations in Diameter
D	HLR – VLR	MAP
D′	pre-R6 HSS/HLR – 3GPP AAA Server	For WLAN
D′/Gr′	3GPP AAA Server – HLR	For WLAN
Dh	AS – SLF (Subscription Locator Function) LIMS-IWF – SLF	AS inquiries to find the user's HSS in a distributed HSS network, using Diameter
Dw	3GPP AAA Server – SLF	For WLAN
Dx	CSCF – SLF	S-CSCF and I-CSCF requesting SLF to locate HSS when HSS is distributed
Dz	BSF (Bootstrapping Function) – SLF	BSF obtains user's HSS address
E	Between MSC servers	2G Mobile signaling
e1	CNG (Customer Network Gateway) – AMF/ARF	Forwarding signaling from UE and CPE onto the NGN access network
e2	CLF – P-CSCF	Location inquiries by IMS from the NGN access network
e3	CNG (Customer Network Gateway) – CNGCF (CNG Configuration Function)	NGN Customer Gateway configuration information to and from a local configuration server
e4	CLF – RACS	NGN Resource allocation inquiries of location
e5	UAAF – PDBF (Profile Database Function)	Interrogation of the authentication database for the NGN access authentication process
F	MSC server – EIR	Mobile signaling
G	Between VLRs	Mobile signaling
Ga	CDF or a CDR transmitting unit (e.g., GGSN or SGSN) – CGF (CDR receiving functionality)	Charging data collection interface

Ref	Between Elements	Main Functions
Gb	SGSN – BSC	Mobile signaling
Gc	GGSN – HLR	Signaling path
Gd	SMS-GMSC–SGSN,SMS-IWMSC–SGSN	SMS to GPRS
Ge	gprsSSF – gsmSCF	Inter-IN/CAMEL AS
Gf	SGSN – EIR	GPRS identifying equipment numbers
Gi	GPRS – external packet data network	Data connection to other IP networks
Gm	P-CSCF – UE	SIP Session initiation, messages and session management
Gmb	GGSN – BM-SC	Media
Gn	GPRS – GPRS nodes within the same PLMN	Relay of GPRS
Gn/Gp	SGSN – GGSN	Support within Home PLMN
Go	GGSN – PDF	COPS to control QoS
Gp	GSN – external GSN	Signaling to external PLMNs
Gp	Between two GPRS nodes in different PLMNs	Allows support of packet-switched domain services across areas served by the cooperating packet-switched domain PLMNs
Gq	PDF – Application Function (AF)PDF – P-CSCF	P-CSCF/AF inquiries on policies and requesting resources using Diameter
Gq'	IBCF – SPDF, SPDF - AF	TISPAN variation of Gq (Diameter), which is a combination of the requirements over Rq and Ialt carries P-CSCF/AF inquiries on policies and requesting resources using Diameter
Gr	SGSN – HLR	Location inquiries
Gr	pre-R6 HSS/HLR – 3GPP AAA Server	WLAN
Gs	MSC/VLR – SGSN	Interconnect between 2G and 3G
Gx	PCRF – PECF	Allows the PCRF to control Policy Enforcement using PCC rules (e.g., GGSN for Flow Based Charging)Based on Diameter
H	HLR – AuC	Location/Authentication
I	MSC – its associated GCR	TDM
Ia	RACS – I-BGF	Convey resource management instructions for gateway control using H.248
Ib	IWF – IBCF	Interworking with non-SIP IP networks
Ic	IBCF to external IBCF	Convey cross-border control in SIP
Iq	P-CSCF – IMS Access Gateway	Release and convey address information to or from the access node

Ref	Between Elements	Main Functions
ISC	CSCF – AS	S-CSCF communication with application servers, using extended SIP
Ix	IMS ALG/IBCF – TrGW	Providing NAT and NA(P)T-PT for IP Interconnect
Lc	MLC – gsmSCF (CAMEL interface)	Mobile location information for Mobile CAMEL AS
Le	GMLC – External LCS Client (usually AS)	Location queries of the Mobile Location Center
Lg	Gateway MLC – VMSC, GMLC – MSC Server, GMLC – SGSN (gateway MLC interface)	Roaming messaging 2G or 3G
Lh	Gateway MLC – HLR (HLR interface)	2G interface HLR interrogation onward to IMS
Li	E-CSCF (Emergency) – Location Server (e.g., LRF)	Emergency calls
Lr	GMLC – GMLC	Signaling relay onward to IMS
Ma	I-CSCF – AS	AS interrogating route and serving servers
Mb	Media	Media and signaling path in IP
Mc	MGW – (G)MSC server	Signaling for a Gateway Mobile Switch Center or a softswitch to control a media gateway toward 2G
Mg	MGCF – CSCF	MGCF converting ISUP to SIP to connect with I-CSCF
Mi	S-CSCF – BGCF	S-CSCF forwarding SIP messages to Breakout Gateway for another IP network
Mj	BGCF – MGCF	Border Gateway and Media Controller in the same IMS network negotiating media path in SIP
Mk	BGCF/ALG – BGCF/ALG	Exchange of SIP messages to route sessions to external IP networks
Mm	CSCF/ALG – Another Multimedia IP network	S-CSCF and I-CSCF communicating with other IP networks
Mn	MGCF – IMS-MGW	Media gateway controller messages to the selected Media Gateway for media flow to CSN, using H.248
Mp	MRFC – MRFP	Media controller commands to the Media Processor, using H.248
Mr	S-CSCF – MRFC	Session controller asking for Media services from a Media Server, using SIP
Mw	CSCF – CSCF	Session management between P-CSCF, I-CSCF, and S-CSCF, using SIP
Mx	CSCF/BGCF – IMS ALG and IBCF	SIP messages bound for IP interconnection

Ref	*Between Elements*	*Main Functions*
Nb	CS-MGW – CS-MGW	TDM media
Nc	MSC Server – GMSC Server	Toward IMS via gateway MSC
Pen	Presence Network Agent – Presence Server	Presence information
Peu	Presence User Agent – Presence Server	Presence information
Pex	Presence External Agent – Presence Server	Presence information
Ph	HSS/HLR – Presence Network Agent	Like Sh, Diameter
Pi	S-CSCF – Presence Network Agent	Same as ISC, SIP
Pw	Watcher applications – Presentity Presence Proxy	Presence information
Px	Presentity Presence Proxy – HSS	Presence information
R	Non-ISDN-compatible TE and MT.	Typically this reference point supports a standard serial interface.
Re	A-RACF – RCEF (Resource Control Emulation Function)	Providing QoS policy rules and control QoS gating on the media transport gateways, using Diameter
Rf	IMS Network Entity – CCF	Offline charging data transfer, Diameter
Rf	IMS Network Entity – CCF	Offline Charging
Ro	AS/MRFC – ECF	Information exchange between AS and the HSS data store, for online charging data, Diameter
Rp	GUP server – HSS	Data retrieval from HSS, which is one of the GUP data repositories
Rq	GUP – AS	Data management by AS via the generic interface
Rr	Toward an external account recharging server	Online charging
Rx	AF – CRF/PCRF	Allowing AF to determine rules for Flow Based ChargingBased on Diameter
Sh	AS – HSS	Transfer of User and apps data over Diameter protocol
Si	HSS – CAMEL IM-SSF	Allows IN CAMEL/MAP application data to be stored and retrieved on HSS
Ua	NAF (application function) – UE	NAF secured interface, including certificate from PKI portal
Ub	UE – BSF	UE obtaining keys, HTTP Digest AKA
Um	Mobile Station (MS) – GSM fixed network part (PLMN)	Provides packet-switched services over the radio to the MSThe MT part of the MS is used to access the packet-switched services in GSM through this interface

Ref	Between Elements	Main Functions
Ut	UE – AS	Endpoint interface directly to AS for service parameter management etc. using HTTP XCAP
Uu	Mobile Station (MS) – PLMN	In UMTS, provides PS services over the radio to the MT in the MS.
Wa	WLAN Access Network – 3GPP AAA Server/Proxy	WLAN charging and control signaling
Wd	3GPP AAA Proxy – 3GPP AAA Server	WLAN charging and control signaling
Wf	Offline Charging System – 3GPP AAA Server/Proxy	WLAN charging and control signaling
Wg	3GPP AAA Server/Proxy – PDG/WAG	WLAN Authentication
Wi	Packet Data Gateway (PDG) – external IP network	Data/media
Wm	Packet Data Gateway (PDG) – 3GPP AAA Server or 3GPP AAA proxy	WLAN signaling for Authentication
Wn	WLAN Access Network – WLAN Access Gateway (WAG)	Access
Wo	3GPP AAA Server – OCS	Online charging of WLAN communications
Wp	WLAN Access Gateway (WAG) – Packet Data Gateway (PDG)	WLAN media
Wu	WLAN UE – Packet Data Gateway (PDG)	Media for WLAN UE
Ww	WLAN UE – WLAN Access Network	WLAN access
Wx	HSS – 3GPP AAA Server	User information correlated to WLAN information
Wy	PDG – Online Charging System (OCS)	WLAN online charging
Wz	PDG – Offline Charging System (OFCS)	WLAN offline charging
Za	Between SEGs belonging to different networks or security domains	Inter-network security
Zb	Between SEGs and NEsbetween NEs	Internal security within the same network or security domain
Zh	BSF – HSS	BSF to fetch user credentials
Zn	NAF (Network Application Function) – BSF	NAF to fetch security keys from BSF

Appendix E

Terms and Definitions

Accounting: The process of apportioning charges between the Home Environment, Serving Network, and Subscriber.

Advice of Charge (AoC): Real-time display of the network utilization charges incurred by the Mobile Station (MS). The charges are displayed in the form of charging units. If a unit price is stored by the terminal, then the display also can include the equivalent charge in the Home currency.

ALG (Application-Level Gateway): Gateway that allows communication between disparate address realms or between different IP versions, when certain applications carry network addresses in the payloads, such as SIP/SDP. The ALG is application specific, as it needs to understand the protocol dialog.

Algorithm for Cryptography: Mathematical formulae and data manipulation procedure that enable masking data by "hashing" it, to achieve communication security and privacy. Such algorithms (e.g., MD5, SHA-1, and HMAC) vary in their level of complexity and the protection that they provide.

ASN.1: A formal notation used for describing data transmitted by telecommunications protocols, regardless of language implementation and physical representation of these data, whatever the application, whether complex or very simple.

Applet: A small program intended not to be run on its own, but rather to be embedded inside another application.

Authentication: A property by which the correct identity of an entity or party is established with a required assurance. The party being authenticated could be a user, subscriber, server, home environment, or serving network.

Authorization: The act of determining whether the requesting entity (user, server, application, etc.) should be granted access to a particular resource. Authorization is the action of determining what a previously authenticated entity is allowed to do. Authentication, which verified that the entity is what it says it is, must precede authorization.

Authorized QoS: The maximum QoS (Quality-of-Service) authorized for a service data flow. In case of an aggregation of multiple service data flows within one IP-CAN bearer (e.g., for GPRS a PDP context), the combination of the "Authorized QoS" information of the individual service data flows is the "Authorized QoS" for the IP-CAN bearer.

Backbone Network: The Backbone Network lies between the UE access node and the core network service node.

Baseline Implementation Capabilities: Set of implementation capabilities, in each technical domain, required to enable a UE to support the required baseline capabilities.

Bearer: An information transmission path of defined capacity, delay and bit error rate, etc. Bearer capability is a transmission function that the UE requests. Bearer service is a type of telecommunications service that provides the capability to transmit signals between access points.

Best-Effort QoS: The lowest of all QoS traffic classes used when QoS is not guaranteed. The bearer network delivers the QoS that can be achieved without any special effort.

Best-Effort Service: A service model that provides minimal or no performance guarantees, allowing an unspecified variance in the measured performance criteria. The Internet is an example of Best-Effort Service.

Billing: The functions utilizing charging data to produce bills for payment. Such bills can be pre-paid or post-paid, printed or online. In case of pre-paid, the billed amount may have been deducted already from the account, but the billing statement still must be produced for accounting purposes. User Billing is not subject to standardization.

Billing Domain: This Billing Domain is considered outside IMS, UMTS, NGN, and the Service environment. It includes billing mediation and bill production, and may be associated with other CRM and management systems.

Binding: The association between a service data flow and the IP-CAN bearer (for GPRS the PDP context) transporting that service data flow. Also associations between line identity and an IP address in user attachment to NGN. Binding mechanism is the method for creating, modifying, and deleting bindings.

Broadband: Data transmission or receipt at a higher speed than available over a traditional phone line. This usually refers to services delivered by NGN and 3G.

Broadcast: A mechanism for delivering the same content to a number of clients at the same time. A service attribute is unidirectional distribution to all users.

CAC (Connection Admission Control): A set of measures taken by the network to balance between the QoS requirements of a new connection request and the current network utilization without affecting the Grade-of-Service of already-established and existing connections.

Call or Call Session: A logical association between two or more users (this could be connection oriented or connectionless), in real- time (i.e., excludes man-to-machine or machine-to-machine sessions).

Camped on a cell: The UE is in idle mode after completing the cell selection or reselection process and a cell is chosen. The UE monitors system information and (in most cases) paging information. Note that the services may be limited, and that the PLMN may not be aware of the existence of the UE within the chosen cell.

Capability Class: A piece of information that indicates general 3GPP System mobile station characteristics (e.g., supported radio interfaces) for the interest of the network.

Cell: Radio network object that can be uniquely identified by a UE from an identification beacon that is broadcast over a geographical area from one UTRAN Access Point.

Chargeable event: Activity utilizing network infrastructure for which the network operator wants to charge. This can be:
- User-to-user communication, (e.g., a single call, a data communication session, or a short message)

- User-to-network communication (e.g., service profile administration)
- Inter-network communication (e.g., transferring calls, signaling, or short messages)
- Mobility (e.g., roaming or inter-system handover)

Charging: The function of collecting and collating information related to chargeable events, sessions, or metered media flows, which will be used in the Billing System. Charging predominantly deals with usage values rather than prices. Charging can include non-session-related services, (e.g., transmitting user status). Charging information is recorded dynamically, and the collection of such records from many network nodes is standardized.

Charging Data Record (CDR): A formatted collection of information about a chargeable event (e.g., time of call set-up, duration of the call, amount of data transferred, etc.) for use in billing and accounting. To charge each party for a chargeable event, a separate CDR is generated. More than one CDR can be generated for a single chargeable event (e.g., because of its long duration or because more than one party is to be charged); CDRs are produced by many Network Elements.

Charging key: Information used by the online and offline charging system for rating purposes.

Charging rule: A set of rules and parameters, including the service data flow filters and the charging key, for a single service data flow.

Cipher key: A code used in conjunction with a security algorithm to encode and decode user and signaling data.

Connection lifetime: Connection-oriented bearer service lifetime is the period of time between the establishment and the release of the connection.

Connection mode: The type of association between two points as required by the Bearer service for the transfer of information. A Bearer service is either connection-oriented or connectionless. In a connection-oriented mode, the connection — a logical association — must be established between the source and the destination entities before information can be exchanged between them.

Connectionless (for a bearer service): In a connectionless bearer, no connection is established beforehand between the source and the destination entities. The source and destination network addresses must be specified in each message. Transferred information cannot have a guaranteed ordered delivery. Connectionless bearer services lifetime is reduced to the transport of one message. Connectionless service is a service that allows the transfer of information among service users without the need for end-to-end call establishment procedures.

Control channel: A logical channel that carries system control information.

Conversational service: An interactive service that provides for bi-directional communication by means of real-time end-to-end information transfer from user to user.

Core network: An architectural term relating to the part of the network that is independent of the connection technology of the terminal (e.g., radio, wired).

Domain: The highest-level group of physical entities. Part of a communication network that provides network resources using a certain bearer technology.

Donor network: The subscription network from which a number is ported in the porting process of a Number Portability service.

Downlink: Unidirectional radio link for the transmission of signals from a network access point of the User Equipment in the general direction.

DTMF (Dual Tone Multiple Frequency): A system that uses combinations of two tones to signify digits. Commonly known as Touch-Tone. DTMF tones are used, for example, to convey to a CSN switch user input from a legacy phone.

Dynamic charging rule: Charging rule where some of the data within the charging rule (e.g., filter information) is assigned via real-time analysis, which can use dynamic application-derived criteria.

E.164: An ITU-T recommendation that defines the structure of numbering for telecommunications in PSTN, enabling global routing of calls. E.164 numbers have a maximum of 15 digits and can include a "+" prefix denoting variable country breakout code.

Early Media: Refers to media that is exchanged before a particular session is accepted by the called party. The media can be announcement, ring-back tones, or push-media.

Element Management Functions: Set of functions for management of network elements on an individual basis.

ENUM/DNS: Electronic Number Mapping is a set of procedures that allows interpreting the Telephony E.164 numbering format via an Internet IP addressing schemes. This involves a conversion formula to generate an IP Address from the E.164 number and lookup tables to obtain NAPTR (Network Address Pointer Resource) records.

Ethernet: Physical LAN technology consisting of standardized cables and interfaces running at 10 Mbps.

FBC Policy Functions: The policy charging rules can be configured in such a way to allow performing Flow Based Charging (FBC) for a certain usage that allows or disallows traffic to pass.

Federated Identity Management (FIM): The use of agreements, standards, and technologies to make identity and entitlements portable across autonomous identity domains.

Fully Qualified Domain Name (FQDN): Domain name that identifies a specific server, expressed with a final dot (e.g., mydomain.my_name.com.). FQDN expressions do not need a lookup query to identify the routing path.

Fully Qualified Partial CDR (FQPC): A type of "Partial CDR" that contains a complete set of the CDR fields but covers only part of the session period. The FQPC includes all the mandatory and conditional fields as well as the operator's specially provisioned fields to be included in the CDR. The first Partial CDR must be a FQPC.

Gating control: The process of blocking or allowing packets, belonging to a service data flow, to pass through to the desired endpoint.

GPRS IP-CAN: This IP-CAN incorporates GPRS over GERAN and UTRAN. GPRS is packet-switched bearer and radio services for both GSM and UMTS systems.

Handover: While moving out of reach of one network and into another, the service is made available by transferring control from one network to another. Ordinary handover can be detectable by the user as an interruption of the media exchange, but the connection should not be halted. Handover can occur between the same type of networks (e.g., two 3G networks), different types of access technology (e.g., WiFi and GSM), within the same operator, or across business borders.

Hard handover: This is a category of handover procedures in which all the old radio links in the User Equipment (UE) are abandoned before the new radio links are established.

Home Network: The network belonging to the operator who owns the user account or subscription. In the case of Virtual Network Operators (VNOs), this is the network that provides connection to the VNO servers.

Hotspot: Public locations, such as airports or hotels where WLAN services have been deployed.

IM CN Subsystem (IMS): IP Multimedia Core Network Subsystem comprises all CN (Core Network) elements for the provision of IP multimedia applications over IP multimedia sessions. This is commonly referred to as IMS.

IMEI (International Mobile Station Equipment Identity): A unique number allocated to each MS (Mobile Station) equipment in the PLMN and is unconditionally implemented by the MS manufacturer.

Interactive service: A service that provides the means for bi-directional exchange of information between users. Services that are between a user and a server are not deemed as interactive.

Inter-Administration or Inter-Operator Accounting: Terms that refer to the exchange of charging information that occurs between operators, service providers, and access network providers (e.g., WLAN) and intermediate network carriers. The information also includes apportioning costs between the service or network providers who are involved in the service delivery.

Interworking WLAN (I-WLAN): A WLAN that can interwork with 3GPP systems, where authentication uses 3GPP existing facilities.

IP flow: A unidirectional flow of IP packets with the same source IP address and port number and the same destination IP address and port number and the same transport protocol. Port numbers are only applicable if used by the transport protocol.

IP Multimedia application: An application that handles one or more media simultaneously such as speech, audio, video, and data (e.g., chat text, shared whiteboard) in a synchronized way from the user's point of view. A multimedia application can involve multiple parties, multiple connections, and the addition or deletion of resources within a single IP multimedia session. A user can invoke concurrent IP multimedia applications in an IP multimedia session.

IP network connection: The unique UE association with an IP network, using an IP-CAN (GPRS, xDSL) and an allocated IP address at the traffic plane.

IP-CAN (IP-Connectivity Access Network): The collection of network entities and interfaces that provides the underlying IP transport connectivity between the UE and the core network or backbone entities. An example IP-CAN is GPRS. An IP-CAN session can incorporate one or more IP-CAN bearers.

IP-CAN bearer (IP-Connectivity Access Network bearer): An IP transmission path of defined capacity, delay, and bit error rate. It is the data communications bearer provided by the IP-CAN. When using GPRS, the IP-CAN bearers are provided by PDP Contexts.

ISIM, (IMS SIM): A client application residing on the UICC that provides access to IP Multimedia Services (IMS).

Jitter: Delay variations on signaling networks, often used as a performance metric.

Key pair: A matching of private and public keys. If a block of data is encrypted using the private key, the public key from the pair can be used to decrypt it. The private key is never divulged to any other party but the public key is available, (e.g., in a certificate).

Latency: Delay associated with network devices — typically measured in microseconds for switches and milliseconds for bridges and controllers.

Local service: A service that can be provided by a network as a local service to inbound roamers and to local subscribers of this network. Such services do not depend on the Home Network User Profile that may not be available for the Visited Network.

Messaging service: An interactive service that offers user-to-user communication between individual users via storage units with store-and-forward capability, mailbox management, and message handling, (e.g., information editing, processing, and conversion functions).

Metadata: "Data about data" in a standard format to describe record content, location, quality, context, and other characteristics.

Mobile termination (MT): The component of the mobile station that supports functions specific to management of the radio interface (Um interface).

Mobility: The ability for the user to communicate while moving, independently of the location. In contrast with "Nomadism" (log in at different points in the network), full Mobility allows seamless handover of connections between network nodes while the user is on the move at some speed, walking or motoring.

Mobility (Generalized Mobility): The ITU-T defines Generalized Mobility as "The ability for the user, or other mobile entities, to communicate and access services irrespective of changes of the location or technical environment. The degree of service availability can depend on several factors, including the Access Network capabilities and the service level agreements between the user's Home Network and the Visited Network (if applicable). Mobility includes the ability of telecommunication with or without service continuity."

Mobility Management: A relation between the mobile station and the UTRAN that is used to set up, maintain, and release the various physical channels.

Multicast service: A service in which a transmission from a single source is relayed to multiple subscribers currently located within a geographical area. The service parameters contain a group identifier indicating whether it should reach all subscribers or only the subset of subscribers belonging to a specific multicast group. By contrast, broadcast is a relay to anyone in the catchment area.

Multimedia service: Services that handle several types of media such as audio and video in a synchronized way from the user's point of view. A multimedia service may involve multiple parties, multiple connections, and the addition or deletion of resources and users within a single communication session.

Multipoint: More than two network terminations.

NAPTR (Naming Authority Pointer – DNS Resource Record): A record type that uses regular expressions to encode a set of rules that define a range of matching servers. RFC 3403 defines this type of record. The ENUM NAPTR is a specific use of this capability.

NA(P)T (Network Address and Port Translation): A method by which IP addresses are mapped from one group to another, transparently to end users. This enables any network address and TCP/UDP ports to be translated into a single network address and its TCP/UDP ports.

NAT (Network Address Translation): Network Address Translation, or a generic term that includes different types of it (i.e., NAT-PT and NA(P)T).

Near-real-time: Services that are delivered at nearly real-time (one minute), but not in real-time (less than one second). For example, billing information can be generated, in less than one minute but a voice call is srequired to be delivered in less than one second.

Network termination: This can be a user equipment or a server that may become a party to a network connection.

Nomadic Operating Mode: Mode of operation where the terminal is transportable but being operated while stationary. Nomadic Terminal does not have full mobile capabilities but would normally be expected to roam between different points of attachment to the network, both wireless and wired.

Nomadism: This is referred to as Mobility with Service Discontinuity; that is, the service stops and is restarted again at the new entry point. It is defined by ITU-T as the "Ability of the user to change his network access point…while the user's service session is completely stopped and then started again, i.e., it is not possible to have session continuity or handover. It is assumed that normal usage pattern is that users shut down their service session before changing to another access point." Note that to be nomadic, there is no requirement to be static or "tethered" (wired). This allows for wireless LAN scenarios on PDAs and laptops.

Nonce: A value that is used only once or that is never repeated within the same cryptographic context. In general, a nonce can be predictable (e.g., a counter) or unpredictable (e.g., a random value). Because some cryptographic properties can depend on the randomness of the nonce, attention should be paid to whether or not a nonce is required to be random.

Number Portability: A service where the provision of numbers that can be dialed is independent of the Home environment or Serving network. Also refers to a regulatory function that permits users to keep their number when changing their subscription carrier within the same country. The donor network forwards communication to the recipient network.

Offline charging: A charging process in which charging information does not affect the service rendered and the information is produced after the event. This mode is not time-critical, but timely information flow may still be needed, that is, in near-real-time (response less than one minute).

Online Charging: A charging process in which charging information can affect, in real-time, the service rendered and therefore directly interacts with the session or service control. A popular example of Online Charging is Pre-pay and instantaneous credit control, which must be exercised dynamically before or during the session. Real-time charging information also may be available to the user, with response time of less than one second.

Open group: A group that does not have a predefined set of members. Any user can participate in an open group.

Open Service Access: Concept for introducing a vendor-independent means of offering new services.

Originating Network: The network where the calling party is located.

Packet: An information unit identified by a label at Layer 3 of the OSI Reference Model. A network protocol data unit (NPDU).

Packet Data Protocol (PDP): Any protocol that transmits data as discrete units known as packets (e.g., IP or X.25).

Packet flow and media flow: This describes a specific user data flow carried through the Traffic Plane. An IP flow is one instance of packet flows, using an IP bearer. Media flow describes a generic flow of Service Data Units (SDUs), of which packet flow is one instance.

Partial CDR: A type of CDR that provides information on part of a subscriber session. A long session can be covered by several partial CDRs. Two formats are considered for Partial CDRs: (1) record containing all of the necessary fields, i.e., Fully Qualified Partial CDR (FQPC); and (2) record with a Reduced Partial CDR (RPC) that does not contain duplicate data but only what is relevant to the part of the session that it describes. The Starting Partial CDR (the FQPC) must have all the data fields to identify the session, but the RPC can contain only mandatory fields.

PCC (Policy and Charging Control): The system of rules and functions, in collaboration with the UE, bearer nodes, and session control functions, that allows the network to differentiate between types of sessions for the purpose of QoS and Charging.

PCC rule: An information set that enables the detection of a service data flow and provides parameters for policy control and charging control.

PDP Context (Packet Data Protocol Context): Packet Data Protocol Context, where context means one or more signaling streams plus one or more media streams, together with their origination and destination information.

Personal mobility (User Mobility): Users can use multiple terminals, of the same type or different equipment, and still receive the same services and same stored data. Users gain access to their own services when they are authenticated by the network, verified by prestored User Profiles that are kept on the network data repository. Examples of this in traditional networks are logging into e-mail via in Internet café or to a telephone Personal Account via PIN (personal identification number).

Port: A software port that is a virtual data connection on a physical channel that allows software logic to control dynamic data exchange directly, instead of using temporary storage location. The most common of these are TCP and UDP ports, which are used to exchange data between computers on the Internet. Ports are often protocol specific and are said to be "listening" to particular transmission.

Portability: In contrast to handover, a basic level of Service Mobility is achieved by lugging equipment to another point and reconnecting to the network there. The user's service environment is made available but there is no service continuity while moving between points. This usually involves restarting the system in the new location. While portability usually refers to the equipment, users are said to be "nomadic."

Post-paid Billing: Billing arrangement between the subscriber and the operator or service providers where the subscriber periodically receives bills for services that the user already has used. If the operator wishes to limit the credit during the post-pay period, an online credit control is required.

Pre-paid Billing: Billing arrangement between the subscriber and the operator or service providers where the user deposits a sum of money in advance and replenishes it periodically. The cost of network or service usage is deducted from the credited amount. A pre-paid service is not synonymous with online charging, but to get dynamic credit control, online charging is required.

QoS (Quality-of-Service): The collective effect of service performances that determines the degree of satisfaction of a user of a service. It is characterized by the combined aspects of performance factors applicable to all services, such as service operability, accessibility, retaining the service reliably, and integrity of the transferred information.

QoS profile: A QoS profile comprises a number of QoS parameters. A QoS profile is associated with each QoS session. The QoS profile defines the performance expectations placed on the bearer network.

QoS session lifetime: A session with a particular QoS setting that lasts for the lifetime of the media flow (e.g., the GPRS PDP context). This is the period between the opening and closing of a network connection whose characteristics are defined by a QoS profile.

Radio interface: The "radio interface" is the tetherless interface between User Equipment and a GSM, WLAN, WiMax, or UTRAN access point.

Realm: A Network Realm generally refers to the operator's total packet network, which may include several domains. It takes the form of an Internet domain name (e.g., "operator. com"). The Realm is the string in the NAI (Network Address Identifier) that immediately follows the "@" character. NAI realm names are required to be unique.

Reduced Partial CDR (RPC): A partial CDR that only provides mandatory fields and information regarding changes in the session parameters relative to the previous partial CDR. For example, location information is not repeated in RPC if the subscriber equipment did not change location. This optimizes information flow of charging details.

Reference point: A conceptual point at the conjunction of two non-overlapping functional groups. A reference point defines an interface for a particular interaction between the two functional elements.

Regular Expression (regex): Regular expression (or regex) allows matching character strings with a template (i.e., a pattern that provides a set of syntax rules and conditions). This is more flexible than a wildcard expression, which merely identifies a range of values in certain positions. The regex can define more complex patterns in a concise way, rather than give a full list of all alternatives. It provides alternative characters expressed within parentheses, and quantification (i.e., repetitions, using "?," "+," and "–," including zero occurrence).

Retrieval service: An interactive service that provides the capability to access information stored in database centers. The information will be sent to the user on demand only. The information is retrieved on an individual basis at the time at which an information flow sequence is to start under the control of the user.

Roaming: The ability of users to access services while outside their subscribed Home Network, that is, by using an access point of a Visited Network. This is usually supported by a roaming agreement between the respective network operators.

Seamless handover: A handover service without any perceivable interruption is said to be "seamless."

Secured Packet: An information unit in a flow on top of which the required security has been applied. An Application Message is transformed according to a chosen Transport layer and chosen level of security into one or more Secured Packets.

Security Association (SA) for IPSec: A unidirectional logical connection created for security purposes. All traffic traversing an SA is provided with the same security protection. The SA itself is a set of parameters to define the security protection between two entities. An IPSec SA includes the cryptographic algorithms, the keys, the duration of the keys, and other parameters for a channel.

Service data flow: Aggregate set of packet flows. A service data flow can be more granular than a PDP context in the case of GPRS.

Service data flow filter: A set of filter parameters used to identify one or more of the packet flows constituting a service data flow.

Service Data Unit (SDU): An amount of information whose identity is preserved when transferred between peer entities in an upper OSI layer and which is not interpreted by the supporting entities in the lower layer (ITU-T).

Service Execution Environment: A platform on which an application or program is authorized to perform a number of functionalities. Examples of Service Execution Environments are the user equipment, integrated circuit card, or a network server.

Service identifier: An identifier for a service. The service identifier can designate an end-user service, a part of an end-user service, or a group of services.

Service Mobility: Concerns the status of services for a Mobile or Nomadic user. Essentially, it refers to the availability of the same service (delivered from the Home Network) to the user, regardless of the access point.

Session (IP Multimedia): An IP multimedia session is a set of multimedia senders and receivers and the data streams flowing from senders to receivers. IP Multimedia Sessions are supported by the IP Multimedia CN Subsystem (IMS) and are enabled by IP connectivity bearers (e.g., GPRS as a bearer). A user can invoke concurrent IP Multimedia Sessions.

Settlement: Payment of amounts resulting from the accounting process submitted to peer business (e.g., between operators).

Shared Channel: A resource (transport channel or physical channel) that can be shared dynamically between several UEs.

SIM application toolkit procedures: The portion of the communication protocol between the ME and the UICC that enables applications on the UICC to send commands to the ME.

SIM code: Code that, when combined with the network and NS codes, refers to a unique SIM. The code is provided by the digits 8 to 15 of the IMSI.

Soft handover: A category of handover procedures where the radio links are added and abandoned in such manner that the UE always keeps at least one radio link to the UTRAN. *See* **hard handover**.

Subscribed QoS: The level of QoS to which the subscriber is entitled. The network will not grant a QoS greater than the subscribed level. The QoS profile parameters are held in the HSS or locally configured on serving network elements. An end user may have several QoS subscriptions. The end user cannot directly modify the QoS subscription profile data.

Subscriber: An entity with one or more users, engaged by contact with a Service Provider. The Subscriber is allowed to subscribe and unsubscribe to services, to register a user or a list of users authorized to enjoy these services, and to set restrictions of use for these users. A Subscription describes the commercial relationship between the Subscriber and the Service Provider.

Supplementary service: A service that modifies or supplements a basic telecommunication service. Consequently, it cannot be offered to a user as a stand-alone service. It must be offered together with or in association with a basic telecommunication service. The same supplementary service can be common to a number of basic telecommunication services.

Terminal: A device that "terminates" the network signaling and acts as an endpoint for media flow in the "user traffic plane." A terminal must be capable of providing access to network services. In 3GPP, this is a device into which a UICC can be inserted. In NGN, this can be a laptop, a phone, or a CPE gateway.

Terminal Mobility: Terminals are capable of mobility when they allow registration to local networks from different locations. This requires that the terminal is able to discover a local network to connect to, and the network to identify and locate that terminal for incoming traffic. Terminal Mobility is independent of the user and concerns the identification of the hardware/firmware rather than the user. Examples are mobile handsets with SIM cards.

TPF: The Traffic Plane Function refers to the bearer functions that carry the media to the user terminal.

Transparent data: Data that is understood syntactically but not semantically by a server. For example, it is data that an AS can store in the HSS to support its service logic, using it as a repository, but the HSS function is unaware of the contents.

Tunneling: Technology enabling one network to send its data via another network's connection by encapsulating a network protocol within packets carried by another protocol. The encapsulated data travels transparently through the connecting networks.

UICC Universal IC card: A physically secure device, an IC card (or "smart card"), that can be inserted and removed from the terminal equipment. It can contain one or more applications. Such applications can be a USIM (for UMTS) or ISIM (for IMS)

Universal Subscriber Identity Module (USIM): An application residing on the UICC used for accessing services provided by mobile networks, on which the application is able to register with the appropriate security.

Universal Terrestrial Radio Access Network (UTRAN): UTRAN identifies part of the network that consists of RNCs and Node Bs, with the Iu and Uu interfaces between them, that enables a UMTS connection to be made.

Uplink: An "uplink" is a unidirectional link for the transmission of signals from the UE to a base station, from a mobile station to a mobile base station, or from a mobile base station to a base station. An "uplink" can also be used generally to denote NGN service data flow from a terminal towards a network access node, and to the Core Network.

User Equipment (UE): UE is a user device allowing a user to access network services. In NGN, this can be, for example, a residential gateway, laptop, or phone. In 3G Mobile, it is a mobile handset.

User Profile: User Profile is the set of information necessary to provide a user with a consistent, personalized service environment, irrespective of the user's location or the terminal used.

Virtual Home Environment: A concept for personal service environment portability across network boundaries and between terminals.

Visited Network: The local network that the roaming user can connect to while away from the Home Network. Users can roam into many Visited Networks but have only one Home Network.

WLAN Direct IP Access: Access to a locally connected IP network (e.g., Internet) from the WLAN, where Authentication and Authorization are performed through the 3GPP Access System but no session control is required. This is in contrast to access to 3GPP/IMS services from the Home Network.

WLAN User Equipment (WLAN UE): UE (equipped with UICC card including (U)SIM) utilized by a subscriber capable of accessing a WLAN. WLAN UE can include entities whose configuration, operation, and software environment are not under the exclusive control of the 3GPP system operator, such as a laptop computer or PDA with a WLAN card.

Index

H

I

T

U

V

W

X

T - #0310 - 101024 - C0 - 254/178/28 [30] - CB - 9780849392504 - Gloss Lamination